QUANTUM MECHANICS
FOR SCIENTISTS AND ENGINEERS

QUANTUM MECHANICS
FOR SCIENTISTS AND ENGINEERS

Harish Parthasarathy
Professor
Electronics & Communication Engineering
Netaji Subhas Institute of Technology (NSIT)
New Delhi, Delhi-110078

CRC Press is an imprint of the
Taylor & Francis Group, an **informa** business

First published 2022
by CRC Press
2 Park Square, Milton Park, Abingdon, Oxon, OX14 4RN

and by CRC Press
6000 Broken Sound Parkway NW, Suite 300, Boca Raton, FL 33487-2742

CRC Press is an imprint of Informa UK Limited

British Library Cataloguing-in-Publication Data
A catalogue record for this book is available from the British Library

Library of Congress Cataloging-in-Publication Data
A catalog record has been requested

ISBN: 978-1-032-11764-5 (hbk)
ISBN: 978-1-003-22143-2 (ebk)

DOI: 10.1201/9781003221432

Preface

This book covers the entire span of modern quantum mechanics including quantum field theory and more recent developments in quantum stochastics. The book starts with the Bohr-Sommerfeld correspondence principle which was the first model developed by Bohr to explain Rutherford's picture of the atom with electrons revolving around the nucleus without losing energy in the form of electromagnetic radiation (produced by accelerating particles according to the classical Maxwell theory) which would otherwise cause the energy of the electron to continuously decrease and collapse to the nucleus leading to the instability of all matter. Bohr explained this by postulating that in a stationary state the electron does not lose or gain energy and further that the angular momentum of the electron which is the line integral of the momentum around one closed orbit is quantized in integral multiples of $h/2\pi$ where h is Planck's constant. Sommerfeld's generalization of this principle of Bohr to arbitrary action-angle varables is mentioned here so that the line integral of the action variable with respect to the angle variable over one complete circuit is an integral multiple of $h/2\pi$. The book then continues with the evolution of the Schrodinger equation from the Planck and De-Broglie hypothesis then proceeds to the matrix mechanics of Heisenberg and its equivalence with the Schrodinger wave mechanics along the lines outlined by Dirac in his celebrated book "The principles of quantum mechanics". In the process, we also discuss the interaction picture of Dirac which is something midway between the Schrodinger and the Heisenberg pictures. Max Born's intepretation of the wave function in terms of probabilities is also discussed. The superposition principle which is at the heart of quantum mechanics is discussed with illustrations from the Young double slit experiments for interference of photons and likewise, the Stern-Gerlach experiments that demonstrate the fact that electrons can sometimes behave like waves thus confirming De-Broglie's wave-particle duality. We then discuss the effect of an electromagnetic field on an atom using time dependent perturbation theory and also the Zeeman effect which predicts the splitting of the degenerate levels of an atom in a constant magnetic field caused by the interaction of spin with the magnetic field. This leads us to the Pauli equation which is the Schrodinger equation for a two component wave function with spin-magnetic field interaction accounted for. A thorough discussion of a time independent quantum system perturbed by time varying potential using the Dyson series is presented which is later on used in the book to calculate scattering amplitudes for electrons, positrons and photons upto any order in perturbation theory. Time independent perturbation theory is also introduced as a tool for calculating approximately the shift in the energy levels and the energy eigenstates of a quantum system when perturbed by a potential. The problem of perturbation of degenerate states is also taken up by showing that if the unperturbed energy E is degenerate with a degeneracy of d, then on perturbation, it will split into d energy levels and whose eigenstates and energy eigenvalues are determined by the eigenvalues and eigenvectors of the "secular matrix" obtained by considering the matrix elements of the perturbing potential with respect to an orthonormal basis of the d dimensional vector space of all eigenfunctions of the unperturbed Hamiltonian having energy E. Using the Cauchy-Schwarz inequality, we give Weyl's proof of the Heisenberg uncertainty principle in its most general form. Three exactly solvable problems in quantum mechanics are presented in full detail, namely derivation the the energy eigenfunctions and energy eigenvalues of the Hamiltonian corresponding to (a) a particle in a box, (b) the harmonic oscillator and (3) the non-relativistic Hydrdogen atom. The last two are based on solving ordinary differential equations using the power series method and for the Hydrogen atom case, the full derivation of the spherical harmonics as eigenfunctions of the angular momentum square operator is presented in full detail. Later on, we show how the spherical harmonics determined all the irreducible reprsentations of the rotation group $SO(3)$. This derivation gives an alternate method to obtain all the irreducible representations of the rotation group in comparison to the method of Wigner and Weyl who constructed the irreducible representations of $SU(2)$ using polynomials in two complex variables. The book contains some major results in group representation theory like the Frobenius- Mackey's theory for constructing irreducible representations of the semidirect product of a normal Abelian group N with another subgroup H. This theory was applied by Wigner to construct the irreducible representations of the Poincare group which is generated by Lorentz transformations and space-time translations. It is also known as the little group method where one starts with a character of the Abelian group N, constructs its orbit under the dual action of H, then constructs the stability subgroup of H for which the chosen character is a fixed point, and then using an irreducible representation of this little group H_0 and the chosen character χ_0, one constructs an irreducible representation of the semidirect product $G_0 = N \otimes_s H_0$ of N and H_0 and by the induced representation theory, the representation of $G = N \times_s H$ is constructed using the Mackey method. Irreduciblity of such induced representations is proved and easier methods of describing such induced representations using functions on the character orbit with values in the vector space on which the irreducible representation of the little group acts are provided. The reader is referred to the excellent book by K.R.Parthasarathy on this subject (KRP:Mathematical Foundations of

Quantum Mechanics, Hindustan Book Agency). After the discussion of non-relativistic quantum mechanics using the Schrodinger method, we describe Feynman's path integral approach to non-relativisti quantum mechanics. Feynman in his Ph.D thesis showed that solution to the Schrodinger equation for the evolving wave function under a Hamiltonian can equivalently be represented as a path integral of $exp(2\pi iS[q]/h)$ over all paths q joining (t_1, q_1) and (t_2, q_2) where $S[q]$ is the action of the path $t \to q(t)$ with the prescribed end points. This action is obtained as the integral $\int_{t_1}^{t_2} L(t, q(t), q'(t))dt$ of the Lagrangian L of the particle or system of particles with the Lagrangian related to the Hamiltonian used in the Schrodinger equation via the Legendre transformation. Thus, Feynman's approach is that along each path q there is a complex amplitude $exp(2\pi iS[q]/h)$ for the particle/system of particles to go from (t_1, q_1) to (t_2, q_2) and the total probability amplitude to go from the initial point to the second in the given time is obtained by summing these amplitudes over all paths. This is an intuitively beautiful picture of presenting quantum mechanics because it displays the quantum interference along different paths and in addition, it clearly shows that in the limit when Planck's constant converges to zero, the rapid oscillations of the phase $2\pi S[q]/h$ for small variations the path q cause cancellations of the probability amplitudes except for paths in the vicinity of the classical path q_* where the variation of the action vanishes, ie where $\delta S[q_*] = 0$. A path q_* satisfing this condition satisfied the classical Euler-Lagrange equation of classical mechanics which is equivalent to Newtonian classical mechanics. The mathematical problems with Feynman's approach have till today defied mathematicians since the constant of proportionality in the FPI is infinite and in addition, the path integral is something like the integral of $exp(-ix^2/2)$ over $\mathbb{R}$ where $i = \sqrt{-1}$ and Feynman's idea is equivalent to replacing this integral by $\sqrt{2\pi i}$. It gives the correct results predicted by Schrodinger's equation but why it gives is not clear since the integral of $exp(-ix^2/2)$ is oscillatory and hence does not converge. The Feynman path integral is the same as Kac formula for $u(t, x) = \mathbb{E}[exp(\int_0^t V(x + B(s))ds)f(x + B(t))]$ where $B(.)$ is Brownian motion but with t replaced by it, ie "complex time". More general forms of Kac formula can be derived when $B(.)$ is an arbitrary diffusion process or even any Markov process with a generator K. These ideas have been discussed in this book. After all these discussions, we come to relativistic quantum mechanics whose wave equation was first correctly proposed by Dirac. Before Dirac, we had the Klein-Gordon equation based on replacing the kinetic energy K by $i\partial/\partial t + eV(t, r)$ and the mometum P by $-i\nabla + eA(t, r)$ in the Einstein energy-momentum relation

$$K^2 - c^2P^2 - m^2c^4 = 0$$

and treating this as an operator acting on a wave function $\psi(t, r)$ set to zero. Earlier, Sommerfeld had solved the corresponding stationary state equation for the Hydrogen atom obtaining thereby relativistic corrections to the $-1/n^2$ spectrum of Schrodinger. However, Dirac observed two major flaws with this equation. One, an evolution equation of quantum mechanics has to be first order in time, since if the evolution operator is $T(t)$, then $T(t+s) = T(t).T(s), t, s \geq 0$ and hence $T(t) = exp(-itH)$ for some selfadjoint operator H. Which is the same as saying that $T'(t) = -iHT(t)$. Of course, we also have $T''(t) = -H^2T(t)$ but that would lead to a superpostion of two solutions $exp(\pm itH)$ and unitarity of the evolution would be broken. One way to rectify this is to consider the wave operator as

$$K - c\sqrt{P^2 + m^2c^2}$$

where $K = i\partial/\partial t + eV(t, r)$ and P replaced by $-i\nabla + eA(t, r)$ but the resulting equation could not be solved for any potential due to ∇^2 occuring inside a square root. The other major problem with such an equation is that it is highly asymmetric in the space and time indices whereas special relativity requires that space and time should be treated on the same footing. Thus, Dirac proposed the factorization of the quadratic form $E^2 - c^2(P_x^2 + P_y^2 + P_z^2) - m^2c^4$ into linear factors by the introducing of non-commuting 4×4 matrices known today as the famous Dirac γ matrices. The Dirac equation was solved by Dirac for the Coulomb potential thus obtaining negative as well as positive energy states, the positive energy states appear in a continuum and Dirac interpreted this as corresponding to a new kind of particle the positron having the same mass but opposite charge as that of an electron. Dirac proved that his relativistic wave equation for the electron is invariant under a representation of the Lorentz group, namely the "Dirac representation" and he also proved that under conjugation followed by application of an appropriate unitary matrix, the resulting wave function also satisfies the Dirac wave equation but with the charge $-e$ replaced by $+e$ and no change in mass. This phenomenon is called charge conjugation and led Dirac to immediately conclude that this charge conjugated wave equation describes the motion of a positron. The discovery of the Positron by Dirac and its subsequent detection in an accelerator by Anderson came as a total surprise to the physics community since it led to the concept of antimatter, ie every elementary particle should have a corresponding antiparticle and that when a particle interacts with its antiparticle then the two may get annihilated to produce just a γ ray photon. The probabilities of such processes and other scattering processes where later on calculated systematically by Feynman in the process of developing quantum electrodynamics. This discovery of antimatter by Dirac must definitely rank as one of the greatest achievements in physics since the time of Newton. The discovery of the Dirac equation also led Dirac to conclude that the spin of the electron is a relativistic effect and he demonstrated that by replacing the momentum operator P by $P + eA$ and the energy E by $E + eV$ where A is the magnetic vector potential and V is the electric potential, and premultiplying the resulting wave equation by its "reversed version" , one obtains the original Klein-Gordon equation of the particle in an electromagnetic field plus terms that describe the interaction of the electric and magnetic field with the spin of the electron. Earlier, Pauli had artificially introduced spin terms into the non-relativistic Schrodinger equation to demonstrate the Zeeman splitting of the energy levels but Dirac's approach led naturally to terms involving the interaction of the electronic spin with the external electromagnetic field. Moreover, for the development of relativistic quantum field theory, ie a quantum field theory that respects Lorentz invariance, it is the Dirac and Maxwell equations that must be used rather than the Schrodinger and Maxwell equations. In quantum field theory, wave functions become operator fields satisfying either commutation or anticommutation relations depending on whether the wave functions describe Bosons or Fermions. The process of replacing wave functions by operator valued fields is called second quantization and from a mathematical viewpoint, this can be achieved by considering multiple tensor products of Hilbert spaces and operators acting on such spaces. If the tensor product is the usual one, the operator fields describe distinguishable particles satisfying the Maxwell-Boltzmann statistics, if the tensor products are symmetrized, then they describe Bosons which are indistinguishable particles having integer spins and that any state can be filled by any number of such particles while if the tensor products are antisymmetrized, then they describe Fermions which are indistinguishable particles having half integral spins and any state can have only either no particle or one particle also known as the Pauli exlcusion principle. In the quantum field theory, for any kind of particle, we can have operators that create, annihilate or conserve the number of particles at any point in space-time or at any value of the four momentum. This leads us to a calculus of such operators and using the Pauli exclusion principle for Fermions, we can show that the Fermionic creation and annihilation operator fields satisfy canonical anticommutation relations while using the analogy of Bosons with the Harmonic oscillator algebra of creation and annihilation operators, we can show that the Bosonic creation and annihilation operator fields satisfy canonical commutation relations. The relation between spin and statistics is also one of the most striking features of quantum mechanics, namely that if we maximize the entropy of a system of particles subject to distinguishability of the particles and no restriction on the number of particles in a state, we get the Maxwell-Boltzmann distribution while if we maximize the entropy subject to indistinguishability and Pauli exclusion, we get the Fermi-Dirac statistics for Fermions and finally if we maximize the entropy subject to indistinguishability no Pauli exclusion, we get the Bose-Einstein statistics. The last of these, was first discovered by the great Indian physicist Satyendranath Bose who derived this statistics and showed that Planck's law of black-body radiation for photons can be obtained from such statistics. Einstein communicated Bose's paper to the most prestigious physics journal at that time "Annalen-der-physik" after translating it into German and adding footnotes explaining the significance of Bose's work at that time when only Planck's law of black-body radiation was avaiable in the quantum theory and further that no rigorous derivation of this law from physical and mathematical principles was available. Our next theme in this book is the development of quantum electrodynamics. This involves first writing down the energy of the electromagnetic field in terms of the electric potential and magnetic vector potential in the frequency domain and then observing that for such a free electromagnetic field, the vector potential satisfies the wave equation which in the spatial frequency domain, is simply a sequence of second order in time ordinary differential equations for harmonic oscillators whose characteristic frequency is simply $c|k|$ where k is the wave vector. The Lagrangian of the field is simply a quadratic functional of $A(t,k)$ and $\partial A(t,k)/\partial t$ and and their complex conjugates and by applying the Euler-Lagrange equations to this Lagrangian, we get the same equation as mentioned above, namely a sequence of harmonic oscillator ode's. We are thus led to conclude that the electromagnetic field is simply a continuous ensemble of harmonic oscillators. Application of the Legendre transformation to this Lagrangian gives us the Hamiltonian density of the field as a sum of position field square and momentum field square where the position field is the magnetic vector potential $A(t,r)$. By postulating the canonical commutation relations between the position and momentum fields in the form

$$[A^{\mu}(t,r),\pi_{\nu}(t,r')] = i\delta^{\mu}_{\nu}\delta^{3}(r-0r')$$

we are led into trouble since the form of the Lagrangian density of the electromagnetic field $F_{\mu\nu}F^{\mu\nu}$ implies that certain constraints on the position and momentum fields like for example since the Lagrangian does not depend on $A^{0}_{,0}$,the corresponding momentum $\pi_0 = 0$ which is inconsistent with the canonical commutation relations. Further, by writing down the Euler-Lagrange equation of motion with a matter-field interaction Lagragian denstiy $J_{\mu}A^{\mu}$, we are lead to a relationship between the spatial momenta $\pi_r, r = 1,2,3$ and and J^{μ} involving only spatial derivatives of π_r. It should be noted that the current field J^{μ} is a matter field described in terms of the Dirac operator wave function:

$$J^{\mu} = -e\psi^{*}(t,r)\gamma^{0}\gamma^{\mu}\psi(t,r)$$

These constraints are therefore either primary ie, arising from the the structure of the Lagrangian or secondary, ie, arising from the Euler-Lagrange equations of motion. These constraints are incompatible with the canonical commutation relations. To rectify this situation, Dirac proposed a new type of bracket expressed in terms of the Poisson/Lie bracket and taking the constraints into account. He showed that this new bracket, known today as the Dirac bracket is conststent with the constraints and also yields the correct equations of motion for the unconstrained observables. Using this he developed the first operator theoretic version of quantum electrodynamics. Feynman then applied his path integral formalism to fields like the electromagnetic four potential and the Dirac four component wave function and showed how one can derive from the path integral formalism, perturbative corrections to the electron and photon propagators arising from interaction between the matter field in the form of Dirac currents and the electromagnetic four potential. He gave a beautiful diagrammatic method for calculating the scattering matrix elements to any order in perturbation theory. The resulting formulae can directly be derived by considering an initial state $|i>$ of photons, electrons and positrons having definite four momenta and spin/polarizations and likewise a final state $|f>$. These initial and final states can be obtained by acting on the vacuum appropriate creation operators, then writing the interaction terms in the total Hamiltonian between photon field and electron current as integrals expanding the time ordered exponential $U(\infty,-i\infty) = T\{exp(-i\int_{-\infty}^{\infty} H(t')dt')\}$ in the interaction picture as a power series in the interaction energy $-\int J^\mu(x)A^\mu(x)d^3x$ where $J^\mu(x) = -e\psi^*(x)\gamma^0\gamma^\mu\psi(x)$, then expressing $\psi(x)$ and $A^\mu(x)$ in terms of creation and annihilation operators of the electron-positron and photon field and then using the standard commutation relations, evaluate the scattering matrix elements $< f|U(\infty,-infty)|i>$ in the interaction picture. This operator formalism was developed by Schwinger-Tomonaga and Dyson while Feynman developed the path integral approach to such calculations with finally Dyson proving the equivalence of the theories of Feynman and Schwigner-Tomonaga. The importance of the propagator computation in quantum field theory has been highlighted in this book. Namely, if $\phi(x)$ is a collection of fields and the action for such a collection of fields is expressed as

$$S[\phi] = S_Q[\phi] + \epsilon S_{NQ}[\phi]$$

where S_Q is a quadratic functional and S_{NQ} is a cubic and higher degree functional, the latter coming from interactions, then the scattering matrix using FPI can be expressed as

$$S[\phi_\infty,\phi_{-\infty}] = \int exp(iS[\phi])D\phi$$

$$= \int ex(iS_Q[\phi])(1 + i\epsilon S_{NQ}[\phi] - \epsilon^2 S_{NQ}[\phi]^2/2 + ...)D\phi$$

It is clear that each term in this infinite series is equivalent to calculating the moments of an infinite dimensional Gaussian distribution with complex variance and we know from basic probability theory, that even higher order moments of a Gaussian vector can be expressed as sums over products of the second order moments. The second order moment appearing here is the propagator of the field:

$$D_\phi[x,y] = \int exp(iS_Q[\phi])\phi(x)\phi(y)D\phi$$

These aspects of quantum electrodynamics have been covered in this book. We give explicit computations of the photon propagator in different gauges (Feynman, Landau and Coulomb gauges) and also for the electron propagator. These expressions are derived using both the methods, first the operator theoretic method and second, using the Feynman path integral for quadratic Lagrangians combined with the standard formulas for the second order moments of a multivariate Gaussian distribution. We also present modern developments in quantum field theory, starting with the standard Yang-Mills group theoretic generalization of the Dirac or Klein-Gordon equations for matter fields interacting with the photon field. In this generalization, we assume that the wave function takes values in $\mathbb{C}^4 \otimes \mathbb{C}^N$ and a subgroup G of the unitary group $U(N)$ acts on this wave function. We introduce a Lie algebra theoretic covariant derivative which is an ordinary four gradient plus a connection field which takes values in the Lie algebra of the group G. This gauge covariant derivative acts on the matter field wave function and if the matter field wave function undergoes a local (ie space-time dependent) transformation $g(x) \in G$, then accordingly the gauge connection field in the covariant derivative has to undergo a certain transformation so that the covariant derivative of the matter field transforms simply by multiplication by $g(x)$. This ensures that the Lagrangian density constructed out of functions of the matter field and its covariant derivative will be invariant under local G-transformations of the matter field and the gauge field provided that the Lagrangian is a G-invariant function of its agruments. Owing to the non-commutativity of the gauge fields, the gauge field tensor defined as the commutator of the covariant derivatives contains an extra quadratic nonlinear term in the gauge potentials apart from its four dimensional curl. This is a characteristic feature of non-Abelian gauge theories in contrast to the case of the electromagnetic field where the gauge group is $U(1)$. In the electromagnetic field case, the matter field is the Dirac field since the gauge group is $U(1)$, the local group transformation element $g(x)$ is simply a modulus one complex number depending on the space-time variables. The corresponding gauge transformation of the gauge field potentials, namely, the electromagnetic four potential simplifies to the standard Lorentz gauge transformation $A_\mu \to A_\mu + \partial_\mu\phi$. Based on the Yang-Mills non-Abelian generalization of electromagnetism, Salam,Weinberg and Glashow were able to unify the electromagnetic forces and the weak forces, calling it the electro-weak theory. Their idea was to start with a non-massive vector gauge field Lagrangian corresponding to the weak forces in the nucleus, coupled to the matter field of Leptons (which include electrons) and then add symmetry breaking terms to this Lagrangian by coupling a scalar field to the Leptons. Symmetry breaking occurs when the scalar field is in its ground state which is a constant and this leads to mass terms involving quadratic non-derivative terms of the electronic field as well as to the gauge Bosonic fields which are components of the vector gauge field. In this context, we give a brief review of symmetry breaking both spontaneous and local versions. Symmetry breaking can occur when Hamiltonian/Lagrangian is invariant under a group G and the ground state is degenerate. If the Hamiltonian is invariant under G, then the subspace of ground states is invariant under G but any given ground state may not be invariant under G. If we take such a ground state and look upon the quantum state as this ground state plus a quantum perturbation, then the resulting Lagrangian will not be invariant under G since the ground state is not, but will be invariant under a subgroup H of G also called a broken subgroup. This sort of symmetry breaking produces massless particles called Goldstone Bosons. We can demonstrate this fact that symmetry breaking produces massless particles much better and in a more generalized framework using groupt theory. To do so, we first express the wave function of the system in terms of a unbroken H part and a broken G/H part. The unbroken part transforms according to H and the broken part can be viewed as a field with values in the coset space G/H. When the Lagrangian is expressed in terms of these components, it turns out that it does not contain any non-derivative quadratic components of the broken part while it contains non-derivative quadratic components of the unbroken part. This demonstrates that the broken part describes massless Goldstone Bosons while the unbroken part of the wave function describes massive particles. We also show that the G-symmetry of a Lagrangian can be broken by adding perturbative terms that are not G-invariant. Sometimes symmetry breaking can lead to massless particles being massive as in the electroweak theory of Salam, Weinberg and Glashow. This happens because of the coupling of the gauge fields to a scalar field. The gauge fields initially are massless but the coupling to the scalar field followed by evaluation of the Lagrangian in the ground state of the scalar field causes terms in the Lagrangian involving quadratic non-derivative terms of the gauge field to be present. These terms cause the gauge fields to become massive. The electroweak theory is an example of such a situation which gives masses to the gauge fields other than the electromagnetic field. All this is quantum field theory from the physical standpoint. We next explain how certain aspects of quantum field theory including the theory of quantum noise can be developed using rigorous mathematics, ie, functional analysis. In particular, we show how certain major stochastic processes in classical probability theory like the Brownian motion and Poisson processes are special cases of quantum stochastic processes, ie, a family of non-commuting observables in a special kind of Hilbert space, namely the Boson Fock space when viewed in specific states. The notion of a quantum probability space as a triplet $(\mathcal{H}, P(\mathcal{H}), \rho)$ where $\mathcal{H}$ is a Hilbert space, $P(\mathcal{H})$ is the lattice of orthogonal projection operators in $\mathcal{H}$ and ρ is a state in $\mathcal{H}$, ie, a positive semidefinite unit trace operator in $\mathcal{H}$ is introduced as compared to a classical probability space $(\Omega, \mathcal{F}, P)$. After doing so, we construct the Boson Fock space which can describe an arbitrary number of Bosons using the symmetric tensor product of a particular Hilbert space. The Boson Fock space is the substratum for constructing basic quantum noise processes, ie, noncommuting family of operators which specialize to Brownian motion and Poisson processes when the state is appropriately chosen. The Boson Fock space or noise Bath space is coupled to the system Hilbert space via a tensor product. Then we follow the marvellous approach of R.L.Hudson and K.R.Parthasarathy of construting the creation, annihilation and conservation operator fields in the Boson Fock space. We show via physical arguments that the creation and annihilation operator fields can also be viewed fromt the standpoint of an infinite sequence of Harmonic oscillators, by constructing the creation and annihilation operator for each oscillator and defining the coherent vectors in terms of a superposion of the energy eigenstates of these oscillators and proving that the coherent states are eigenstates of the annihilation operator field now constructed as superpostions of the annihilation operators for the different oscillators. The creation operator is the adjoint of the annihilation operator and turns out to have the same action on coherent vectors as the complex derivative of the latter with respect to the complex numbers used to construct the coherent vector from the oscillator energy eigenstates. In the work of Hudson and Parthasarathy, coherent vectors in the Boson Fock space were constructed as a weighted direct sum of multiple tensor products of a fixed vector in the Hilbert space with itself and the creation, conservation and annihilation operators were defined using the generators of one parameter unitary groups derived from the Weyl operator by restricting to translation and unitary rotation. The Weyl operator in the Hudson-Parthasrathy (HP) theory is itself described by its action on coherent vectors. Coherent vectors without normalization in the HP theory were called exponential vectors. This approach to the construction of the basic noise fields is highly mathematical and we provide in this book some physical insight into this correspondence by making an isomorphism between the coherent vectors of the HP theory and coherent vectors constructed using eignstates of harmonic oscillators. Further, in our book, we construct the unitary rotation operators needed for constructing the conservation process by resorting to quadratic forms of the creation and annihilation operators of the harmonic oscillators. The theory of quantum stochastic calculus developed by Hudson and Parthasarathy introduces time dependent creation, annihilation and conservation processes which statisfy a quantum Ito formula for products of time differentials of these processes. The classical Ito formula for Brownian motion and the Poisson process is shown to arise as a special commutative version of the quantum Ito formula of HP. The HP theory says more, namely that in the general noncommutative case, the products of creation and annihilation operator differentials with conservation differentials need not be zero. This unlike the classical probability case where if B is Brownian motion and N is Poisson process, then $dB.dN = 0$. Our analogy of the HP theory with the quantum harmonic oscillator theory provides us with a direct route to quantum optics as described in the celebrated book by Mandel and Wolf on optical coherence and quantum optics. The idea is to first set up the famous Glauber-Sudarshan non-orthogonal resolution of the identity operator in the Boson Fock space as a complex integral of the coherent states $|e(u)><e(u)|$, then to express the Hamiltonian of the system interacting with a photon bath as the sum of the system Hamiltonian which consists of spin matrices interacting with a constant classical magnetic field, the bath photon field Hamiltonian expressed as a quadratic form in the creation and annihilation operators and an interaction Hamiltonian expressed as a time varying linear combination of the products(tensor) between the atomic spin observables and the bath annihilation and creation variables, then assume that the density of $\rho(t)$ of the system and bath can be expanded as a Glauber-Sudarshan integral:

$$\rho(t) = \int \rho_A(t,u) \otimes |e(u)><e(u)|du$$

where $\rho_A(t,u)$ is a finite dimensional matrix (of the same order as the spin observables of the atomic system). Finally, we substitute this expansion into the quantum equation of motion

$$i\rho'(t) = [H_A + H_F + H_{AF}(t), \rho(t)]$$

where H_A is the atomic Hamiltonian, H_F is the bath field Hamiltonian and $H_{AF}(t)$ is the atomic-bath field interaction Hamiltonian. Using properties of creation and annihilation operators acting on the exponential/coherent vectors and integration by parts, we then derive a partial differential equation for $\rho_A(t,u)$ which may be called the fundamental equation of quantum optics. After these discussions, we proceed to one of the most modern techniques in time dependent quantum measurement theory. The time dependent HP theory, ie quantum stochastic calculus also has a nice physical interpretation. When the average values of the creation, and annihilation processes in a coherent state are calculated, we get time integrals of the product of the coherent state defining vector with the creation/annihilation process defining vector upto time t. This result can be interpreted physically by saying that the annihilation (creation) process acts on a coherent state and yields the total amplitude and phase of photons annihilated (created) upto time t taking into account the relative polarization of the photons in the coherent state with respect to that of the annihilation (creation) process defining vector. On the other hand, the conservation process average in a coherent state gives a time integral of a quadratic product of the coherent state defining vector upto time t which has the interpretation as being the number of photons in the state present upto time t. The conservation process in the HP calculus has the Poisson process interpretation in classical probability theory just as the creation and annihilation process are linked to Brownian motion. The whole subject of quantum probability and quantum stochastic processes can be viewed as an linear-algebra-functional analytic approach to probability theory with generalizations to the non-commutative case. As an important application of the HP stochastic calculus, we present the celebrated work of V.P.Belavkin on quantum filtering. To do so, we first note that in quantum mechanics two or more observables may not be simultaneously measurable when they do not commute but when they commute, they can be simultaneously diagonalized and hence simultaneously measured. Measurement on a quantum system causes the state of the system to collapse to a state dictated by the outcome of the measurement or of the measurement is made by a set of projection valued operators (pvm) or more generallyn by a set of positive operators (povm), then the state of the system collapses to a state formed by superposing the collapsed states corresponding to each measurement outcome. Belavkin proposed a scheme of constructing a filtration on an Abelian Von-Neumann algebra of observables that satisfies the non-demolition property, ie, the algebra generated by the elements of the filtration upto time t is Abelian and also commutes with the states of the HP noisy Schrodinger equation at times $s \geq t$. Such measurements, he called non-demolition measurements. The HP noisy Schrodinger equation determines a unitary evolution in the joint system and bath space $\mathfrak{h} \otimes \Gamma_s(L^2(\mathbb{R}_+) \otimes \mathbb{C}^d)$. The dynamics of the unitary evolution is dictated by the system Hamiltonian, the fundamental creation, annihilation and conservation processes of the HP quantum stochastic calculus which are operators in the bath space and connect to the system dynamics via system operators. There is also a quantum Ito correction term to the system Hamiltonian in the form of an additive skew Hermitian operator that ensures unitarity of the evolution. It is known that if one computes the Heisenberg dynamics of a system observable using these unitary evolution operators, then the standard Heisenberg equations of motion are obtained along with noise correction terms which and the system observables after finite time t evolves to an observable in the tensor product of the system Hilbert space and the noise bath space. Specifically, if $U(t)$ denotes the evolution operator of the HP noisy Schrodinger equation and X is a system observable, then after time t it evolves to $j_t(X) = U(t)^* XU(t)$ which satisfies the noisy Heisenberg equations of motio. Belavkin proposed that if we take an input noise process $Y_{in}(t)$ which is a commuting family of operators in the bath space and define the output noise process $Y_{out}(t) = U(t)^* Y_{in}(t)U(t)$, then by virtue of the fact that the unitarity of $U(t)$ depends only on system operators which commute with bath operators, it follows that $Y_{out}(t) = U(T)^* Y_{in}(t)U(T)$ for all $T \geq t$ from which it is easy to see that $Y_{out}(t)$ commutes with $Y_{out}(s)$ for all $t, s \geq 0$ and further that $Y_{out}(t)$ commutes with $j_s(X)$ for all $s \geq t$. This commutativity enables us to jointly measure $j_t(X), Y_{out}(s), s \leq t$ and hence define the conditional expectation $\pi_t(X) = \mathbb{E}(j_t(X)|Y_{out}(s), s \leq t)$. The family $\pi_t(X), t \geq 0$ of operators forms an Abelian family since $\pi_t(X)$ is a function of $Y_{out}(s), s \leq t$ and the latter form a commuting family of operators. Belavkin derived a stochstic differential equation for $\pi_t(X)$ driven by the process $Y_{out}(.)$ and its derivation was greatly simplified using quantum versions of the Kallianpur-Striebel formula and other methods based on orthogonality properties of the conditional expectation estimate by John Gough and Kostler. The resulting equation for $\pi_t(X)$ is the fundamental quantum filtering equation and is the non-commutative generalization of Kushner's equation for classical non-linear filtering. These aspects have been discussed in this book. It should be noted that although this version of the filtering equations takes place in the observable domain, it could be directly transformed to the density domain. This is analogous to the situation of classical filtering where we can describe the evolution of the conditional moments of the state or more generally of the conditional expectation of any function of the state at time t given measurements upto time t or equivalently of the conditional probability density of the state at time t given measurements upto time t. To obtain the Belavkin stochastic differential equation for the conditoinal density of the state at time t given output measurements upto time t, we must simply note that we can write $\pi_t(X) = Tr(\rho_t X)$ where ρ_t can be viewed as a density matrix in the system Hilbert space that is a function of the output measurement process upto time t, or equivalently, simply as a classical random process with values in the space of system space density operators. Classical random because the measurement operators commute. We substitute $\pi_t(X) = Tr(\rho_t X)$ into the Belavkin observable version of the filtering equation and using the arbitrariness of X, we derive a classical stochastic differential equation for the system state valued classical random process ρ_t driven by $Y_{out}(t)$. Such equations are called "Stochastic Schrodinger equations" (Luc Bouten, Ph.D thesis on quantum optics filtering and control). We present a generalization of Belavkin's work by first constructing a family of p commuting inputmeasurement processes which are expressible as linear combinations of the creation, annihilation and conservation processes. Such processes have independent increments in coherent states and the quantum Ito's formula leads in general to the fact that any integer power of the differentials of such processes is non zero just as in the classical case $(dN)^k = dN, k = 1, 2, ...$ where $N(.)$ is a Poisson process. The Belavkin filter is constructed by assuming it to have the form

$$d\pi_t(X) = F_t(X)dt + \sum_{k\geq 1, m=1,2,...,p} G_{mkt}(X)(dY_{mout}(t))^k$$

with $F_t(X) and G_{mkt}(X)$ being measurable with respect to the algebra generated by the output measurments upto time t. The coefficients $F_t(X), G_{mkt}(X)$ are determined by applying the quantum Ito formula to the orthogonality equation

$$\mathbb{E}((j_t(X) - \pi_t(X))C_t) = 0$$

where

$$dC_t = \sum_{m,k} f_{mk}(t)C_t(dY_{mout}(t))^k, t \geq 0, C_0 = 1$$

and using arbitrariness of the complex valued functions $f_{mk}(t)$. It should be noted that in the absence of any measurement, ie, when the system evolves according to the HP noisy Schrodinger equation, we can take a system observable X, obtain its noisy evolution $j_t(X)$ and then choose a pure states of the form $|f_k \otimes \phi(u) >, k = 1, 2$ of the system and the bath where $|f_k >$ is a system state and $|\phi(u) >$ is a bath coherent state, then compute the matrix elements $< f_1 \otimes \phi(u)|j_t(X)|f_2 \otimes \phi(u) >$ and write down the evolution of this quantity. More generally, we can compute the system state at time t defined via the duality equation

$$Tr(\rho_s(t)X) = Tr((\rho_s(0) \otimes |\phi(u) < \phi(u)|)j_t(X))$$

or equivalently, as

$$\rho_s(t) = Tr_B(U(t)(\rho_s(0) \otimes |\phi(u) >< \phi(u)|)U(t)^*)$$

or equivalently, as

$$< f_1|\rho_s(t)|f_2 >= Tr(\rho_s(t)|f_2 >< f_1|) =$$
$$Tr((\rho_s(0) \otimes |\phi(u) >< \phi(u)|)j_t(|f_2 >< f_1|))$$

where $|f_k >, k = 1, 2$ are system states. The resulting differential equation for the system state $\rho_s(t)$ is the generalized GKSL equation (Gorini, Kossakowski, Sudarshan, Lindblad). If $u = 0$, ie, the bath is in vaccuum, then the ordinary GKSL equation is obtained. Using the dual of the GKSL equation, we get a description of the evolution of system observables when corrupted by bath noise in such a way that the system observable always remains a system observable, ie, averaging out over the bath noise variables is being performed at each time. The dual GKSL can be used to describe non-Hamiltonian quantum dissipative systems, for example, damped quantum harmonic oscillators or lossy quantum transmission lines. We then introduce the notion of quantum control of the Belavkin stochastic Schrodinger equation in the sense of Luc Bouten. This involves taking non-demolition Belavkin measurements on a quantum system evolving according to the HP noisy Schrodinger equation, then considering at each time point t, the Belavkin filtered and controlled state $\rho_c(t)$, applying the Belavkin filter evolution by making a non-demolition measurement $dY(t)$ in time $[t, t+dt]$ to obtain the Belavkin filtered state at time $t+dt$ as

$$\rho(t+dt) = \rho_c(t) + F_t(\rho_c(t))dt + G_t(\rho_c(t))dY(t)$$

and then applying a control unitary $U^c_{t,t+dt} = exp(iZdY(t))$ in the time interval $[t, t+dt]$ to the Belavkin filter output $\rho(t+dt)$ to get the controlled state at time $t+dt$ as

$$\rho_c(t+dt) = U^c_{t,t+dt}\rho(t+dt)U^{c*}_{t,t+dt}$$

Here, Z is a system observable that commutes with $dY(t)$. More precisely, Z should be replaced by $U(t+dt)^*ZU(t+dt)$ to make it commute with $dY(t)$. We can then show by application of the quantum Ito formula that the system operator Z can be selected so that the evolution from $\rho_c(t)$ to $\rho_c(t+dt)$ is such that a major portion of the noise in the Belavkin filter is removed and the evolution of $\rho_c(t)$ nearly follows that of the noiseless Belavkin equation, ie, the GKSL equation. The next topic discussed in this book is that of designing quantum gates using scattering theory experiments. The basic idea is to realize a unitary quantum gate using the quantum scattering matrix. The system consists of a free projectile having Hamiltonian $H_0 = P^2/2m$ and the interaction potential energy between the projectile and the scattering centre is $V(Q)$ so that the total Hamiltonian of the projectile interacting with the scattering centre is $H = H_0 + V$. The projectile arrives from the infinite past $(t \to -\infty)$ from the input free particle state $|\phi_{in}>$. This free state evolves according to H_0, ie, at time t this free particle state is $exp(-itH_0)|\phi_{in}>$. After interacting with the scatterer, it goes to the input scattered state $|\psi_{in}>$ which evolves according to H, ie, at time t, this input scattered state is $exp(-itH)|\psi_{in}>$. It follows that these two states coincide in the remote past, ie,

$$lim_{t\to-\infty}(exp(-itH)|\psi_{in}> - exp(-itH_0)|\phi_{in}>) = 0$$

from which, we easily deduce that

$$|\psi_{in}>= \Omega_-|\phi_{in}>$$

where

$$\Omega_- = slimexp(itH).exp(-itH_0)$$

After interacting with the scatterer for a sufficient long time, we ask the question, what is the probability amplitude of the projectile being in the free particle state $|\phi_{out}>$?. To evaluate this, we must first determine the scattered state, ie, the output scattered state $|\psi_{out}>$ that develops as $t \to \infty$ to $|\phi_{out}>$. It is clear that the condition required for this is that

$$lim_{t\to\infty}(exp(-itH)|\psi_{out}> - exp(-itH_0)|\phi_{out}>) = 0$$

or equivalently,

$$|\psi_{out}>= \Omega_+|\phi_{out}>$$

where

$$\Omega_+ = slim_{t\to\infty}exp(itH).exp(-itH_0)$$

The domains of Ω_- and Ω_+ will not in general be the entire Hilbert space $L^2(\mathbb{R}^3)$. In fact, determining the domains of definition of $\Omega_\pm$ is a hard problem in operator theory and nice treatments of this delicate problem have been given in the masterful monographs of T.Kato (Perturbation theory for linear operators) and W.O.Amrein (Hilbert space methods in quantum mechanics). The scattering matrix S is an operator that maps free input particle states to free output particle states so that the scattering amplitude for the process $|\phi_{in}>\to|\phi_{out}>$ is given by

$$<\phi_{out}|S|\phi_{in}>=<\psi_{out}|\psi_{in}>=$$

$$<\phi_{out}|\Omega_+^*\Omega_-|\phi_{in}>$$

or equivalently,

$$S = \Omega_+^*\Omega_-$$

We define

$$R = S - I$$

and derive a formula for the operator R in the from

$$<\lambda,\omega'|R|\lambda,\omega>=<\lambda,\omega'|V - V(H-\lambda)^{-1}V|\lambda,\omega>$$

where λ is the energy of the total system or equivalently of the free projectile coming from $t = -\infty$. The energy is conserved during the scattering process. The state $|\lambda, \omega >$ represents the incident free projectile having energy $P^2/2m = \lambda$ and with incoming momentum P being directed along the direction $\omega \in S^2$ and likewise $|\lambda, \omega' >$ represents the free projectile state having energy λ and outgoing momentum P being directed along $\omega' \in S^2$. It should be noted that the scattering operator S conserves the energy since it commutes with H_0. This follows from the relations

$$\Omega_- exp(-itH_0) = exp(itH)\Omega_-,$$

$$\Omega_+ exp(-itH_0) = exp(itH)\Omega_+$$

for all $t \in \mathbb{R}$, from which follows

$$exp(itH_0)\Omega_+^*\Omega_- exp(-itH_0) = \Omega_+^*\Omega_-$$

which gives formally, on differentiating w.r.t t at $t = 0$,

$$[H_0, S] = 0$$

The design of the quantum gate is based on choosing a potential V so that for a given energy λ, the matrix $((<\lambda, \omega'|R|\lambda, \omega >))_{\omega,\omega'}$ is as close as possilble to $U_g - I$ where U_g is a given $N \times N$ unitary matrix and N is the number of discretized directions ω chosen on the unit sphere. Another topic of importance discussed in this book is a rough idea about how a unified quantum field theory can be developed encompassing gravitation, electromagnetism, the electron-positron field of matter and if possible other Yang-Mills fields. The theory developed relies heavily on identifying the correct connection for the covariant derivative of spinor fields in the presence of a curved space-time metric. The construction of this gravitational connection for Dirac spinors can be achieved in terms of the tetrad components of the gravitational metric and the Lorentz generators in the Dirac spinor representation. This construction can be found in Weinberg's book (Gravitation and Cosmology: Principles and Applications of the General Theory of Relativity). This connection can also be used for Yang Mills field. Once this has been done, it is an easy matter to construct the Lagrangian density of the Dirac or Yang-Mills field in curved space-time. This Lagrangian density is added to the Einstein-Hilbert Lagrangian density of the gravitational field and to the Lagrangian density of the electromagentic and Yang-Mills gauge fields, and to the Lagrangian density of scalar Higgs field taking once again into account the gravitational connection using the Tetrad. This total Lagrangian density is a function of the Dirac wave function, the Yang-Mills wave function, the tetrad components of the gravitational metric and the gauge fields, namely the electromagnetic four potential, ie, the $U(1)$ gauge fields for the Dirac equation and the non-Abelian $SU(N)$ gauge fields for the Yang-Mills equations. The corresponding four dimensional action integral is then constructed and Feynman diagrammatic rules are formulated using the path integral approach for determining the probability amplitude for scattering/absorption-emission processes involving gravitons, photons, electrons, positrons, and particles associated with the Yang-Mills fields. The basic idea for calculating these amplitudes is to use Feynman's trick: Identify the parts S_q of the action S that are quadratic in the fields retain them in the exponent $exp(iS)$. The cubic and higher degree terms of the fields appearing in S denoted by S_c are considered as perturbations thus writing down

$$exp(iS) = exp(iS_q)(1 + iS_c - S_c^2/2 - iS_c^3/6 + ...)$$

and the path integral is evaluated using the basic Wick theorem which roughly states that higher moments of a Gaussian distribution can be decomposed into a sum over products of second order moments, ie, propagators. Other schemes of quantization of all the fields can be found in the textbook by Thiemann (Modern canonical quantum general relativity). Finally, we discuss the modern theory of Supersymmetry which is one of the mathematically successful attempts to unify the various field theories like electromagnetism, Dirac's relativistic quantum mechanics, the scalar field theories of Klein-Gordon and Higgs, the Yang-Mills gauge theories and even general relativity. The idea here is to introduce four anticommuting Majorana Fermionic variables θ and to define a superfield $S[x, \theta]$ as a function of both the bosonic space-time coordinates $x = (x^\mu)$ and the four Fermionic variables $\theta = (\theta^a)$. When the superfield is expanded in powers of θ, all terms involving five or more $\theta's$ vanish owing to the anticommutativity of the four $\theta's$. Hence, the superfield S is a fourth degree polynomial in θ and the coefficients of $\theta^a, \theta^a\theta^b, \theta^a\theta^b\theta^c$, $\theta^0\theta^1\theta^2\theta^3, 0 \le a < b < c \le 4$ are functions of x only. Then following Salam and Stratadhee, we introduce supersymmetric generators $L_a, \bar{L}_a a = 0, 1, 2, 3$ that are vector field vector field in super-space (x, θ). These generators satisfy the standard anticommutation relations required for supersymmetry generators, ie, their anticommutators are bosonic generators:

$$\{L_a, \bar{L}_b\} = \gamma^\mu_{ab}\partial/\partial x^\mu$$

The general form of these generators can be obtained using standard group theoretic arguments: Define the composition of superspace variables as

$$(x^\mu, \theta^a).(x^{\mu'}, \theta^{a'}) = (x^\mu + x\mu' + \theta^T\Gamma^\mu\theta, \theta^a + \theta^{a'})$$

Then, the supersymmetry generators (which are supervector fields) are calculated by expressing

$$\frac{\partial}{\partial x^{\mu'}} f(x,\theta).(x',\theta'))$$

and

$$\frac{\partial}{\partial \theta^{a'}} f((x,\theta).(x',\theta'))$$

evaluated at $x' = 0, \theta' = 0$ in terms of $\frac{\partial}{\partial x^{\mu}}$ and $\frac{\partial}{\partial \theta^{a}}$. The supersymmetry generators when acting on a superfield induce transformations on the component superfield and it is noted that only the coefficient of the fourth degree term in θ suffers a change that is a total differential. In fact, if the coefficient of the fourth degree in θ is split into a sum of two components, one we call the D component and the other we write as a constant times the D'Alembertian acting on the scalar superfield component, then the D component suffers a change that is a perfect divergence. Hence, the four dimensional space-time integral of the D component remains invariant under a supersymmetry transformation and hence this integral can be used as a candidate action for a supersymmetric field theory. It is also noted that if we impose conditions that one part of the θ^3 component, namely the λ component (called the gaugino) and the D component is zero, and further if the gauge component $V_\mu(x)$ which appear as coefficients of the θ^2 portions is a perfect gauge, ie, a perfect divergence, then after a supersymmetry transformation again λ and D vanish. Hence we obtain a subclass of the class of all superfields, namely the Chiral superfields which can be expressed as the sum of a left Chiral and a right Chiral superfield. Chiral fields are characterized by the property that their λ and D components vanish and the V_μ component is a perfect gauge. We further note following the exposition of Wienberg that a left Chiral superfield can be expressed as a function of x^μ_+ and θ_L only and likewise, a right Chiral superfield can be expressed as a function of x^μ_- and θ_R where $x^\mu_\pm = x^\mu \pm \theta_L^T \epsilon \gamma^\mu \theta_R$ and $\theta_L = (1+\gamma_5)\theta/2, \theta_R = (1-\gamma_5)\theta/2$ are respectively the left chiral and right chiral projections of the Fermionic parameter θ. There are just two linearly independent left chiral Fermionic parameters and likewise two linearly independent right chiral Fermionic parameters. This means that cubic and higher degree terms in θ_L vanish and likewise cubic and higher terms in θ_R vanish. Further, left chiral superfields are characterized by the property that certain "right" superdifferential operators D_R acting on these superfields give zero and right chiral superfields are characterized by the property that certain "left" superdifferential operators D_L acting on these superfields give zero. Here D_R and D_L are defined as respectively the right chiral and left chiral projections (mutliplication by $(1+\gamma_5)/2$ and $(1-\gamma_5)/2$) of a super vector field D obtained by changing a sign in the expression of the supersymmetry generators. The proofs of these facts follow by showing first that $D_R x^\mu_+ = 0, D_L x^\mu_- = 0$ and then noting that $D_R(anyleftsuperfield) = 0, D_L(anyrightsuperfield) = 0$ since left superfields are functions of x_+, θ_L only and right superfields are functions of x_-, θ_R only. Further it is readily shown that the coefficient of θ_L^2 (called the F-term) in a left chiral superfield $\Phi_L(x,\theta)$ suffers change by a perfect divergence under a supersymmetry transformation and hence its space-time integral is a candiadate supersymmetric action. After this, we come to the crucial point: A kinematic Lagrangian density should canonically by a quadratic function of component superfields just as kinetic energies are quadratic functions of momenta/velocities and potential energies of harmonic oscillators are quadratic functions of positions, or equivalently, the Klein-Gordon Lagrangian density is a quadratic function of the space-time derivatives of the field minus a mass term which is a quadratic function of the field itself. Now, if we start with a supefield $S[x,\theta]$ and consider the D-component of the superfield $S[x,\theta]^* S[x,\theta]$, then we get terms that are bilienar in the D and C terms but no term that is quadratic or higher in the D term. Likewise, we get terms that are bilinear in the λ and ω terms (λ is a cubic component and ω is a first degree component) but no terms that are quadratic or higher in the λ term. Now if we build our action from these terms, then standard Gaussian path integral considerations show that we must evaluate our path integrals by setting the variational derivatives of the action with respect to the various component fields to zero. But the variational derivative of the above mentioned action w.r.t. the D term of S^*S is the C term and likewise, the variational derivative of S^*S w.r.t the λ term is the ω terms. So we are led to the disastrous consequence that $C = 0$ and $\omega = 0$, ie, we cannot have scalar fields like the Klein-Gordon field or Fermionic fields like the Dirac field. To rectify this problem, we assume that $D = 0, \lambda = 0$ in S while constructing the action $[S^*S]_D$. In other, words, to construct matter field Lagrangians, we take our basic matter superfield S as a Chiral field. For a given superfield $S[x,\theta]$, the component superfields $C(x)$ (the scalar field) (zeroth power of θ coefficient), the Ferminoic field $\omega(x)$ (first power of θ coefficient) and the other fields $M(x), N(x)$ (which are coefficients of the second power of θ) constitute the matter fields and the other components $V_\mu(x)$ (one set of components of the second power of θ) which is also called the gauge field, $\lambda(x)$ (one part of the Fermion field appearing as a coefficient of the third power of θ) also called the gaugino field (and interpreted as the superpartner of the gauge field) and the auxiliary field $D(x)$ appearing as a coefficient of the fourth power of θ constitute the gauge part of the superfield. To get a gauge invariant theory of supefields, first we must form out of a left Chiral superfield $\Phi[x,\theta]$ (built out of the matter fields C, ω, M, N) and a matrix $\Gamma[x,\theta]$ built out of the gauge superfields V_μ, λ, D the D component $[\Phi^*\Gamma\Phi]_D$ of $\Phi^*\Gamma\Phi$ (which will of course be supersymmetry invariant since it is the D component of a superfield) and the transformation law of the matter part Φ and the gauge part Γ under a gauge transformation is defined in such a way so that $[\Phi^*\Gamma\Phi]_D$ remains invariant. This forms the part of the total Lagrangian density that describes matter like the scalar field, the Dirac Fermionic field of electrons and positrons etc., and the interaction of the matter fields with the gauge fields. Finally, another superfield $W[x,\theta]$ is constructed out of the gauge and auxiliary components V_μ, λ, D such that $W[x,\theta]$ is a left Chiral field (and hence its F-component is supersymmetry invariant) and its F component has the form

$$c_1 F_{\mu\nu}F^{\mu\nu} + c_1\bar{\lambda}^T\gamma^\mu D_\mu\lambda + c_3 D^2$$

which guarantees gauge invariance of the F-part of W. Here, $iF_{\mu\nu}[=\partial_\mu + iV_\mu, \partial_\nu + iV_\nu]$ where in an Abelian gauge theory, V_μ is simply a function of x so that $F_{\mu\nu} = V_{\nu,\mu} - V_{\mu,\nu}$ is the electromagnetic field of photons and $\bar{\lambda}^T D_\mu\lambda$ is the sum of $\bar{\lambda}^T\gamma^\mu\lambda_{,\mu}$ which is like a kinematic part of the Dirac Lagrangian density and $\bar{\lambda}^T i\gamma^\mu\lambda V_\mu$ which is like the interaction Lagrangian between the Dirac field of Fermions and the photon electromagnetic four potential field V_μ. In a non-Abelian gauge theory, V_μ is a Lie algebra valued function of x and then $[V_\mu, V_\nu] \neq 0$. It follows then that the structure constants of this Lie algebra will enter into the picture while defining and $F_{\mu\nu}, D_\mu\lambda$. In this case, the above gauge Lagrangian density will describe particles arising in the Yang-Mills theories like Electroweak theories etc. When the matter and matter-gauge interaction Lagrangian $[\Phi^*\Gamma\Phi]_D$ is added to the above gauge Lagrangian, we obtain a Lagrangian density that is both supersymmetry and gauge invariant. In such a theory, if path integrals are evaluated with respect to certain auxiliary fields, we obtain equations like the Dirac relativistic wave equation with mass dependent on the scalar field, the Maxwell photon equations, the Yang-Mills equations and even gravitational fields can be included by more additions to the superfield. In short, supersymmetry provides the ideal ground for unifying all the known theories into a single framework, studying interactions between the particles associated with each theory and finally quantizing such a unified field theory using the Feynman path integral. We do not give all the details here for they can be found in the masterpiece of Wienberg (Supersymmetry, Cambridge University Press).

Table of Contents

Chapter 1
Wave and Matrix Mechanics of Schrodinger, Heisenberg and Dirac

[1] The De-Broglie Duality of particle and wave properties of matter. A plane wave in one dimension is expressed by the following complex amplitude:

$$\psi(t,x) = A.exp(i(kx - \omega t))$$

where $k = 2\pi/\lambda$ with λ as the wavelength and $\nu = \omega/2\pi$ is the frequency. According to De-Broglie, we can associate a particle having momentum $p = h/\lambda = hk/2\pi$ with such a wave where h is Planck,s constant. According to Planck, we can associate a quantum of energy $E = h\nu = h\omega/2\pi$ with such a wave. Thus, we have

$$(ih/2\pi)\frac{\partial\psi}{\partial t} = (h\omega/2\pi)\psi, = E\psi$$

$$(-ih\nabla/2\pi)\psi = (hk/2\pi)\psi = p\psi$$

In the presence of an external potential $V(x)$, E should be taken as $p^2/2m + V(x)$ and hence,

$$E\psi = (p^2/2m + V)\psi = (-h^2\nabla^2/8\pi^2 m + V(x))\psi$$

Although the above plane wave ψ cannot satisfy this equation for general $V(x)$, we assume that the actual De-Broglie matter wave ψ associated with the particle is a solution to the above equation.. This is called the one dimensional Schrodinger equation. In short, by putting together De-Broglie's matter-wave duality principle and Planck's quantum hypothesis, we have heuristically derived the Schrodinger wave equation.

[2] Bohr's correspondence principle.

Bohr's correspondence principle is that associated with each observable in classical mechanics is an observable in quantum mechanics that follows certain quantization axioms which can be used to explain the energy spectra of atoms and quantum oscillators. For example, consider an electron of charge $-e$ and mass m moving in a circular orbit of radius r around the nucleus. Its mometum is $p = mv$ and the correspondence principle implies that $\int_\Gamma pdq = nh, n \in \mathbb{Z}$ where q is the position coordinate and the lhs is the line integral around a complete orbit of the electron. Thus, we derive

$$2\pi pr = nh, p = nh/2\pi r, n \in \mathbb{Z}$$

Thi is the same as

$$mvr = nh/2\pi$$

According to the centripetal force law that keeps a particle in an orbit,

$$mv^2/r = KZe^2/r^2, K = 1/4\pi\epsilon_0$$

Thus, we get

$$n^2h^2/4\pi^2 mr^3 = KZe^2/r^2$$

or

$$r = n^2h^2/4\pi^2 mKZe^2$$

and

$$v^2 = KZe^2/mr = 4\pi^2K^2Z^2e^4/n^2h^2$$

and finally, we get the energy spectrum of the particle:

$$E == E_n = p^2/2m - KZe^2/r = mv^2/2 - Ze^2/r = KZe^2/2r - KZe^2/r = -KZe^2/2r$$

$$= -2\pi^2mK^2Z^2e^4/n^2h^2, n = 1, 2, ...$$

This spectrum first derived by Bohr, agreed with experiments.

[3] Bohr-Sommerfeld's quantization rules.

If q is a canonical position coordinate vector and p the canonical momentum vector, then for cyclic motion, the Bohr-Sommerfeld's quantization rules state that

$$\int_\Gamma p.dq = nh, n \in \mathbb{Z}$$

where Γ is a closed loop. A special case of this rule was applied earlier by Bohr to derive the spectrum of the Hydrogen atom. For example, if the system is described by action-angle variables I, θ, then we get

$$\int_0^{2\pi} I(\theta)d\theta = nh, n \in \mathbb{Z}$$

Sommerfeld applied this to relativistic quantum mechanics according to which the Hamiltonian of the particle is given by

$$H(q,p) = c\sqrt{p^2 + m^2c^2} - eV(q)$$

Writing $H = E$ and solving for p, we get

$$p = \pm[(E + eV(q))^2 + c^2p^2 + m^2c^4]^{1/2} = \pm p(q,E)$$

If p given by this expression vanishes at $q = q_1, q_2$, then the Bohr-Sommerfeld quantization rule states that the energy spectrum of this relativistic particle is given by

$$2\int_{q_1}^{q_2} p(q,E)dq = nh, n \in \mathbb{Z}$$

Sommerfeld used this to derive the relativistic spectrum of the Hydrogen atom which improves upon the non-relativistic spectrum given by Bohr.

[4] The principle of superposition of wave functions and its application to the Young double slit diffraction experiment.

In quantum mechanics, a pure state in the position representation is described by a complex valued wave function $\psi(x), x \in \mathbb{R}^3$. Given two such wave functions $\psi_1(x), \psi_2(x)$, we can construct another wave function $\psi(x) = c_1\psi_1(x) + c_2\psi_2(x)$, where c_1, c_2 are two complex numbers. $|\psi_k(x)|^2$ is proportional to the intensity of the wave ψ_k at $x, k = 1, 2$ and $|\psi(x)|^2 = |c_1|^2|\psi_1(x)|^2 + |c_2|^2|\psi_2(x)|^2 + 2Re(\bar{c}_1c_2\bar{\psi}_1(x)\psi_2(x))$ is proportional to the intensity of the wave ψ at x. In particular, we can take $c_1 = c_2 = 1$, then the intensity of ψ is the intensity of ψ_1 plus the intensity of ψ_2 plus an interference term $2Re(\bar{\psi}_1(x)\psi_2(x))$. This last term is a purely quantum mechanical effect. This fact has been marvellously illustrated by Feynman using the Young double slit experiment: We may two slits in a cardboard sheet and place an electron gun behind this sheet. On the other side of the sheet is a screen than can record the impact of electrons. If the first slit is open and the second closed, then the electron intensity pattern on the screen shows a maximum value at the portion of the screen directly in front of the first slit and decays down as the distance from the first slit increases on both the sides. Likewise, if slit one is closed and slit two is open, then the electron intensity on the screen in front of slit two shows a maximum. We denote the former intensity pattern by $I_1(x)$ and the latter intensity pattern by $I_2(x)$. Now, when both the slits are open, if the electron were to behave as classical particles, we should expect the intensity pattern on the screen to be $I_1(x) + I_2(x)$, ie, maxima at both the points on the screen in front of the first and second slit respectively. But this is not what we observe. Instead we observe a maximum on the screen at a point in front of the middle between the two slits, ie at a point on the screen that is equidistant from the two slits. Further, as we move away from this intensity maximum point, the intensity shows a sinusoidal variation. This can be explained only by the presence of the above interference term. For example, if $\psi_1(x) = A_1.exp(ik_1x), \psi_2(x) = A_2.exp(ik_2x)$ where x is the distance on the screen from the central point P that is equidistant from both the slits, then the intensity pattern on the screen when both the slits are open is given by

$$I(x) = |\psi_1(x) + \psi_2(x)|^2 = |A_1|^2 + |A_2|^2 + 2|A_1||A_2|cos((k_1 - k_2)x + \phi)$$

where

$$\phi = arg(A_1) - arg(A_2)$$

It follows that if $\phi = 0$ (which will be the case when the two slits are equidistant from the electron gun, then $I(x)$ will show a maximum at $x = 0$ and its spatial variation will be sinusoidal with the distance between two successive maxima or between two successive minima being given by $2\pi/|k_1 - k_2|$. Thus we are forced to conclude that electrons exhibit wave behaviour at some time which is completely in accord with the De-Broglie matter-wave duality principle. The Heisenberg uncertainty principle can also be explained using this setup. The difference between the two electron momenta parallel to the screen is $\Delta p = h|k_1 - k_2|/2\pi$ according to De-Broglie and the position measurement uncertainty is the distance between an intensity maximum and an intensity miniumum on the screen, ie, $\Delta x = \pi/|k_1 - k_2|$. Thus,

$$\Delta p.\Delta x \approx h/2$$

This means that if we attempt to measure the momentum difference accurately then the position measurement will become less accurate and vice versa.

[5] Schrodinger's wave mechanics and Heisenberg's matrix mechanics.

In wave mechanics, the state of a quantum system at time t is defined by a normalized vector $|\psi(t)>$ in a Hilbert space $\mathcal{H}$ and this vector satisfies a "Schrodinger wave equation":

$$i\frac{d|\psi(t)>}{dt} = H(t)|\psi(t)>$$

where $H(t)$ is a possibly time varying self-adjoint operator in $\mathcal{H}$. If $H(t) = H$ is time independent, then the solution can be expressed as

$$|\psi(t)>= exp(-itH)|\psi(0)>$$

where $exp(-itH)$ is a unitary operator in $\mathcal{H}$ and may be defined via resolvents:

$$exp(-itH) = lim_{n\to\infty}(1 + itH/n)^{-n}$$

We note that even if H is an unbounded operator, this definition of the exponential may make sense using the theory of resolvents and spectra. (A complex number z is said to be in the resolvent set $\rho(H)$ of H if $(z - H)^{-1}$ exists and is a bounded operator. The complement of $\rho(H)$ denoted by $\sigma(H)$ is called the spectrum of the operator H. Since here H is Hermitian, we have the spectral representation

$$H = d\int_{\mathbb{R}} xE(dx)$$

so that

$$exp(-itH) = \int_{\mathbb{R}} exp(-itx)E(dx)$$

and

$$(1 + itH/n)^{-n} = \int (1 + itx/n)^{-n}E(dx)$$

so for any $|f>\in \mathcal{H}$, we have

$$\| (1 + itH/n)^{-n}f - exp(-itH)f \|^2 = \int_{\mathbb{R}} |(1 + itx/n)^{-n} - exp(-itx)|^2 < f|E(dx)|f>$$

which converges by the dominated convergence theorem to zero as $n \to \infty$. We note that if T is any densely defined operator in $\mathcal{H}$, bounded or unbounded, and if for $z \in \rho(T)$, we have an inequality of the form

$$\| (z - T)^{-1} \| \leq f(z)$$

then

$$\| (1 + itT/n)^{-n} \| \leq \| (1 + itT/n)^{-1} \|^n \leq nf(in/t)/|t|$$

and this may remain bounded as $n \to \infty$. On the other hand, $(1 + itT/n)^n$ is unbounded if T is unbounded and hence we cannot defined $exp(itT)$ as the limit of this as $n \to \infty$ (Reference: T.Kato, Perturbation theory for linear operators, Springer).

In wave mechanics, the observables like position, momentum, angular momentum, energy etc. are represented by self-adjoint operators in the Hilbert space $\mathcal{H}$ and these do not vary with time while the state $|\psi(t)>$ varies with time. Hence, if X is an observable, then its average at time t is given in the Schrodinger wave mechanics picture by $<\psi(t)|X|\psi(t)>$. On the other hand, in the Heisenberg matrix mechanics picture, observables change with time while the state remains the same and hence, the average of X at time t in the Heisenberg picture is $<\psi(0)|X(t)|\psi(0)>$. Since the physics must be the same no matter which model we use, we must have

$$<\psi(t)|X|\psi(t)>=<\psi(0)|X(t)|\psi(0)> ---(1)$$

and differentiating this equation w.r.t. time gives

$$i<\psi(t)|[H(t), X]|\psi(t)>=<\psi(0)|X'(t)|\psi(0)>$$

Writing

$$|\psi(t)>= U(t)|\psi(0)>$$

then gives us

$$X'(t) = iU(t)^*[H(t), X]U(t) = [iU(t)^*H(t)U(t), X(t)]$$

We could also directly write (1) as

$$< \psi(0)|U(t)^*XU(t)|\psi(0) >=< \psi(0)|X(t)|\psi(0) >$$

and hence

$$X(t) = U(t)^*XU(t)$$

from which we obtain on differentiation,

$$X'(t) = iU(t)^*[H(t), X]U(t) = i[U(t)^*H(t)U(t), X(t)]$$

This is Heisenberg's equation of matrix mechanics. In particular, if $H(t) = H$ does not vary with time, we get $U(t)^*HU(t)^* = H$ and Heisenberg's equation of matrix mechanics becomes

$$X'(t) = i[H, X(t)]$$

with solution,

$$X(t) = exp(itH).X.exp(-itH)$$

[6] Dirac's replacement of the Poisson bracket by the quantum Lie bracket.
The Poisson bracket between two observables $u(q,p)$ and $v(q,p)$ satisfies

$$\{u, vw\} = \{u, v\}w + v\{u, w\}$$

and likewise for the first argument. If we agree that an analogous bracket [.] exists for non-commutative quantum observables with the same ordering preserved, then we must have

$$[uv, wz] = [uv, w]z + w[uv, z] = u[v, w]z + [u, w]vz + wu[v, z] + w[u, z]v$$

on the one hand and on the other hand,

$$[uv, wz] = [u, wz]v + u[v, wz] = [u, w]zv + w[u, z]v + u[v, w]z + uw[v, z]$$

Equating these two expressions gives

$$[u, w]vz + wu[v, z] = [u, w]zv + uw[v, z]$$

Hence

$$[u, w](vz - zv) = (uw - wu)[v, z]$$

It follows from the arbitrariness of the four observables u, v, w, z that the quantum bracket [.] should have the form

$$[u, w] = c(uw - wu)$$

where c is a complex constant. In other words, the quantum bracket that replaces the classical Poisson bracket must be proportional to the Lie bracket.

[7] Duality between the Schrodinger and Heisenberg mechanics based on Dirac's idea.
In Schrodinger's wave mechanics, the state of the system at any time t is described by a density operator $\rho(t)$ in a Hilbert space $\mathcal{H}$. In other words, $\rho(t) \geq 0, Tr(\rho(t)) = 1$. Further, it satisfies the Schrodinger-Liouville- Von-Neumann equation of motion:

$$i\rho'(t) = [H(t), \rho(t)]$$

This follows by writing the spectral decomposition of $\rho(t)$ as

$$\rho(t) = \sum_k |\psi_k(t) > p_k < \psi_k(t)|$$

in diagonal form with $p_k's$ constant and $\sum_k p_k = 1, p_k \geq 0$. The $\psi_k(t)'s$ satisfy the Schrodinger wave equation

$$i|\psi_k'(t) >= H(t)|\psi_k(t) >$$

and hence

$$-i < \psi_k'(t)| =< \psi_k(t)|H(t)$$

so that

$$i\rho'(t) = \sum_k (i|\psi_k'(t) > p_k < \psi_k(t)| + i|\psi_k(t > p_k < \psi_k'(t)|)$$

$$= \sum_k (H(t)|\psi_k(t) > p_k < \psi_k(t)| - |\psi_k(t)p_k < \psi_k(t)|H(t))$$

$$= [H(t), \rho(t)]$$

We can write its solution as

$$\rho(t) = U(t)\rho(0)U(t)^*$$

where $U(t)$ is a unitary operator satisfying the Schrodinger equation

$$U'(t) = -iH(t)U(t)$$

In this picture, observables do not vary with time. Thus, the average value of an observable X at time t is given by

$$< X > (t) = Tr(\rho(t)X) = \sum_k p_k < \psi_k(t)|X|\psi_k(t) >$$

In the Heisenberg picture, the observable X changes at time t to $X(t)$ while the state $\rho(0)$ does not change. Hence, in this picture the average of the observable at time t is

$$< X > (t) = Tr(\rho(0)X(t))$$

and the two pictures must give the same average. Thus,

$$Tr(\rho(0)X(t)) = Tr(\rho(t)X) = Tr(U(t)\rho(0)U(t)^*X) = Tr(\rho(0)U(t)^*XU(t))$$

and hence from the arbitrariness of $\rho(0)$, it follows that dynamics of observables in the Heisenberg picture must be given by

$$X(t) = U(t)^*XU(t)$$

and hence

$$X'(t) = iU(t)^*(H(t)X - XH(t))U(t) = i[\tilde{H}(t), X(t)]$$

where

$$\tilde{H}(t) = U(t)^*H(t)U(t)$$

In particular, if $H(t) = H$ is time independent, then $U(t) = exp(-itH)$ and we get

$$X'(t) = i[H, X(t)]$$

for the Heisenberg dynamics and

$$i\rho'(t) = [H, \rho(t)]$$

for the Schrodinger dynamics.

[8] Quantum dynamics in Dirac's interaction picture.
Suppose

$$H(t) = H_0 + V(t)$$

is the system Hamiltonian. The Schrodinger evolution operator $U(t)$ satisfies

$$iU'(t) = H(t)U(t)$$

We write

$$U(t) = U_0(t)W(t), U_0(t) = exp(-itH_0)$$

Then, substituting this into the above Schrodinger evolution equation gives

$$iW'(t) = \tilde{V}(t)W(t), \tilde{V}(t) = U_0(t)^*V(t)U_0(t)$$

This leads us to the Dirac interaction picture where a state $|\psi>$ evolves according to the Hamiltonian $\tilde{V}(t)$ while an observable X evolves according to the Hamiltonian H_0. This keeps the physics invariant since if X is an observable and $|\psi>$ is the state at time 0, then in the Dirac interaction picture, the average of the observable in the state at time t is given by (The subscript i stands for evolution of states or observables in the interaction picture. Thus $|\psi_i(t)>= W(t)|\psi<, X_i(t) = U_0(t)^* X U_0(t))$.

$$< \psi_i(t)|X_i(t)|\psi_i(t) >=< \psi|W(t)^*(U_0(t)^* X U_0(t))W(t)|\psi >$$

where

$$W(t) = T\{exp(-i\int_0^t \tilde{V}(\tau)d\tau)\}$$

Now, as a check

$$\frac{d}{dt}(U_0(t)W(t)) = U_0'(t)W(t) + U_0(t)W'(t) =$$

$$-iH_0U_0(t)W(t) + U_0(t)(-i\tilde{V}(t))W(t)$$

$$= -iH_0U_0(t)W(t) - iU_0(t)U_0(t)^*V(t)U_0(t)W(t)$$

$$= -iH_0U_0(t)W(t) - iV(t)U_0(t)W(t) = -iH(t)U_0(t)W(t)$$

In other words, $U_0(t)\tilde{U}(t)$ satisfies the same equation as $U(t)$ and hence must equal $U(t)$. Thus, $U_0(t)W(t) = U(t)$ as expected and this shows that

$$< \psi_i(t)|X_i(t)|\psi_i(t) >=< \psi|U(t)^* X U(t)|\psi >$$

which is in agreement with the Schrodinger and Heisenberg pictures.

[9] The Pauli equation: Incorporating spin in the Schrodinger wave equation in the presence of a magnetic field.
We describe how the Pauli equation can be used to calculate the average electric dipole moment and average magnetic dipole moment of an atom in an external electromagnetic field. This calculation enables us to calculate the electric polarization field and the magnetization field of a material. It will be a consequence of this calculation that the average electric dipole moment and hence polarization depends on both the external electric and magnetic fields and likewise for the average magnetic dipole moment and magnetization.

The electric dipole moment observable is $-er$, $r = (x, y, z)$ being the electron position relative to the nucleus. The magnetic dipole moment observable is $\mu = (e/2m)(L+g\sigma/2)$ where $L = (L_x, L_y, L_z)$ is the angular momentum observable vector and $\sigma = (\sigma_x, \sigma_y, \sigma_z)$ is the vector of Pauli spin matrices. A heuristic justification of this fact

[10] The Zeeman effect: Let H_0 be the Hamiltonian of an atom that commutes with the orbital and spin angular momentum operators $(L_x, L_y, L_z) = L$ (ie, H_0 is rotation invariant as happens when $H_0 = p^2/2m_0 + V(r)$ where V depends only on the radial coordinate) and $(\sigma_x, \sigma_y, \sigma_z) = \sigma$. Then the eigenstates of the Hamiltonian operator

$$H = H_0 + e(L + g\sigma, B)/2m_0$$

where B is a constant magnetic can be calculated as follows. We may assume without loss of generality that B is along the z axis. Then,

$$H = H_0 + (eB/2m)(L_z + g\sigma_z)$$

and since L^2, L_z, σ_z, H_0 mutually commute, the energy eigenstates of H are labelled by four quantum numbers $|n, l, m, s>$ where $l(l+1)$ is an eigenvalue of L^2, m is an eigenvalue of L_z and s is an eigenvalue of σ_z. If $E(n, l)$ is an energy eigenvalue of H_0 ie of H without the magnetic field, then this eigenvalue has a degeneracy of $2(2l+1)$ corresponding to the $2l+1$ eigenvalues of L_z for a given eigenvalue $l(l+1)$ (ie,+ $m = -l, -l+1, ..., l-1, l$)of L^2 and the two eigenvalues $\pm 1/2$ of σ_z. When the magnetic field is turned on, these $2(2l+1)$ degenerate eigenstates split into the the same number of non-degenerate eigenstates with eigenvalues $E(n, l, m, s) = E(n, l) + (eB/2m_0)(m + gs), m = -l, -l+1, ..., l-1, l, s = \pm 1/2$. The eigenstate of H corresponding to the eigenvalue $E(n, l, m, s)$ is denoted by $|n, l, m, s>$.

Chapter 2
Exactly Solvable Schrodinger Problems, Perturbation Theory, Quantum Gate Design

[11a] The spectrum of the Hydrogen atom.
This is described by the stationary Schrodinger equation

$$\nabla^2\psi(r,\theta,\phi)+2m(E+e^2/r)\psi(r,\theta,\phi)=0$$

Writing

$$\psi(r,\theta,\phi)=R(r)Y_{lm}(\theta,\phi)$$

where Y_{lm} are the spherical harmonics, we get using

$$\nabla^2=r^{-2}\frac{\partial}{\partial r}r^2\frac{\partial}{partial r}-L^2/r^2$$

where L^2 is the squared angular momentum operator:

$$L^2=-\frac{1}{sin(\theta)}\frac{\partial}{\partial\theta}sin(\theta)\frac{\partial}{\partial\theta}-\frac{1}{sin^2(\theta)}\frac{\partial^2}{\partial\phi^2}$$

Y_{lm} is an eigenfunction of both the commuting operators L^2 and L_z:

$$L=r\times p=-ir,L^2=-(r\times\nabla)^2$$

$$L_zY_{lm}=mY_{lm},L^2Y_{lm}=l(l+1)Y_{lm}$$

Exercise: Verify that L^2 given by the above differential expression coincides with

$$-(r\times\nabla)^2=(yp_z-zp_y)^2+(zp_x-xp_z)^2+(xp_y-yp_x)^2,$$

$$p_x=-i\partial_x,p_y=-i\partial_y,p_z=-i\partial_z$$

The radial equation for $R(r)$ thus becomes

$$R''(r)+2R'(r)/r-l(l+1)R(r)/r^2+2m(E+e^2/r)R(r)=0$$

or equivalently,

$$r^2R''(r)+2rR'(r)+2m(Er^2+e^2r-l(l+1)/2m)R(r)=0$$

As $r\to\infty$, this equation approximates to

$$r^2R''(r)+2mEr^2R(r)\approx 0$$

or equivalently,

$$R''(r)=-2mER(r)$$

Since for bound states, E must be negative, it follows that as $r\to\infty$, we have $R(r)\approx C.exp(-\alpha r),\alpha=\sqrt{-2mE}$. So, writing the exact solution as

$$R(r)=f(r)exp(-\alpha r),\alpha=\sqrt{-2mE}$$

we get

$$R'=(f'-\alpha f)exp(-\alpha r),R''=(f''-2\alpha f'+\alpha^2f)exp(-\alpha r)$$

The Schrodinger equation now becomes

$$r^2(f''-2\alpha f'+\alpha^2f)+2r(f'-\alpha f)+(-\alpha^2r^2+2me^2r-l(l+1))f=0$$

or

$$r^2f''+2r(1-\alpha r)f'+(2me^2r-l(l+1))f=0$$

We solve this equation by the power series method. Substitute

$$f(r)=\sum_{n\geq 0}c(n)r^{n+s}$$

Then

$$\sum_{n\geq 0}(n+s)(n+s-1)c(n)r^{n+s}+2\sum_{n\geq 0}(n+s)c(n)r^{n+s}-2\alpha\sum_{n\geq 0}(n+s)c(n)r^{n+s+1}$$

$$+2me^2\sum_{n\geq 0}c(n)r^{n+s+1}-l(l+1)\sum_{n\geq 0}c(n)r^{n+s}=0$$

or equivalently,

$$\sum_{n\geq 0}((n+s)(n+s+1)-l(l+1))c(n)r^{n+s}$$

$$+\sum(2me^2 - 2(n+s)\alpha)c(n)r^{n+s+1} = 0$$

Equating coefficients of equal powers of r gives

$$(s(s+1) - l(l+1))c(0) = 0,$$

and for all $n \geq 1$,

$$((n+s)(n+s+1) - l(l+1))c(n) + (2me^2 - 2(n+s-1)\alpha)c(n-1) = 0$$

We must assume $c(0) \neq 0$ for otherwise, this recursion would imply that $c(n) = 0 \forall n \geq 0$ which would mean that the wave function vanishes. Then, we get

$$s(s+1) = l(l+1)$$

so $s = l$ since $s = -l-1$ would give a singularity at $r = 0$ and the wave function would fail to be square integrable when $l > 0$.

Further, for n large, ie, $n \rightarrow \infty$, the above difference equation asymptotically is equivalent to

$$c(n) \approx 2\alpha c(n-1)/n$$

or equivalently,

$$c(n) \approx (2\alpha)^n/n!$$

and hence

$$f(r) = \sum_n c(n)r^{n+l} \approx r^l exp(2\alpha r)$$

so

$$f(r)exp(-\alpha r) \approx r^l exp(\alpha r)$$

which is not square integrable, in fact, it diverges exponentially as $r \rightarrow \infty$. The only way out is that $c(n) = 0$ for all $n \geq n_0$ for some $n_0 \geq 1$. Let n_0 be the smallest such integer. Then we get from the above recursion by putting $n = n_0$, and $c(n_0 - 1) \neq 0$,

$$me^2 = (n_0 + s - 1)\alpha = (n_0 + l - 1)\alpha$$

Thus,

$$E = -me^4/2(n_0 + l - 1)^2$$

Thus, we get the result that the possible energy levels of the hydrogen atom are

$$E_n = -me^4/2n^2, n = max(1, l), l+1, l+2, ...,$$

This result was first derived rigorously by Schrodinger in this way although it was earlier obtained by Niels Bohr using adhoc arguments like the correspondence principle.

[11b] The spectrum of particle in a $3-D$ box. This is described by the stationary state Schrodinger equation

$$-\nabla^2\psi(r)/2m = E\psi(r), r = (x, y, z) \in [0, a] \times [0, b] \times [0, c]$$

with boundary conditions that $\psi(r)$ vanishes on the boundary, ie when either $x = 0, a$ or $y = 0, b$ or $z = 0, c$. The solution to this boundary valued problem is obtained by separation of variables and the result is the set of orthonormal wave functions

$$\psi_{nmk}(r) = ((2/a)(2/b)(2/c))^{1/2} sin(n\pi x/a)sin(m\pi y/b)sin(l\pi z/c), n, m, k = 1, 2, 3, ...$$

with the corresponding energy eigenvalues

$$E = \pi(n^2/a^2 + m^2/b^2 + k^2/c^2)^{1/2}$$

[11c] The spectrum of a quantum harmonic oscillator.
A quantum Harmonic oscillator has the Hamiltonian

$$H = p^2/2m + m\omega^2 q^2/2$$

where

$$[q,p] = i$$

We define the annihilation and creation operators by

$$a = (p - im\omega q)/\sqrt{2m}, a^* = (p + im\omega q)/\sqrt{2m}$$

Then,

$$a^*a = p^2/2m + m\omega^2 q^2/2 - \omega/2 = H - \omega/2, aa^* = H + \omega/2$$

Thus,

$$[a, a^*] = \omega$$

Now let $|E>$ be any normalized eigenstate of H with eigenvalue E. Then we have

$$0 \leq < E|a^*a|E >=< E|H - \omega/2|E >= E - \omega/2$$

Hence

$$E \geq \omega/2$$

ie, the minimum energy level of the oscillator is $\omega/2$. This is attained when and only when $a|E >= 0$. In other words, we have

$$a|\omega/2 >= 0$$

and hence

$$(d/dq + m\omega q) < q|\omega/2 >= 0$$

or equivalently,

$$< q|\omega/2 >= C.exp(-m\omega^2 q^2/2)$$

with C being the normalizing constant chosen so that

$$1 =< \omega/2|\omega/2 >= \int_{\mathbb{R}} | < q|\omega/2 > |^2 = |C|^2 \int_{\mathbb{R}} exp(-m\omega^2 q^2)dq$$

$$= |C|^2 = (\pi/m)^{1/2}\omega^{-1}$$

so we may take

$$C = (\pi/m\omega^2)^{1/4}$$

[12] Time independent perturbation theory: Calculation of the energy levels and eigenfunctions of non-degenerate and degenerate systems using perturbation theory.

The perturbed Hamiltonian is

$$H = H_0 + \sum_{k=1}^{\infty} \epsilon^k V_k$$

The stationary state wave functions for H are expanded as

$$|\psi >= |\psi_0 > + \sum_{k \geq 1} \epsilon^k |\psi_k >$$

and the corresponding perturbed energy level is also expanded as a power series:

$$E = E_0 + \sum_{k \geq 1} \epsilon^k E_k$$

Substituting these expressions into the eigenvalue problem

$$H|\psi >= E|\psi >$$

and equating coefficients of $\epsilon^k, k = 0, 1, 2, ...$ gives us the series

$$H_0|\psi_0 >= E_0|\psi_0 >,$$

$$(H_0 - E_0)|\psi_k> + \sum_{r=1}^{k} V_r|\psi_{k-r}> + \sum_{r=1}^{k} E_r|\psi_{k-r}>, k \geq 1$$

For each eigenvalue $E_0 = E_0(m)$ of H_0, let $|m, r>, r = 1, 2, ..., d(m)$ denote an orthonormal basis of eigenstates of $\mathcal{N}(H_0 - E_0(m))$. Thus, the eigenvalue $E_0(m)$ has a degeneracy of $d(m)$. So we can write

$$|\psi_0> = \sum_{r=1}^{d(m)} c(m, r)|m, r>$$

for $E_0 = E_0(m)$. The $O(\epsilon)$ equation

$$(H_0 - E_0(m))|\psi_1> + V_1|\psi_0> = E_1|\psi_0> - - -(1)$$

then gives on premultiplying by $< m, r|$,

$$\sum_r c(m, r) < m, k|V_1|m, r> = E_1 c(m, r)\delta[k - r]$$

and hence, it follows that the for the unperturbed energy level $E_0(m)$, the possible first order perturbations $E_1 = E_1(m, s), s = 1, 2, ..., d(m)$ to the energy levels are given by the eigenvalues of the $d(m) \times d(m)$ secular matrix $((< m, k|V_1|m, r>))_{1 \leq k, r \leq d(m)}$. Further, if $c_s(m) = ((c_s(m, r)))_{r=1}^{d(m)}$ is the eigenvector of this secular matrix corresponding to the eigenvalue $E_1(m, s)$, then then the corresponding unperturbed state is given by $\sum_{r=1}^{d(m)} c_s(m, r)|m, r>$. We note that the constants $c_s(m, r)$ can be chosen so that the states $\sum_{r=1}^{d(m)} c_s(m, r)|m, r>, s = 1, 2, ..., d(m)$ form an orthonormal basis for the $d(m)$ dimensional vector space $\mathcal{N}(H_0 - E_0(m))$. Further, we get from (1) by premultiplying by $< l, r|$ for $l \neq m$,

$$(E_0(l) - E_0(m)) < l, r|\psi_1> + < l, r|V_1|\psi_0> = E_1 < l, r|\psi_0> = 0$$

so that

$$< l, r|\psi_1> = \frac{< l, r|V_1|\psi_0>}{E_0(m) - E_0(l)}$$

which implies that the unperturbed state $|\psi_0> = \sum_r c_s(m, r)|m, r>$ gets perturbed to

$$|\psi_0> + \epsilon|\psi_1> + O(\epsilon^2)$$

where

$$|\psi_1> = \sum_{l \neq m, r=1,2,...,d(m)} |l, r>< l, r|V_1|\psi_0> /(E_0(m) - E_0(l))$$

and

$$< l, r|V_1|\psi_0> = \sum_{k=1}^{d(m)} c_s(m, k) < l, r|V_1|m, k>$$

We now calculate the second order pertrubation to the energy levels and the the corresponding eigenfunctions.

[13] Time dependent perturbation theory: Calculation of the transition probabilities of a quantum system from one stationary state of the unperturbed system to another stationary state in the presence of a time varying interaction potential.

The perturbed Hamiltonian has the form

$$H(t) = H_0 + \sum_{m=1}^{\infty} \epsilon^m V_m(t)$$

The Schrodinger evolution operator $U(t)$ of this system is expanded as a perturbation series:

$$U(t) = U_0(t) + \sum_{m \geq 1} \epsilon^m U_m(t)$$

Substituting this into the evolution equation

$$iU'(t) = H(t)U(t)$$

and equating equal powers of the perturbation parameter ϵ gives us the sequence of differential equations:

$$iU_0'(t) = H_0U_0(t),$$

$$iU_m'(t) - H_0U_m(t) = \sum_{k=1}^{m} V_k(t)U_{m-k}(t), m \geq 1$$

Thus

$$U_0(t) = exp(-itH_0),$$

$$U_m(t) = -i\sum_{k=1}^{m}\int_0^t U_0(t-s)V_k(s)U_{m-k}(s)ds, m \geq 1 - - - (1)$$

Let now $|n>$ be an eigenstate of the unperturbed system with energy $E(n)$:

$$H_0|n>= E(n)|n>$$

Then when the perturbation is switched on, the transition probability amplitude from state $|n>$ to state $|m>$ in time $[0,T]$ with $m \neq n$ is given by

$$<m|U(T)|n>= -i\sum_{k=1}^{r}\epsilon^k <m|U_k(T)|n> +O(\epsilon^{r+1})$$

where $U_1(T),....,U_r(T)$ are successively determined from (1). For example,

$$U_1(t) = -i\int_0^t U_0(t-s)V_1(s)U_0(s)ds,$$

$$U_2(t) = -i\int_0^t U_0(t-s)V_1(s)U_1(s)ds - i\int_0^t U_0(t-s)V_2(s)U_0(t-s)ds$$

$$= -\int_{0<s'<s<t} U_0(t-s)V_1(s)U_0(s-s')V_1(s')U_0(s')ds'ds - i\int_0^t U_0(t-s)V_2(s)U_0(t-s)ds$$

etc. In particular since $U_0(t)|n>= exp(-iE(n)t)|n> \forall n$, it follows that the transition probability from $|n>$ to $|m>$ for $m \neq n$ in time T is given by

$$P_T(n \to m) = \epsilon^2| <m|U_1(T)|n> |^2 + O(\epsilon^3) = \epsilon^2|\int_0^T exp(-i(E(m)(t-s) - E(n)s)) <m|V_1(s)|n> ds|^2 + O(\epsilon^3)$$

$$= \epsilon^2|\int_0^T exp(i(E(m) - E(n))s) <m|V_1(s)|n> ds|^2 + O(\epsilon^3)$$

In the limit as $T \to \infty$, this probability converges to

$$\epsilon^2|\hat{V}_1(E(m) - E(n))|^2 + O(\epsilon^3)$$

where $\hat{V}_1(\omega) = \int_0^\infty V_1(t)exp(i\omega t)dt$ is the one sided Fourier transform of V_1.

[14] The full Dyson series for the evolution operator of a quantum system in the presence of a time varying potential.

[15] The transition probabilities in the presence of a stochastically time varying potential.

If the potentials $V_k(t), k = 1,2,...$ appearing in section (13) are operator valued stochastic processes, then the various perturbation terms $U_k(t), k = 1,2,...$ in the expansion of the evolution operator will also be operator valued stochastic processes and in fact, $U_k(t)$ will be a sum of terms, each term of which is a multilinear functional of $V_r(.), r = 1,2,...$ with $V_{r_j}(.)$ appearing m_j times such that $\sum_j m_jr_j = k$. The transition probability from state $|n>$ to state $|m>$ in time T will then be given by

$$P_T(n \to m) = \mathbb{E}[<m|U(T)|n> |^2] = \mathbb{E}[(|sum_{k\geq 1}\epsilon^k <m|U_k(T)|n> |^2]$$

where the expectation is taken w.r.t the probability distribution of the random potentials $V_k(.), k = 1,2,....$

[16] Basics of quantum gates and their realization using perturbed quantum systems.

If H_0 is the unperturbed Hamiltonian and we apply a perturbation time varying potential

$$V(t) = \sum_{k=1}^{p} f_k(t)V_k$$

where the $f_k(t)'s$ are control functions, then the solution to the Schrodinger evolution unitary operator $U(t)$ defined by

$$U'(t) = -iH(t)U(t), H(t) = H_0 + \delta V(t)$$

upto $O(\delta^m)$ is given by

$$U(t) = U_0(t)W(t), U_0(t) = exp(-itH_0),$$

$$W(T) = W(T,f) = I + \sum_{1\leq r\leq m, 1\leq k_1,...,k_r\leq p} (-i)^r\delta^r \int_{0<t_r<t_{r-1}<...<t_1<T} f_{k_1}(t_1)...f_{k_r}(t_r)\tilde{V}_{k_1}(t_1)...\tilde{V}_{k_r}(t_r)dt_1...dt_r + O(\delta^m)$$

where

$$\tilde{V}_k(t) = U_0(-t)V_kU_0(t)$$

We can thus formulate the problem of designing a unitary gate U_g by minimizing the error energy

$$\| U_g - I - \sum_{1\leq r\leq m, 1\leq k_1,...,k_r\leq p} (-i)^r\delta^r \int_{0<t_r<t_{r-1}<...<t_1<T} f_{k_1}(t_1)...f_{k_r}(t_r)\tilde{V}_{k_1}(t_1)...\tilde{V}_{k_r}(t_r)dt_1...dt_r \|^2 + O(\delta^m)$$

w.r.t the control functions $f_k(t), k = 1, 2, ..., p, 0 \leq t \leq T$
using standard variational calculus. The particular case when $r = 1$ amounts to approximating $W(T, f)$ by

$$I - i\sum_{k=1}^{p}\int_0^T f_k(t)\tilde{V}_k(t)dt$$

and if H_g is the Hermitian generator of U_g, ie, $U_g = exp(-iH_g)$, or equivalently, $H_g = ilog(U_g)$, then we can design the gate by choosing the functions f_k over $[0, T]$ so that

$$\| H_g - \sum_{k=1}^{p}\int_0^T f_k(t)\tilde{V}_k(t)dt \|^2$$

is a minimum. This is called the matching generator technique [Kr.Gautam et.al].

[17] Bounded and unbounded linear operators in a Hilbert space.

If $\mathcal{H}$ is a Hilbert space and T is a linear operator defined in a dense domain $D(T)$ of $\mathcal{H}$, then we say that T is bounded if $K = sup_{x\neq 0, \in D(T)} \| Tx \| / \| x \| < \infty$. In that case, we define $\| T \| = K$. In this case, it is clear that T admits a unique extension to the whole of $\mathcal{H}$, ie, given any $x \in \mathcal{H}$, we choose a sequence $x_n \in D(T)$ such that $x_n \to x$ and then since $\| Tx_n - Tx_m \| \leq \| T \| . \| x_n - x_m \|$, it follows that Tx_n is a Cauchy sequence in $\mathcal{H}$ and hence converges to some $y \in \mathcal{H}$. We then define $Tx = y$. This definition is unique because given any other sequence $z_n \in D(T), z_n \to x$, we have

$$\| Tz_n - Tx_n \| \leq \| T \| . \| z_n - x_n \| \to 0$$

since

$$\| z_n - x_n \| \leq \| z_n - x \| . \| x_n - x \|$$

If T is unbounded, ie $K = \infty$, then we cannot in general extend T to the whole of $\mathcal{H}$. For example, consider the position operator x acting in $L^2(\mathbb{R})$ by multiplication. It is clear that $f(x) \in L^2(\mathbb{R})$ does not imply that $xf(x) \in L^2(\mathbb{R})$. The domain $D(x)$ of x is defined by the set of all $f \in L^2(\mathbb{R})$ for which $xf \in L^2(\mathbb{R})$. Equivalently,

$$D(x) = \{f : \int_{\mathbb{R}} (1 + x^2)|f(x)|^2 dx < \infty\}$$

$D(x)$ is dense in $L^2(\mathbb{R})$. Exercise:Prove this.

A densely defined linear operator T in $\mathcal{H}$ is said to be symmetric if $T \subset T^*$, ie, if $D(T) \subset D(T^*)$ and the restriction of T^* to $D(T)$ equals T. The adjoint T^* of T is defined as follows. First $D(T^*)$ consists of all $g \in \mathcal{H}$ for which $f \in D(T)$ implies that $< g, Tf > = < T^*(g), f >$ for some $T^*(g) \in \mathcal{H}$. In that case $T^*g = T^*(g)$. If $T = T^*$, ie, if T is symmetric and $D(T) = D(T^*)$, then we say that T is self-adjoint/Hermitian. Consider for example the position operator x in $L^2(\mathbb{R})$. Given $f \in D(x)$, we have $g \in D(x^*)$ iff $< g, xf >=< h, f >$ for some $h \in L^2(\mathbb{R})$. Since

$$(\int |x\bar{g}(x)f(x)|dx)^2 = (\int_{\mathbb{R}} |\bar{g}(x)xf(x)|dx)^2 \leq (\int \| |g(x)|^2 dx).(\int x^2|f(x)|^2 dx), g \in L^2(\mathbb{R})$$

and $\int x^2|f(x)|^2 dx < \infty$, it is clear that $D(x^*)$

$$D(x^*) = D(x)$$

ie, x is self-adjoint. Likewise, the momentum operator $p = -id/dx$ is selfadjoint in $L^2(\mathbb{R})$ since it is the same as FxF^{-1} where F is the Fourier transform in $L^2(\mathbb{R})$ which is a unitary operator.

[18] The spectral theorem for compact normal and bounded and unbounded self-adjoint operators in a Hilbert space.

The spectral theorem for a Hermitian operator H (also called self-adjoint operator) in a Hilbert space $\mathcal{H}$ represents H as a "spectral integral" (defined in the operator norm):

$$H = \int_{\mathbb{R}} xdE(x)$$

where $x \rightarrow E(x)$ from $\mathbb{R}$ into $P(\mathcal{H})$ (the lattice of orthogonal projections in $\mathcal{H}$) is a strongly right continuous map with the property that it is non-decreasing ie $R(E(x)) \subset R(E(y))$ for $x \leq y$ and $slim_{x\rightarrow-\infty}E(x) = 0, slim_{x\rightarrow\infty}E(x) = I$. By $slim$ we mean strong limit, ie if $T_n, n = 1, 2, ...$ is a family of linear operators in $\mathcal{H}$ we say that $slimT_n = T$ if for all $x \in \mathcal{H}$, $T_n x \rightarrow Tx$, ie $\| T_n x - Tx \| \rightarrow 0$. We say that $limT_n = T$ (ie convergence in the operator norm) if $\| T_n - T \| \rightarrow 0$, ie $sup_{x\neq 0, \|x\|\leq 1} \| T_n x - Tx \| \rightarrow 0$ and we write $wlimT_n = T$ if for all $x, y \in mathcal H, < x, (T_n - T)y > \rightarrow 0$. operator convergence implies strong convergence implies weak convergence but the converse is not true.

Exercise: Give examples when weak convergence of operators holds but strong convergence does not hold and when strong convegence holds but operator convergence does not hold.

Remark: In a finite dimensional vector space, all norms are equivalent by the Bolzano-Weierstrass theorem (which states that a bounded sequence in $\mathbb{R}^n, n < \infty$ has a convergent subsequence. Use this fact to show that in a finite dimensional Hilbert space, all the three types of convergence discussed above are equivalent.

Remark: If $|e_n >, n = 1, 2, ...$ is an orthonormal sequence in a separable Hilbert space $\mathcal{H}$, then by Bessel's inequality, for any $|f > \in \mathcal{H}$, we have

$$\| f \|^2 \geq \sum_n | < f|e_n > |^2$$

and this implies $< f|e_n > \rightarrow 0, n \rightarrow \infty$, ie the sequence $|e_n >$ converges weakly to zero. However since $\| e_n - e_m \| = \sqrt{2}, n \neq m$, the sequence $|e_n >$ does not converge strongly and in fact, no subsequence of this sequence can converge strongly. Now define the operators $T_n = |e_n >< e_n|, n = 1, 2, ...$ Then $< f|T_n|g >=< f|e_n >< e_n|g > \rightarrow 0$, hence $wlimT_n = 0$. Further

$$\| T_n f \|^2 = | < e_n|f > |^2 \rightarrow 0$$

hence $slimT_n = 0$. Further, $\| T_n \| = 1$ and hence T_n does not converge in the operator norm.

Assuming the spectral theorem stated above, we can define for any Borel set B in $\mathbb{R}$ the orthogonal projection $E(B) = \int_B dE(x)$. This means that for all $f, g \in \mathcal{H}, < f|E(B)|g >= \int_B d < f|E(x)|g >$ where $d < f|E(x)|g >$ defines a signed measure for each f, g. Note that the function $x \rightarrow < f|E(x)|g >$ has bounded variation since

$$|d < f|E(x)|g > | = | < f|dE(x)|g > | \leq \| f \| < g|dE(x)|g >^{1/2} = \| f \| \| dE(x)g \|$$

$$= \| f \| .(d(\| E(x)g \|)^2)^{1/2}$$

where we use the fact that $(dE(x))^2 = dE(x) = dE(x)^*$ and that for $a < b$,

$$\| (E(b) - E(a))g \|^2 = \| E(b)g \|^2 - \| E(a)g \|^2$$

We also have for $< f|f >= 1$,

$$|d < f|E(x)|g > |^2 = | < f|dE(x)|g > |^2 \leq < dE(x)g|dE(x)g > = \| dE(x)g \|^2$$

so that

$$|d < f|E(x)|g > | \leq \| dE(x)g \|$$

Note that $x \rightarrow \parallel E(x)g \parallel$ is a non-decreasing function bounded above by $\parallel g \parallel$ and hence is a function of bounded variation or more precisely, $\int_{\mathbb{R}} d \parallel E(x)g \parallel = \parallel g \parallel < \infty$. We have further, for $< f|f >= 1$ that

$$| < f|E(x)|g > | \leq \parallel E(x)g \parallel$$

and also

$$| < f|E(y)|g > - < f|E(x)|g > | = | < f|(E(y) - E(x))|g > | \leq \parallel (E(y) - E(x))g \parallel$$

Now $x \rightarrow \parallel E(x)g \parallel$ is a function of bounded variation and $x \rightarrow < g|E(x)|g >= \parallel E(x)g \parallel^2$ is also a function of bounded variation since it is a non-decreasing function bounded above by $\parallel g \parallel^2$. Now,

$$4Re(< f|E(x)|g >) =< f + g|E(x)|f + g > - < f - g|E(x)|f - g >$$

implying that $x \rightarrow Re(< f|E(x)|g >)$ is of bounded varation an likewise $x \rightarrow Im(< f|E(x)|g >)$ is also of bounded variation obtained by replacing f with $i.f$. Hence $x \rightarrow < f|E(x)|g >$ has bounded variation which enables us to define spectral integrals weakly using Riemann-Stieltjes integrals. Further, if

$$H = \int_{\mathbb{R}} x dE(x)$$

then for $c \in \mathbb{R}$, the projection P_c onto $R((H - c)_+)$ is given using the formula

$$(H - c)_+ = \int_{x \geq c} (x - c)dE(x)$$

as

$$P_c = E(c)$$

This formula enables us to constructively describe the spectral measure associated with a Hermitian operator. In fact, in the proof of the spectral theorem, we define $P_c = f(H) = (H - c)_+$ by taking $f(x) = max(x - c, 0) = (|x - c| + (x - c))/2$ and then proving that $c \rightarrow P_c$ is a spectral measure such that $\int c dP_c = H$. See T.Kato, "Perturbation theory for linear operators", Springer, for details.

Remark: If z is any complex number with $Im(z) \neq 0$, then for Hermitian H,

$$(H - z)^{-1} = \int (x - z)^{-1} dE(x)$$

so that the spectral norm of $(H - z)^{-1}$ is given by

$$\parallel (H - z)^{-1} \parallel = max(|x - z)^{-1} : x \in \mathbb{R}\} = |Im(z)|^{-1} < \infty$$

In paricular, $(H - z)^{-1}$ is a bounded operator, ie $z \in \rho(H)$ where $\rho(H)$ is the resolvent set of H. More generally, it can be shown that if H is a closed symmetric operator, then $\pm i$ belong to $\rho(H)$ iff H is Hermitian.

Remark: If T is an operator in a Hilbert space, its graph is defined by

$$Gr(T) = \{(x, Tx) : x \in D(T)\}$$

T is said to be closed if $Gr(T)$ is closed, ie if $x_n \in D(T), x_n \rightarrow x, Tx_n \rightarrow y$ imply $x \in D(T)$ and $y = Tx$. An operator T is said to be closable if the closure $Cl(Gr(T))$ of $Gr(T)$ is also the graph of some operator $\tilde{T}$. It is easy to see that this happens iff $x_n \in D(T), x_n \rightarrow x, Tx_n \rightarrow 0$ imply $x = 0$. Then we say that T is closable with $\tilde{T}$ being the closure of T. A operator T in $\mathcal{H}$ is said to be essentially selfadjoint if T is closable with its closure $\tilde{T}$ being selfadjoint. If $\mathcal{E}$ is a subset of $D(T)$ such that $\{(x, Tx) : x \in D(T)\}$ is dense in $Gr(T)$, then $\mathcal{E}$ is called a core for T. It is easy to see that this happens iff for any $x \in D(T)$ there exists a sequence $x_n \in \mathcal{E}$ such that $\parallel x_n - x \parallel^2 + \parallel Tx_n - Tx \parallel^2 \rightarrow 0$, or equivalently, iff for any $x \in D(T)$, $< (x, T(x)), (z, T(z)) >= 0$ for all $z \in \mathcal{E}$ implies $x = 0$.

[19] The general theory of Events, states and observables in the quantum theory.

The basic axiomatic set up that defines a quantum probability space is the triplet $(\mathcal{H}, P(\mathcal{H}), \rho)$ where $\mathcal{H}$ is a Hilbert space, $P(\mathcal{H})$ is the lattice of orthogonal projections in $\mathcal{H}$ and ρ is a state in $\mathcal{H}$, ie, a positive definite matrix of unit trace. This is to be compared with a classical probability space defined by the triplet $(\Omega, \mathcal{F}, P)$ where Ω is the sample space, $\mathcal{F}$ is a $\sigma - algebra$ of subsets of Ω and P is a probability measure on $\mathcal{F}$. The elements of $\mathcal{F}$ are called events in classical probability and likewise the elements of $P(\mathcal{H})$, namely the orthogonal projections are the events in quantum probability. The elements of Ω in classical probability are called the elementary outcomes and likewise the elements of

$\mathcal{H}$ after normalizing are the elementary outcomes in quantum probability. Just as an event E in classical probability is a set of elementary outcomes, likewise in quantum probability, an event P which is an orthogonal projection can be viewed as an aggregated of elements of $\mathcal{H}$ using the spectral diagonalization method:

$$P = \sum_{a \in I} |\psi_a >< \psi_a||, |\psi_a > \in \mathcal{H}, < \psi_a|\psi_b >= \delta_{ab}$$

P is viewed as the aggregate of $|\psi_a >, a \in I$. Finally if μ is any measure on $P(\mathcal{H})$, ie, $\mu : P(\mathcal{H}) \to [0,1]$ is a map such that if $P_n, n = 1, 2, ...$ are pairwise orthogonal orthogonal projections in $\mathcal{H}$, then $\mu(\sum_n P_n) = \sum_n \mu(P_n), \mu(0) = 0, \mu(I) = 1$, then Gleason's theorem asserts that there exists a positive definite operator ρ of unit trace such that

$$\mu(P) = Tr(\rho P), P \in P(\mathcal{H})$$

This proves the generality of the setup of a quantum probability space. An observable in the quantum probability space $(\Omega, \mathcal{F}, P)$ is a Hermitian operator X in $\mathcal{H}$. It has spectral representation

$$X = \int x E_X(dx)$$

and the probability that when X is measured, the outcome will fall in the range $(a, b]$ is $Tr(\rho E_X((a, b]))$. In other words, $F_X(x) = Tr(\rho E((-\infty, x]))$ is the probability distribution of X in the state ρ. This is to be compared with the classical setup where an observable X is a random variable in the the probability space $(\Omega, \mathcal{F}, P)$, ie a measurable map $X : (\Omega, \mathcal{F}) \to (\mathbb{R}, \mathcal{B}(\mathbb{R}))$. The probability distribution of X in the "state" P, is

$$F_X(x) = PX^{-1}((-\infty, x])) = P(\{\omega \in \Omega : X(\omega) \le x\})$$

To show that classical random variables are special cases of quantum observables, we define the spectral measure E_X in $L^2(\mathbb{R})$ by

$$E_X(B) = \chi_B, B \in \mathcal{B}(\mathbb{R})$$

ie $E_X(B)$ is multiplication by the indicator of the set B. Then we obviously have

$$X = \int_{\mathbb{R}} x E_X(dx)$$

Then, for any function$f : \mathbb{R} \to \mathbb{R}$, we have

$$f(X) = \int f(x) E_X(dx)$$

and hence

$$(f(X)g)(y) = \int f(x) \chi_{(x, x+dx]}(y) g(y) = f(y) g(y)$$

ie $f(X)$ is the operator of multilication by $f(x)$. The crucial difference between classical and quantum probability is that in the former, all observables commute while in the latter, they need not commute. Thus, if X, Y are two classical random variables defined over the same probability space $(\Omega, \mathcal{F}, P)$, then both have same spectral measure:

$$E_X(B) = E_Y(B) = \chi_B,$$

$$X = \int x E_X(dx), Y = \int y E_Y(dy), E_X = E_Y$$

Hence, we can define the joint probability of X taking values in a Borel set B_1 and Y taking values in a Borel set B_2 as

$$P(X \in B_1, Y \in B_2) = P(X^{-1}(B_1) \cap Y^{-1}(B_2))$$

and in particular the joint probability distribution of the pair (X, Y) is

$$F_{XY}(x, y) = P(X^{-1}((-\infty, x]) \cap Y^{-1}((-\infty, y]))$$

On the other hand, in quantum probability two observables X, Y need not commute, ie $XY - YX \neq 0$. Then the spectal measures of X, Y will differ:

$$X = \int x E_X(dx), Y = \int y E_Y(dy), E_X \neq E_Y$$

and hence, we cannot speak of the event that (X,Y) takes values in $B_1 \times B_2$ in a given state. In fact, we have the basic Heisenberg uncertainty inequality

$$Tr(\rho X^2)Tr(\rho Y^2) \geq Tr(\rho XY)^2 = (Re(Tr(\rho XY))^2 + (Im(Tr(\rho XY))^2$$

$$= (Tr(\rho(XY+YX)))^2/4 + (Tr(\rho(XY-YX)/i))^2/4 \geq (Tr(\rho[X,Y]/i))^2/4$$

In particular, if $[X,Y] = ic$, ie,

$$Tr(\rho X^2).Tr(\rho Y^2) \geq c^2/4$$

Thus, if X is replaced by $X - Tr(\rho X)$ and Y by $Y - Tr(\rho Y)$, we get that the product of the variances of X and Y in any state ρ is always $\geq c^2/4$ and hence we cannot measure them simultaneously. We can measure any single Hermitian function of X, Y by not jointly two or more functions unless they commute. If X, Y commute, then they have the same spectral measure (jointly diagonable) and hence we can represent

$$X = \int xE_X(dx), Y = \int f(x)E_X(dx)$$

for some real valued function f and hence $Tr(\rho E_X(B_1 \bigcap f^{-1}(B_2)))$ is the joint probability of X taking values in B_1 and Y in B_2. If X, Y do not commute, then the quantity

$$\psi_X(t_1,t_2) = Tr(\rho.exp(i(t_1X + t_2Y)))$$

need not be positive definite and hence it is not a characteristic function of any probability measure in $\mathbb{R}^2$. However, for any $a, b \in \mathbb{R}$,

$$\psi(t) = Tr(\rho.exp(it(aX + bY))$$

is positive definite in t and hence the characteristic function of the observable $aX + bY$. Thus, any single real linear combination or more generally, any single Hermitian matrix valued function of (X,Y) is measurable with a definite characteristic function but two different linear combinations may not have a joint characteristic function that is positive definite and hence the 2-D inverse Fourier transform of such a function need not be a probability measure on $\mathbb{R}^2$.

Exercise: Choose a general 2×2 state

$$\rho = \begin{pmatrix} p_1 & z \\ \bar{z} & p_2 \end{pmatrix}$$

so that

$$p_1, p_2 \geq 0, p_1 + p_2 = 1, p_1p_2 - |z|^2 \geq 0$$

These conditions ensure that ρ is a state in $\mathbb{C}^2$, ie, positive definite of unit trace. Take the three Pauli matrices

$$\sigma_1 = \begin{pmatrix} 0 & 1 \\ 1^0 & \end{pmatrix}, \sigma_2 = \begin{pmatrix} 0 & -i \\ i & 0 \end{pmatrix},$$

$$\sigma_3 = \begin{pmatrix} 1 & 0 \\ 0 & -1 \end{pmatrix}$$

Calculate the joint characteristic function of these three observables in the state ρ, ie,

$$\psi(t_1,t_2,t_3) = Tr(\rho.exp(i(t_1\sigma_1 + t_2\sigma_2 + t_3\sigma_3)))$$

and show that there exists a ρ such that ψ is not positive definite.

Chapter 3
Open Quantum Systems, Tensor Products, Fock Space Calculus, Feynman's Path Integral

[20] The evolution of the density operator in the absence of noise.

[21] The Gorini-Kossakowski-Sudarshan-Lindblad (GKSL) equation for noisy quantum systems.

[22] Distinguishable and indistinguishable particles: The Maxwell-Boltzmann, Bose-Einstein and Fermi-Dirac statistics.

[23] The relationship between spin and statistics.

[24] Tensor products of Hilbert spaces: Let $\mathcal{H}_1, \mathcal{H}_2$ be two Hilbert spaces. For $(x_1, x_2), (y_1, y_2) \in \mathcal{H}_1 \times \mathcal{H}_2$, define

$$K((x_1, x_2), (y_1, y_2)) = < x_1, y_1 >_1 < x_2, y_2 >_2$$

By Schur's theorem, K is a positive definite kernel on $\mathcal{H}_1 \times \mathcal{H}_2$. Hence, by Kolmogorov's consistency theorem, there exists a Gaussian process $\{\lambda(x_1, x_2), (x_1, x_2) \in \mathcal{H}_1 \times \mathcal{H}_2\}$ on $\mathcal{H}_1 \times \mathcal{H}_2$ such that

$$\mathbb{E}(\lambda(x_1, x_2)\lambda(y_1, y_2)) = K((x_1, x_2), (y_1, y_2))$$

or equivalently, if $< .,. >$ denotes the standard inner product in the underlying probability space, we can write

$$< \lambda(x_1, x_2), \lambda(y_1, y_2) >= K((x_1, x_2), (y_1, y_2)) = < x_1, y_1 >_1 < x_2, y_2 >_2$$

λ is called the tensor product of $\mathcal{H}_1$ with $\mathcal{H}_2$. It is easy to prove that the maps $x_k \to \lambda(x_1, x_2), k = 1, 2$ are linear. This is proved by expanding

$$\| \lambda(c_1 x + c_2 y, z) - c_1 \lambda(x, z) - c_2 \lambda(y, z) \|^2$$

and likewise for the second argument.

[25] Symmetric and antisymmetric tensor products of Hilbert spaces, the Fock spaces.

Let $\mathcal{H}$ be a finite dimensional Hilbert space of dimension N. Choose an orthonormal basis $\{|e_i >: i = 1, 2, ..., N\}$ for this space. Let $\otimes$ be a tensor product of this Hilbert space with itself as many times as needed. Using the definition of the inner product

$$< u_1 \otimes ... \otimes u_n, v_1 \otimes ... \otimes v_n >= \Pi_{i=1}^n < u_i, v_i >$$

we easily see that

$$\{e_{i_1} \otimes ... \otimes e_{i_n} : 1 \leq i_1, ..., i_n \leq N\}$$

is an orthonormal basis of the Hilbert space $\mathcal{H}^{\otimes n}$ and hence $dim\mathcal{H}^{\otimes n} = N^n$. The state

$$|e_{i_1} \otimes ... \otimes e_{i_n} >$$

corresponds to a state of n distinguishable particles in which the kth particle is in the state $|e_{i_k} >$. Since there may be repetitions of an index in $i_1, ..., i_k$, it follows that more than one particle can occupy a given state. Now suppose that the n particles are indistinguishable but are bosons, ie, more than one particle can occupy a given state. Then a given state of the n-particle system is specified by saying how many particles occupy each distinct state $|e_i >$. Such a state should be invariant under a permutation of the particles, ie, it must be superpositions of vectors in $\mathcal{H}^{\otimes n}$ of the form $\sum_{\sigma \in S_n} |u_{\sigma 1} \otimes ... \otimes u_{\sigma n} >$ where $u_k \in \mathcal{H}$. The set of all such superpositions forms a Hilbert space $\mathcal{H}^{\otimes n}$, called the n-fold symmetric tensor product of $\mathcal{H}$ with itself. An orthonormal basis for this space is $|e_1^{r_1}...e_N^{r_N} >, r_1, ..., r_N \geq 0, r_1 + ... + r_N = n$. where $|e_1^{r_1}...e_N^{r_N} >$ equals the symmetrization of $|e_1^{\otimes r_1} \otimes ... \otimes e_N^{\otimes r_N} >$ multiplied by the normalizing factor $(r_1!...r_N!)^{-1}(n!/(r_1!...r_N!))^{-1/2} = (n!r_1!...r_N!)^{-1/2}$. It is easy to see from this that $dim\mathcal{H}^{\otimes sn}$ is the number of ways in which n indistinguishable balls can be placed in N boxes and this number is the coefficient of x^n in the expansion $(1 + x + x^2 + ...)^N = (1 - x)^{-N}$. We thus get

$$dim\mathcal{H}^{\otimes_s n} = \binom{-N}{n}(-1)^n = \binom{N + n - 1}{n}$$

Finally, we consider the situation of Fermions which are also indistinguishable particles but no state can have more than one particle. Such a state consisting of n particles is a superposition of states of the form $\sum_{\sigma \in S_n} sgn(\sigma) u_{\sigma 1} \otimes ... \otimes u_{\sigma n}$, ie, it is antisymmetric with respect to interchange of particles. The set of all such superpositions forms a subspace of $\mathcal{H}^{\otimes n}$ and is denoted by $\mathcal{H}^{\otimes an}$ and an orthonormal basis for this space is $|e_{i_1} \wedge e_{i_2} \wedge ... \wedge e_{i_n} >, 1 \leq i_1 < ... < i_n \leq N$, where

$$u_1 \wedge ... \wedge u_n = (n!)^{-1/2} \sum_{\sigma \in S_n} sgn(\sigma) u_{\sigma 1} \otimes ... \otimes u_{\sigma n}$$

It is easily seen that if any two of the $u_i's$ are identical, then this n-fold wedge product is zero and this corresponds to the fact that no state have more than one particle, also called the Pauli exclusion principle for Fermions. The above results are sometimes expressed by saying that bosonic wave functions are symmetric under particle interchange while Fermionic wave functions are antisymmetric under particle interchange. We are now in a position to derive the Maxwell-Boltzmann, Bose-Einstein and Fermi-Dirac statistics respectively for distinguishable particles, Bosons and Fermions. Suppose our quantum system has energy levels $E_1, E_2, ...$ with respective degeneracies of $g_1, g_2, ...$ The number of ways of placing n_k distinguishable articles in the kth energy level E_k is $g_k^{n_k}$. Hence, the total number of ways of placing n distinguishable particles in these states such that n_k particles occupy the energy level E_k is

$$\Omega(n_1, n_2, ...) = (\frac{n!}{n_1!n_2!...}(g_1^{n_1}...g_k^{n_k}...)$$

The most probable distribution $\{n_k\}$ is obtained by maximizing $\Omega(n_1, n_2, ...)$ subject to the number and energy constraints $\sum_k n_k = n, \sum_k g_k n_k = E$. Using Stirling's formula for the factorial and incorporating the above constraints using Lagrange multipliers yields the fact that the most probable distribution is obtained by maximizing

$$F[n_1, n_2, ..., a, b] = \sum_k (n_k log(g_k) - n_k log(n_k)) - a(\sum_k n_k - n) - b(\sum_k E_k n_k - E)$$

Setting $\frac{\partial F}{\partial n_k} = 0$ gives

$$log(g_k) - log(n_k) - 1 = a + bE_k$$

or

$$n_k = Cg_k.exp(-bE_k), C = exp(-a-1)$$

This is called the Maxwell-Boltzmann distribution C, b are determined by the conditions $\sum_k n_k = n, \sum_k E_k n_k = E$. Suppose now that we are concerned with bosons, ie, indistinguishable particles with no constraint on the number of particles in each state. The number of ways in which n_k bosons can be distributed in the energy level E_k having g_k boxes is $\binom{n_k+g_k-1}{n_k}$ and hence if no constraint is placed on the number of particles, the most probable distribution $\{n_k\}$ with a constraint on the total energy E is obtained by maximizing

$$F = \sum_k log\binom{n_k + g_k - 1}{n_k} - b(\sum_k E_k n_k - E)$$

or using Stirling' formula and neglecting constants,

$$F = \sum_k ((n_k + g_k)log(n_k + g_k) - n_k log(n_k)) - b(\sum_k E_k n_k - E)$$

Setting $\frac{\partial F}{\partial n_k} = 0$ gives

$$log(n_k + g_k) - log(n_k) - bE_k = 0$$

or

$$1 + g_k/n_k = bE_k, n_k = g_k/(exp(bE_k) - 1)$$

If we put a constraint on the total number of particles n, then the function to be maximized is

$$F = \sum_k ((n_k + g_k)log(n_k + g_k) - n_k log(n_k)) - a(\sum_k n_k - n) - b(\sum_k E_k n_k - E)$$

Setting $\frac{\partial F}{\partial n_k} = 0$ then gives

$$log(n_k + g_k) - log(n_k) - a - bE_k = 0$$

or

$$n_k = \frac{g_k}{exp(a + bE_k) - 1}$$

This is the Bose-Einstein distribution. The constants a, b are determined from the constraints

$$\sum_k n_k = n, \sum_k E_k n_k = E$$

[26] Coherent/exponential vectors in the Fock spaces.

Let $\mathcal{H}_0 = L^2(\mathbb{R})$ and consider the standard position and momentum operators, $q, p = -id/dq$ acting in this space. Construct the annihilation and creation operators a, a^* in this space as

$$a = (q + ip)/\sqrt{2}, a^* = (q - ip)/\sqrt{2}$$

Show that

$$[a, a^*] = 1$$

Now consider copies of $\mathcal{H}_0$ and in the kth copy, let the annihilation and creation operators be denoted by a_k, a_k^*. Thus, in the tensor product of p such copies $\mathcal{H}_0^{\otimes p}$, we have defined operators $a_1, ..., a_p, a_1^*, ..., a_p^*$ satisfying $[a_i, a_j^*] = \delta_{ij}, [a_i, a_j] = 0, [a_i^*, a_j^*] = 0$. Let $|n_1, ..., n_p>$ be the tensor product of the state $|n_1>, ..., |n_p>$ where $|n_k>$ is the state in the kth copy of $\mathcal{H}_0$ such that $a_k|n_k>= \sqrt{n_k}|n_k-1>, a_k^*|n_k>= \sqrt{n_k+1}|n_k+1>$ and $a_k^*a_k|n_k>= n_k|n_k>, a_k a_k^*|n_k>= (n_k+1)|n_k>$. Define for $z \in \mathbb{C}^p$, the state

$$|e(z)>= \sum_{n_1,...,n_p \geq 0} z_1^{n_1}...z_p^{n_p}|n_1, ..., n_p> / \sqrt{n_1!...n_p!}$$

$$= \sum_{n_1,...,n_p \geq 0} z_1^{n_1}...z_p^{n_p} a_1^{*n_1}...a_p^{*n_p}|0> / n_1!...n_p!$$

$$= exp(z.a^*)|0>, z.a^* = z_1 a_1^* + ... + z_p a_p^*$$

Show using

$$< n_1, ..., n_p|m_1, ..., m_p>= \Pi_{k=1}^p \delta[n_k - m_k] = \delta[n-m]$$

that for $z, u \in \mathbb{C}^p$,

$$< e(z)|e(u)>= exp(< z, u>)$$

Now, define for $z \in \mathbb{C}^p, n \geq 0$,

$$\psi(z^{\otimes n}) = \sum_{n_1+...+n_p=n} \frac{\sqrt{n!}}{\sqrt{n_1!...n_p!}} z_1^{n_1}...z_p^{n_p}|n_1, ..., n_p>$$

Then show using the multinomial theorem that for $z, u \in \mathbb{C}^p$,

$$< \psi(z^{\otimes n}), \psi(u^{\otimes n})>=< z, u>^n$$

Now extend the map ψ linearly to the Boson Fock space

$$\Gamma_s(\mathbb{C}^p) = \bigotimes_{n \geq 0} (\mathbb{C}^p)^{\otimes_s n}$$

by defining for

$$|f(z)>= \bigotimes_{n \geq 0} \frac{z^{\otimes n}}{\sqrt{n!}} \in \Gamma_s(\mathbb{C}^p)$$

$$\psi(|f(z)>) = |e(z)>$$

or equivalently,

$$\psi(z^{\otimes n}/\sqrt{n!}) =$$

$$\sum_{n_1+...+n_p=n} \frac{z_1^{n_1}...z_p^{n_p}}{\sqrt{n_1!...n_p!}}|n_1, ..., n_p>, n \geq 0$$

Show that ψ defines a Hilbert space isomorphism between $\Gamma_s(\mathbb{C}^p)$ and $L^2(\mathbb{R}^p)$. This gives a physical interpretation of the coherent states in Boson Fock space in terms of the p-dimensional Harmonic oscillator algebra.

[25] Creation, conservation and annihilation operators in the Boson Fock space.
[26] The general theory of quantum stochastic processes in the sense of Hudson and Parthasarathy.
[27] The quantum Ito formula of Hudson and Parthasarathy.
[28] The general theory of quantum stochastic differential equations.
[29] The Hudson-Parthasarathy noisy Schrodinger equation and the derivation of the GKSL equation from its partial trace.
[30] The Feynman path integral for solving the Schrodinger equation.

Let $X(t), t \geq 0$ be a Markov process with values in $\mathbb{R}$ having generator K, ie

$$\mathbb{E}[\phi(X(t+dt))|X(t) = x] = \phi(x) + K\phi(x)dt + o(dt)$$

Note that by including generalized functions such as the Dirac Delta function and its derivatives, we may represent K as an integral kernel:

$$K\phi(x) = \int K(x, y)phi(y)dy$$

We define

$$u(s, t, x) = \mathbb{E}[f(X(t))exp(\int_s^t V(X(s))ds)|X(s) = x], 0 \leq s \leq t$$

Then an easy application of the Markov property shows that

$$u(s,t,x) = (1+V(x)ds)\mathbb{E}[(u(s+ds,t,X(s+ds))|X(s)=x] + o(ds)$$

$$= (1+V(x)ds)(u(s,t,x)ds + u_{,s}(s,t,x) + ds\int K(x,y)u(s,t,y)dy) + o(ds)$$

and hence,

$$\partial_s u(s,t,x) + V(x)u(s,t,x) + \int K(x,y)u(s,t,y)dy = 0$$

Further,

$$u(t,t,x) = f(x)$$

In particular, if $X(t)$ is Brownian motion, we have

$$K(x,y) = (1/2)\delta''(x-y)$$

and the above becomes

$$\partial_s u(s,t,x) + V(x)u(s,t,x) + (1/2)\partial_x^2 u(s,t,x) = 0$$

which is the Feynman-Kac formula. Since we are assuming that the Markov process is time homogeneous, ie its generator K is time independent, it follows that $u(s,t,x) = u(0,t-s,x)$. We denote this by $v(t-s,x)$ and then we get

$$\partial_t v(t,x) = V(x)v(t,x) + \int K(x,y)v(t,y)dy, t \geq 0, v(0,x) = f(x)$$

Feynman formally replaced t by it, $i = \sqrt{-1}$, ie define $w(t,x) = v(it,x)$ in the above formula to get

$$i\partial_t w(t,x) = -V(x)w(t,x) - \int K(x,y)w(t,y)dy$$

Now replace V by $-V$ to get the generalized Schrodinger equation

$$i\partial_t w(t,x) = V(x)w(t,x) - \int K(x,y)w(t,y)dy, w(0,x) = f(x)$$

where

$$w(t,x) = \mathbb{E}[exp(-\int_0^{it} V(X(s))ds)f(X(it))|X(0)=x]$$

$$= \mathbb{E}[exp(-i\int_0^{t} V(X(is)ds)f(X(it))|X(0)=x]$$

Formally, this formula can be interpreted as follows: $v(t,x)$ is approximated as

$$v(t,x) = \mathbb{E}[exp(-\int_0^{t} V(X(s))ds)f(X(t))|X(0)=x] \approx$$

$$\int exp(\sum_{k=0}^{n}(-V(x_k)) + (log(K))(x_k,x_{k+1}))\delta s_k)f(s_n)\Pi_{0\leq k\leq n}dx_k$$

where

$$0 = s_0 < s_1 < ... < s_n = t$$

and $log(K)$ is the operator logarithm of K (not $log(K(x,y))$).

[31] Comparison between the Hamiltonian (Schrodinger-Heisenberg) and Lagrangian (path integral) approaches to quantum mechanics.

The Schrodinger-Heisenberg approaches to quantum mechanics are Hamiltonian approaches. Feynman proposed an alternative approach to non-relativistic quantum mechanics that is based on the Lagrangian. To see how this proceeds, assume that the Hamiltonian of the system is $H(t,q,p)$. The Lagrangian is then

$$L(t,q,q') = (p,q') - H(t,q,p)$$

with

$$p = \frac{\partial L}{\partial q'}$$

Thus,

$$< t + dt, q''|t, q' >=< t, q''|exp(-iH(t, q, p)dt)|t, q' >$$

We can applying the commutation relations between q and p, assume that in the Hamiltonian $H(t, q, p)$, all the $p's$ appear to the left of all the $q's$. Then we get

$$< t + dt, q''|t, q' >= \int < t, q''|t, p' > dp' < t, p'|exp(-iH(t, q, p)dt)|t, q' >$$

$$= C \int exp(i(q", p'))exp(-iH(t, q', p')dt)|t, q' > dp'$$

$$= C \int exp(i((q'', p') - H(t, q', p')dt)).exp(-i(q', p'))dp'$$

$$= C \int exp(i[(q'' - q', p')/dt - H(t, q', p')]dt))dp'$$

By composing these infinitesimal amplitudes, we get the transition amplitude for finite time as

$$K(t_2, q_2'|t_1, q_1') =< t_2, q_2'|t_1, q_1' >=$$

$$C \int_{q(t_1)=q_1', q(t_2)=q_2'} exp(i \int_{t_1}^{t_2} (p(t), q'(t)) - H(t, q(t), p(t)))dt)\Pi_{t_1 \leq t \leq t_2} dq(t)dp(t)$$

This is indeed the formula for the time evolution operator kernel in the position representation, ie

$$K(t_2, q_2'|t_1, q_1') =< t_2, q_2'|T\{exp(-i \int_{t_1}^{t_2} H(t)dt)\}|t_1, q_1' >$$

where $T\{\}$ is the time ordering operator. In the special case, when the Hamiltonian has the form

$$H(t, q, p) = p^2/2m + V(q)$$

the integral over p becomes a Gaussian integral and therefore it can be replaced by evaluating the action integral at the stationary point, ie at $p(t)$ given by

$$\frac{d}{dp(t)}((p(t), q'(t)) + p^2(t)/2m) = 0$$

ie

$$q'(t) = p(t)/m, p(t) = mq'(t)$$

Thus, in this special case,

$$K(t_2, q_2'|t_1, q_1') = C \int_{q(t_1)=q_1', q(t_2)=q_2'} exp(i \int_{t_1}^{t_2} (mq'(t)^2/2 - V(q(t)))dt)\pi_{t_1 < t < t_2} dq(t)$$

Chapter 4
Quantum Field Theory, Many Particle Systems

[32] The quantum theory of fields: Dirac's relativistic equation for the electron, Quantization of the Klein-Gordon field, quantization of the electromagnetic field, quantization of the Dirac electron-Positron field interacting with a quantum electromagnetic field, The Feynman diagrams for calculating scattering probabilities between electrons, positrons and photons. Dyson's general theory of renormalization. Why is the gravitational field non-renormalizable?

Quantum theory of fields:

[a] Quantization of the Klein-Gordon field: The Lagrangian density for this field is

$$L(\phi(x), \phi_{,\mu}(x) = (1/2)(\partial_\mu \phi)(\partial^\mu \phi) - m^2\phi^2/2$$

$$= (1/2)\phi_{,0}^2 - (1/2)(\nabla\phi)^2 - m^2\phi^2/2$$

We are assuming here that the field is real. The canonical classical field equations are

$$\partial_\mu \frac{\partial L}{\partial \phi_{,\mu}} - \frac{\partial L}{\partial \phi} = 0$$

which give us the Klein-Gordon equation

$$\partial_\mu \partial^\mu \phi + m^2\phi = 0$$

or equivalently,

$$\phi_{,00} - \nabla^2\phi + m^2\phi = 0$$

This has the following interpretation in quantum mechanics (not yet quantum field theory): In the special theory of relativity, the energy momentum relation is

$$E^2 - p^2c^2 - m^2c^4 = 0, p^2 = p_x^2 + p_y^2 + p_z^2$$

According to the standard rules of quantum mechanics, the momentum three vector p is replaced by $-ih\nabla/2\pi$ and the energy E by $(ih/2\pi)\partial/\partial t$ yielding thereby the KG equn.

$$(E^2 - p^2c^2 - m^2c^4)\phi = 0$$

or

$$[-h^2\partial_t^2/4\pi^2 + (h^2c^2/4\pi^2)\nabla^2 - m^2c^4]\phi = 0$$

or

$$[\partial_t^2 - c^2\nabla^2 + 4\pi^2m^2c^4/h^2]\phi = 0$$

which is the KG equation once normalized units are used: $h/2\pi = 1, c = 1$. To canonically quantize this field, we first construct the Hamiltonian density corresponding to the Lagrangian density by performing a Legendre transformation:

$$\pi(x) = \frac{\partial L}{\partial \phi_{,0}} = \phi_{,0}$$

The Hamiltonian density is then

$$\mathcal{H}(\phi, \nabla\phi, pi) = \pi\phi_{,0} - L =$$
$$(1/2)(\pi^2 + (\nabla\phi)^2 + m^2\phi^2)$$

We can check that classical Hamiltonian equations of motion of the field obtained from the Hamiltonian

$$H = \int \mathcal{H} d^3x$$

as

$$\pi_{,0} = -\delta H/\delta\phi, \phi_{,0} = \delta H/\delta\pi$$

with the variational derivative

$$\delta H/\delta\phi = \frac{\partial \mathcal{H}}{\partial \phi} - (\nabla, \frac{\partial \mathcal{H}}{\partial \nabla\phi}),$$

$$\delta H/\delta\pi = \frac{\partial \mathcal{H}}{\partial \pi}$$

yield the correct classical KG field equations.

Remark: The variational derivative of H w.r.t ϕ is obtained by considering the change in H under a small change $\delta\phi$ in ϕ:

$$\delta H = \int \frac{\partial \mathcal{H}}{\partial \phi}\delta\phi + (\frac{\mathcal{H}}{\partial \nabla \phi}, \nabla \delta\phi)d^3x$$

$$= \int [\frac{\partial \mathcal{H}}{\partial \phi} - (\nabla, \frac{\partial H}{\partial \nabla \phi})]\delta\phi d^3x$$

where we have used integration by parts, $\delta\nabla\phi = \nabla\delta\phi$ and the assumption that $\delta\phi(x)$ vanishes when the spatial coordinates go to ∞. The Schrodinger equation for this Hamiltonian reads:

$$[(1/2)\int(-\delta^2/\delta\phi(r)^2 + (\nabla\phi(r))^2 + m^2\phi(r)^2)d^3r]\psi_t(\phi(r) : r \in \mathbb{R}^3) = (ih/2\pi)\frac{\partial}{\partial t}\psi_t(\phi(r) : r \in \mathbb{R}^3)$$

In other words, the second quantized Klein Gordon field is determined by a continuous infinity of quantum Harmonic oscillators. This does not make much sense so we work ih the spatial frequency domain wherein we express the solution the KG equation as a superposition of plane waves:

$$\phi(t, r) = \int [f(k)a(k)exp(-i(\omega(k)t - k.r)) + \bar{f}(k)a(k)^*exp(i(\omega(k)t - k.r))]d^3k$$

$$\omega(k) = (k^2 + m^2)^{1/2}$$

so that

$$\pi(t, r) = \phi_{,0} = -i\int [\omega(k)f(k)a(k)exp(-i(\omega(k)t - k.r)) - \bar{f}(k)a(k)^*exp(i(\omega(k)t - k.r))]d^3k$$

In the second quantized picture, $a(k), a(k)^*$ are operators and $\phi(t, r)$ is also (field) operator. The canonical equal time commutation rules (we are assuming ϕ to be a bosonic field) are

$$[\phi(t, r), \pi(t, r')] = i\delta(r - r')$$

ϕ is called the canonical position field and π the canonical momentum field. In accordance with the commutation rules for creation and annihilation operators of a harmonic oscillator, we assume that

$$[a(k), a(k')^*] = \delta(k - k'), [a(k), a(k')] = [a(k)^*, a(k')^*] = 0$$

Then we get

$$i\delta(r-r') = i\int [\omega(k')f(k)\bar{f}(k')[a(k), a(k')^*]exp(-i(\omega(k)-\omega(k'))t$$
$$+i(k.r-k'.r'))+\omega(k')f(k)\bar{f}(k')[a(k'), a(k)^*]exp(i(\omega(k)-\omega(k'))t-$$

or equivalently,

$$\delta(r-r') = \int [\omega(k')f(k)\bar{f}(k')exp(-i(\omega(k)-\omega(k'))t+i(k-k').r)+\omega(k')f(k)\bar{f}(k')exp(i(\omega(k)$$
$$-\omega(k'))t-i(k-k').r)]\delta(k-k')d^3kd^3k'$$

$$= \int |f(k)|^2\omega(k)(exp(ik.(r - r')) + exp(-ik.(r - r')))d^3k$$

so we require that

$$|f(k)|^2\omega(k) = (1/2)(1/(2\pi)^3) = 1/16\pi^3$$

We may thus take

$$f(k) = (1/4\pi^{3/2})\omega(k)^{-1/2}$$

$a(k)$ is to be interpreted as the annihilation operator of a KG boson having momentum k (spinless boson) and $a(k)^*$ is to be interpreted as teh creation operator of a KG boson. We now compute the second quantized Hamiltonian in terms of the creation and annihilation fields:

$$H = (1/2)\int [\phi_{,0}^2 + (\nabla\phi)^2 + m^2]d^3r$$

$$= \int \omega(k)a(k)^*a(k)$$

Exercise:Verify this formula by performing the integrations w.r.t $d^3k, d^3k'd^3r$. Now to check this, we calculate the Heisenberg equation of motion of the creation and annihilation operators:

$$da(k,t)/dt = i[H, a(k,t)] = i\int \omega(k')[a(k',t)^*, a(k,t)]a(k,t)d^3k'$$

$$= -i\int \omega(k')\delta(k-k')d^3k'a(k,t) = i\omega(k)a(k,t)$$

and hence

$$a(k,t) = a(k)exp(-i\omega(k)t)$$

Likewise,

$$a(k,t)^* = a(k)^*exp(i\omega(k)t)$$

Thus, we get

$$\phi(t,r) = \int (f(k)a(k,t)exp(ik.r) + \bar{f}(k)a(k,t)^*exp(-ik.r))d^3k$$

and hence

$$\nabla^2\phi - m^2\phi = -\int \omega(k)^2(f(k)a(k,t)exp(ik.r) + \bar{f}(k)a(k,t)^*exp(-ik.r))d^3k$$

$$= \phi_{,00}$$

in agreement with the KG equation. Now suppose that we apply an external field $f(x)$ to the KG system. The resulting Lagrangian density is then given by

$$L = L_0 + f(x)V(\phi(x))$$

where $V(\phi(x))$ is some linear/nonlinear function of the KG field. We wish to approximately compute the transition probability of the KG field from the initial state $|\phi_i>$ at time t_1 in which the field is exactly a given function $\phi_i(r), r \in \mathbb{R}^3$ of the spatial variables to the final state $|\phi_f>$ at time t_2 in which the field is exactly another given function $\phi_f(r), r \in \mathbb{R}^3$ of the spatial variables. The corresponding Hamiltonian will then have the form

$$H(t) = \int \mathcal{H}_0(\phi(x), \nabla\phi(x), \pi(x))d^3x - \int f(x)V(\phi(x))d^3x$$

(Note: $x = (t,r), d^3x = d^3r, d^4x = d^3xdt = d^3rdt$) where $\mathcal{H}_0$ is the Hamiltonian density corresponding to the Lagrangian density L_0 (ie obtained by applying the Legendre transform to L_0):

$$H_0 = (1/2)\int (\pi^2 + (\nabla\phi)^2 + m^2\phi^2)d^3x$$

The transition probability amplitude from $|\phi_i>\rightarrow|\phi_f>$ in the time duration $[t_1, t_2]$ can be calculated using the Feynman path integral formula:

$$< \phi_f|S[t_2,t_1]|\phi_i >=$$

$$C\int_{\phi(t_1,.)=\phi_i, \phi(t_2,.)=\phi_f} exp(-(\int_{[t_1,t_2]\times\mathbb{R}^3} Ldtd^3x))\Pi_{r\in\mathbb{R}^3, t\in(t_1,t_2)}d\phi(x)$$

$$= C\int exp(iS_0)(1 + i\int f(x)V(\phi(x))dtd^3x + (i^2/2!)\int f(x)f(x)'\phi(x)\phi(x')dtdt'd^3xd^3x' + ..)\Pi d\phi(x)$$

where

$$S_0 = \int L_0dtd^3x = (1/2)\int (\partial_\mu\phi\partial^\mu\phi - m^2\phi^2)d^td3x$$

By expanding $V(\phi(x)$ as a power series in $\phi(x)$, the computation of the above path integral reduces to computing the moments of a complex infinite dimensional zero mean Gaussian distribution sinc S_0 is a quadratic functional of ϕ. In particular, we note that the odd moments of a symmetric Gaussian distribution are zero and the even moments can be computed by summing the products of the second moments taken over all partitions of the product fields into pairs. Thus, computation of the second moments of such a Gaussian distribution beomes significant, ie,

$$D(x,y) = C\int exp(iS_0)\phi(x)\phi(y)\Pi_{z\in\mathbb{R}^4}d\phi(z)$$

if we are interested in transitions from $t = -\infty$ to $t = +\infty$. From standard methods in quantum mechanics, it is easily seen that

$$D(x,y) =< 0|T\{\phi(x)\phi(y)\}|0 >$$

provided that we use the interaction representation which removes the effect of the unperturbed Hamiltonian H_0. If we use the Schrodinger representation, then we would have to compute D as

$$D(x,y) =< 0|T\{U(\infty,-\infty)\phi(x)\phi(y)\}|0 >=$$

$$< 0|U(\infty,t_x)\phi(x)U(t_x,t_y)\phi(y)U(t_y,-\infty)|0 >$$

assuming $t_x \geq t_y$ and where U is the unperturbed Schrodinger evolution operator. Here $|0>$ is the vacuum state of the field. The function $D(x, y)$ is called the propagator. The complete propagator taking into account interactions is defined as

$$D_c(x, y) = C \int exp(iS)\phi(x)\phi(y)\Pi_z d\phi(z)$$

where

$$S = S_0 + \int f(x)V(\phi(x))d^4x = \int Ld^4x$$

We can write a perturbative expansion for D_c as

$$D_c(x, y) = \int exp(iS_0)(1 + iS_1 + i^2S_1^2/2! + ..)\phi(x)\phi(y)\Pi_z d\phi(z)$$

where

$$S_1 = \int f(x)V(\phi(x))d^4x$$

is the perturbation to the action caused by external field coupling. Even if there is no external field, but there is a small perturbation to the Lagrangian density/Hamiltonian density, the above series expansion can be used to determine the complete propagator It was Feynman's genius to recognize that the various perturbation terms in D_c can be calculated easily using a digrammatic method which could be applied to more complex situations like quantum electrodynamics wherein the quantum fields are the electromgnetic four potential $A_\mu(x)$ and the Dirac four component spinor wave function $\psi(x)$. Let us now formally compute the propagator of the unperturbed KG field:

$$S_0 = \int \phi(x)[(1/2)\partial_\mu\partial^\mu - m^2/2)\delta^4(x - y)]\phi(y)d^4xd^4y$$

$$= \int \phi(x)K(x, y)\phi(y)d^4xd^4y$$

and hence a simple Gaussian second moment evaluation gives

$$D(x, y) = \int exp(iS_0[\phi])\phi(x)\phi(y)\Pi_z d\phi(z)$$

$$= C_1(det(iK))^{-1/2}.K^{-1}(x, y)$$

In other words $D(x, y)$ is proportional to $K^{-1}(x, y)$ where K^{-1} is the inverse Kernel of K:

$$\int K^{-1}(x, y)K(y, z)d^4y = \delta^4(x - z)$$

We can write $K(x, y) = K(x - y)$ and then defining its four dimensional Fourier transform:

$$\hat{K}(p) = \int K(x)exp(-ip.x)d^4x, p.x = p_\mu x^\mu = p^0x^0 - p^1x^1 - p^2x^2 - p^3x^3$$

we get

$$K^{-1}(p) = 1/K(p)$$

Clearly,

$$K(x) = (1/2)\partial_\mu\partial^\mu - m^2/2)\delta^4(x)$$

and hence

$$\hat{K}(p) = (p_\mu p^\mu - m^2)/2$$

Thus,

$$\hat{D}(p) = \frac{C_0}{p^2 - m^2}$$

where

$$p^2 = p_\mu p^\mu = p^{02} - p^{12} - p^{22} - p^{32}$$

Finally,

$$D(x, y) = D(x - y) = C_0/(2\pi)^4 \int \frac{exp(ip.x)}{p^2 - m^2} d^4p$$

The corrected (complete) propagator:

$$D_c(x, y) = \int exp(iS_0[\phi])\phi(x)\phi(y)(1 + i \int f(x)V(\phi(z))d^4z + ...)\Pi_u d\phi(u)$$

Clearly, we can write this in operator kernel notation as $D_c = D + D\Sigma D + D\Sigma D\Sigma D + ...$ using the property of moments of a Gaussian distribution. For example, if $V(\phi) = \phi^4$ and $f = c_0$, then in the Gaussian average of the product $\phi(x)\phi(y)\int f(z)\phi(z)^4 d^4z$, we get the coupling terms $4 < \phi(x)\phi(z) >< \phi(z)^2 >< \phi(z)\phi(y) >$ so if we define $\Sigma(z) = 4f(z)0 < \phi(z)^2 >$, we can write

$$< \phi(x)\phi(y)(\int f(z)\phi(z)^4 d^4z) >= \int D(x-z)\Sigma(z)D(z-y)d^4z$$

Likewise, for the next perturbation term

$$< \phi(x)\phi(y)(\int f(z)\phi(z)^4 d^4z)^2 >=$$

$$\int f(z_1)f(z_2) < \phi(x)\phi(y)\phi^4(z_1)\phi^4(z_2) > d^4z_1 d^4z_2$$

Again, this can be expressed using the Gaussian moments formula as a sum of terms of the form

$$\int f(z_1)f(z_2) < \phi(x)\phi(z_1) >< \phi(z_1)^3\phi(z_2)^3 >< \phi(z_2)\phi(y) > d^4z_1 d^4z_2$$

and

$$\int f(z_1)^2 < \phi(x)\phi(z_1) >< \phi(z_1)^2\phi(z_2)^4 >< \phi(z_1)\phi(y) > d^4z_1 d^4z_2$$

etc. Now, each term $< \phi(z_1)^m\phi(z_2)^m >$ is a product of propagators $D(z_1 - z_2)$ and $D(0)$ so the above general form is valid.

[b] Quantization of the electromagnetic field. The Lagrangian density is

$$L_F = -\frac{1}{4}F_{\mu\nu}F^{\mu\nu}, F_{\mu\nu} = A_{\nu,\mu} - A_{\mu,\nu}$$

Thus,

$$F_{0r} = E_r, F_{12} = -B_3, F_{23} = -B_1, F_{31} = -B_2$$

We get

$$L_F = \frac{1}{2}(E^2 - B^2)$$

as required. We compute the canonical momenta:

$$\pi_r = \frac{\partial L_F}{\partial A^r_{,0}} = -F_{0r}$$

and

$$\pi_0 = \frac{\partial L_F}{\partial A^0_{,0}} = 0$$

This is inconsistent with the canonical commutation relations

$$[A^\mu(t,r), \pi_\nu(t,r')] = i\delta^\mu_\nu\delta^3(r - r')$$

Hence we must use the Dirac brackets which are modifications of the Poisson/Lie bracket when constraints are taken into account. Thie first constraint is

$$\pi_0 = 0$$

and the second constraint is obtained from the equations of motion:

$$\partial_r \frac{\partial L_F}{\partial A^0_{,r}} = -J^0$$

where J^μ the four current density is a matter field unlike the $A^{\mu'}s$ which are the Maxwell fields. These equations can be expressed as

$$\chi_1 = \partial_r\pi_r + J^0 = 0$$

and since time derivatives do not appear here, this equation should be regarded as a constraint, ie, a relationship between the matter field and the electromagnetic field. The above equation is obtained by adding to the field Lagrangian density, the matter-field interaction Lagrangian

$$L_{int} = -J^\mu A_\mu$$

Now, we work in the Coulomb gauge (we are free to impose a gauge condition on the potentials that leaves the actual electric and mangetic field invariant). In this gauge, $divA = 0$, ie,

$$\chi_2 = A^r_{,r} = 0$$

The Maxwell equations in this gauge imply that

$$\nabla^2 A^0 = -J^0$$

which has solution

$$A^0(t,r) = \int \frac{J^0(t,r')}{|r-r'|} d^3r'$$

and since both sides of the above equation are taken at the same time, we can regard A^0 as a matter field. Hence, the quantized electromagnetic field is described by only three position fields $A^r, r = 1,2,3$ and once we impose the Coulomb gauge condition, there are only two degrees of freedom for the position fields. We now calculate the field Hamiltonian when it interacts with matter. The Hamiltonian density is with $L = L_F + L_{int}$,

$$\mathcal{H} = \pi_r A^r_{,0} - L = -F_{0r} A^r_{,0} + (1/4) F_{\mu\nu} F^{\mu\nu} + J^\mu A_\mu$$

$$\frac{1}{2} F_{0r}(A_{r,0} + A_{0,r}) + (1/4) F_{rs} F_{rs} + J^\mu A_\mu$$

Thus, making use of the matter equation $A_{0,rr} = -J^0$, the constraint χ_1 and neglecting a 3-dimensional divergence (which will not affect the total Hamiltonian), we get

$$\begin{aligned}\mathcal{H} &= \frac{1}{2} F_{0r} F_{0r} + F_{0r} A_{0,r} + (1/4) F_{rs} F_{rs} + J^\mu A_\mu \\ &= \pi^2/2 + (\nabla \times A)^2/2 - F_{0r,r} A_0 + J^\mu A_\mu \\ &= (1/2)(\pi^2 + (\nabla \times A)^2) + \pi_{r,r} A^0 + J^\mu A_\mu \\ &= (1/2)(\pi^2 + (\nabla \times A)^2) - J^0 A^0 + J^\mu A_\mu\end{aligned}$$

The term $J^0 A^0$ is a pure matter field while the two terms within the bracket are pure field terms. This simplifies to

$$\mathcal{H} = (1/2)(\pi^2 + (\nabla \times A)^2) - J.A$$

where

$$J.A = J^r A^r$$

There is no pure matter term in this Hamiltonian. We define

$$\pi_\perp = \pi - \nabla A^0$$

ie

$$\pi_{\perp r} = \pi_r - A^0_{,r}$$

Then,

$$div \pi_\perp = \pi_{r,r} - A^0_{,rr} = -J^0 + J^0 = 0$$

Thus $\pi_\perp$ is a solenoidal field. Then, we can express

$$\mathcal{H} = \frac{1}{2}(\pi_\perp^2 + (\nabla \times A)^2) - J.A + (\nabla A^0)^2/2 - - - (1)$$

since the term $(\nabla A^0, \pi_\perp)$ on perfoming a $3 - D$ integration is zero because it is a perfect 3-D divegence:

$$(\nabla A^0, \pi_\perp) = div(A^0 \pi_\perp)$$

(1) is our final form of the Hamiltonian of the electromagnetic field interacting with an external current source.

We note that the last term $(\nabla A^0)^2/2$ is a pure matter field. Hence, if we are bothered only about the electromgnetic field and its interaction with matter, the Hamiltonian density is

$$\mathcal{H}_F = (1/2)(\pi^2 + (\nabla \times A)^2)$$

where we have renamed $\pi_\perp$ as π for convenience of notation. Our constraints are $div\pi = 0, divA = 0$.

Dirac brackets for constraints: Suppose $Q_1, ..., Q_n, P_1, ..., P_n$ are the unconstrained positions and momenta of a system. The constraints are $Q_j = P_j = 0, j = n+1, ..., n+p$. Without loss of generality, we are choosing our constrained variables as new positions and momenta. The Poisson bracket relations are

$$\{f, g\} = \sum_{i=1}^{n+p} f_{,Q_i} g_{,P_i} - f_{,P_i} g_{,Q_i})$$

In particular, we get the contradiction

$$f_{,Q_i} = -f_{,P_i}, \{f, P_i\} = f_{,Q_i}, i > n$$

since $Q_i = P_i = 0, i > n$. In order to rectify this problem, Dirac introduced a new kind of bracket defined as follows: Let

$$\chi_{ij} = \{\eta_i, \eta_j\} = J_{ij}$$

J is the standard symplectic matrix of size $2p \times 2p$. where

$$\eta = [Q_{n+1}, ..., Q_{n+p}, P_{n+1}, ..., P_{n+p}]^T$$

$Q_{n+i}, P_{n+i}, i = 1, 2, ..., p$, ie η_i are functions of $Q_i, P_i, i = 1, 2, ..., n$ and the bracket $\{.,.\}_{eff}$ is calculated using $Q_i, P_i, i = 1, 2, ..., n$ and regarding Q_{n+i}, P_{n+i} as functions of $Q_j, P_j, j \leq n$. The bracket $\{f, g\}_P$ is computed using $Q_i, P_i, i \leq n$ and taking $Q_{n+i} = 0, P_{n+i} = 0$:

$$\{f, g\}_P = \sum_{i=1}^{n} (f_{,Q_i} g_{,P_i} - f_{,P_i} g_{,Q_i})$$

We have

$$C = \chi^{-1} = -J$$

as $2p \times 2p$ matrices. Then, the Dirac bracket is defined as

$$\{f, g\}_D = \{f, g\} + \sum_{i,j} \{f, \eta_i\} J_{ij} \{\eta_j, g\}$$

We see that for $k \leq n$,

$$\{f, Q_k\}_D = \{f, Q_k\} = -f_{,P_k}$$

since

$$\{\eta_j, Q_k\} = 0, k \leq n$$

Note that $\{.,.\}$ is the unconstrained Poisson bracket. Again, we note that

$$\{f, P_k\}_D = \{f, P_k\} = f_{,Q_k}$$

since $\{P_k, \eta_j\} = 0, k \leq n$. Further, for $i, j \geq 1$, we have

$$\{f, \eta_i\}_D = \{f, \eta_i\} + \sum_{k,l} \{f, \eta_k\} J_{kl} \{\eta_l, \eta_i\}$$

$$= \{f, \eta_i\} - \sum_{k,l} \{f, \eta_k\} C_{kl} \chi_{li} = 0$$

since

$$\sum_l C_{kl} \chi_{li} = \delta_{ki}$$

We note that

$$\{f, g\}_D = \{f, g\} - \sum_{i,j} \{f, \eta_i\} J_{ij} \{\eta_j, g\}$$

$$= \sum_{i \leq n} (f_{,Q_i} + \sum_j (f_{,\eta_j} \eta_{j,Q_i}))(g_{,P_i} + \sum_j g_{,\eta_j} \eta_{j,P_i})$$

$$-interchange of f and g$$

$$+\sum_i f_{,\eta_i} J_{ij} g_{,\eta_j} + \sum \{f, \eta_i\} J_{ij} \{\eta_j, g\}$$

$$= \sum_{i \le n} (f_{,Q_i} + \sum_j (f_{,\eta_j} \eta_{j,Q_i}))(g_{,P_i} + \sum_j g_{,\eta_j} \eta_{j,P_i})$$

This formula tells us that the Dirac bracket between two observables is calculated using the Poisson bracket w.r.t. the unconstrained variables and by regarding the constrained variables as functions of the unconstrained variables.

More generally, suppose $Q = (Q_1, ..., Q_n), P = (P_1, ..., P_n)$ are arbitrary canonical coordinates and the constraints are of the form

$$\chi_i(Q, P) = 0, i = 1, 2, ..., r$$

Define the Poisson bracket $\{.,.\}$ as usual w.r.t Q, P, and then define the Dirac bracket

$$\{f, g\}_D = \{f, g\} - \sum_{i,j=1}^{r} \{f, \chi_i\} C_{ij} \{\chi_j, g\}$$

where

$$((C_{ij})) = ((\{\chi_i, \chi_j\}))^{-1}$$

Then,

$$\{f, \chi_k\}_D = \{f, \chi_k\} - \sum_{i,j} \{f, \chi_i\} C_{ij} \{\chi_j, \chi_k\}$$

$$= \{f, \chi_k\} - \sum_i \{f, \chi_i\} \delta_{ik} = 0$$

as required. Further,

$$\{f, Q_k\}_D = -f_{,P_k} - \sum_{i,j} f_{,P_i} C_{ij} \chi_{j,P_k}$$

It is then not hard to show that the rhs is the same as $-d_{P_k} f$, ie, the partial derivative of f w.r.t. P_k where we regard f as an independent function of Q, P and the $\chi_i's$ and then defined

$$d_{P_k} f = f_{,P_k} + \sum_j f_{,\chi_j} \chi_{j,P_k}$$

Now, we evaluate the Dirac bracket between π_i and A^j taking into account the constraints:

$$\chi_1 = \pi_{i,i} = 0, \chi_2 = A_{i,i} = 0$$

We get

$$\{\chi_1(t, r), \chi_2(t, r')\} = i\nabla^2 \delta^3(r - r')$$

The inverse kernel of the rhs is $K(r - r') = i/4\pi|r - r'|$. Further,

$$\{A^m(t, r), \chi_1(t, r')\} = -i\delta_k^m \delta_{,k}^3(r - r') = i\delta_{,m}^3(r - r')$$

$$\{A^m(t, r), \chi_2(t, r')\} = 0,$$

$$\{\chi_1(t, r), \pi_m(t, r')\} = 0,$$

$$\{\chi_2(t, r), \pi_m(t, r')\} = i\delta_m^k \delta_{,k}^3(r - r') = i\delta_{,m}^3(r - r')$$

Hence,

$$\{A^m(t, r), \pi_k(t, r')\}_D = i\delta_k^m \delta^3(r - r') - \int d^3r'' d^3r''' \{A^m(t, r), \chi_1(t, r'')\} K(r'' - r''') \{\chi_2(t, r'''), \pi_k(t, r')\}$$

$$= i\delta_k^m \delta^3(r - r') + \int d^3r'' d^3r''' \delta_{,m}^3(r - r'') K(r'' - r''') \delta_{,k}^3(r''' - r')$$

$$= i\delta_k^m \delta^3(r - r') - \int (\frac{\partial^2}{\partial x^{m''} \partial x^{k'''}} K(r'' - r''')) \delta^3(r - r'') \delta^3(r''' - r') d^3r'' d^3r'''$$

$$= i\delta_k^m \delta^3(r - r') - \frac{\partial^2}{\partial x^m \partial x^{k'}} K(r - r')$$

$$= i\delta_k^m \delta^3(r - r') + K_{,mk}(r - r')$$

where

$$K(r) = i/4\pi|r|$$

Quantum electrodynamics using creation and annihilation operators for photons, electrons and positrons: We work in the Coulomb gauge so that $divA = 0$ and this implies $\nabla^2 A^0 = -J^0$, ie, A^0 is a matter field. The Maxwell wave equation for A in the absence of matter, ie charge and current densities is given by

$$\nabla^2 A - A_{,0} = 0$$

and the general solution to this is

$$A^k(t, r) = \int e^r(K, \sigma)[a(K, \sigma)exp(-i(|K|t - K.r)) + \bar{e}^r(K, \sigma)a(K, sigma)^* exp(i(|K|t - K.r))]d^3K$$

Here, the summation is over $\sigma = 1, 2$ corresponding to only two linearly independent polarizations of the photon, ie, $divA = 0$ implies $\sum_{r=1}^3 K^r e^r(K, \sigma) = 0$. The energy of the electromagnetic field in the Coulomb gauge is

$$H_F = (1/2)\int (E^2 + B^2)d^3x = (1/2)\int [(A_{,t}^2 + (\nabla \times A)^2]d^3x$$

$$= \int 2|K|^2|e(K, \sigma)|^2 a(K, \sigma)^* a(K, \sigma)d^3K$$

once we make use of the fact that $|K \times e(K, \sigma)| = |K||e(K, \sigma)|$. For this to be interpretable as the sum of energies of harmonic oscillators, each oscillator in the spatial frequency domain having energy $|K|$, ie, the frequency of the wave. This means that we must have

$$|e(K, \sigma)| = (2|K|)^{-1/2}$$

in order to ensure that

$$H_F = \int |K|a(K, \sigma)^* a(K, \sigma)d^3K$$

We can cross check this result as follows. Assuming that the $a(K, \sigma)'s$ satisfy the canonical commutation relations:

$$[a(K, \sigma), a(K', \sigma')^*] = \delta^3(K - K')\delta_{\sigma,\sigma'}$$

it follows from the Heisenberg equations of motion that

$$a(t, K, \sigma)_{,t} = i[H_F, a(t, K, \sigma)] = -i|K|a(t, K, \sigma),$$

$$a^*(t, K, \sigma)_{,t} = i[H_F, a(t, K, \sigma)^*] = i|K|a^*(t, K, \sigma)$$

These equations imply

$$a(t, K, \sigma)_{,t} = -|K|^2 a(t, K, \sigma)$$

$$a^*(t, K, \sigma)_{,tt} = -|K|^2 a^*(t, K, \sigma)$$

which are the correct equations for the spatial Fourier transform of the vector potential arrived from the wave equation. Another way to check these commutation relations which we leave as an exercise, is to start with the Lagrangian density

$$L_F = (1/2)(A_{,t})^2 - (1/2)(\nabla \times A)^2$$

so that the momentum density is

$$\pi_k(t, r) = \frac{\partial L_F}{\partial A_{,t}^k} = A_{,t}^k$$

then apply the canonical commutation relations

$$[A^k(t, r), \pi_m(t, r')] = i\delta_m^k \delta^3(r - r')$$

and verifiy that these relations are satisfied by the above Fourier integral representation of A assuming the canonical commutation relations between $a(K, \sigma)$ and $a(K', \sigma')$. We leave this verification as an exercise to the reader.

Now consider the second quantized Dirac field described by the four component ield operators $\psi(x), \psi(x)^*$ where $x = (t, r), t \in \mathbb{R}, r \in \mathbb{R}^3$. In the absence of any classical or quantum electromagnetic field ,ψ satisfies the Dirac equation

$$[i\gamma^\mu \partial_\mu - m]\psi(x) = 0$$

or equivalently,

$$[\gamma^\mu p_\mu - m]\psi = 0, p_\mu = i\partial_\mu$$

The solutions to ψ are plane waves:

$$\psi(x) = \int (u(P,\sigma)a(P,\sigma)exp(-ip.x) + v(P,\sigma)b(P,\sigma)^* exp(ip.x))d^3P$$

where

$$p.x = p_\mu x^\mu = E(P)t - P.r, E(P) = \sqrt{m^2 + P^2}, p = (\pi^\mu) = (E, P), u(P,\sigma), v(P,\sigma) \in \mathbb{C}^4$$

Here, the summation is over $\sigma = \pm 1/2$ corresponding to the fact that Dirac's equation can be expressed as

$$[i\partial_0 - (\alpha, P) - \beta m]\psi(x) = 0, P = -i\nabla$$

and hence if P denotes an ordinary 3-vector (not an operator), then $u(P)exp(-ip.x)$ satisfies the Dirac equation iff

$$[p^0 - (\alpha, P) - \beta m]u(P) = 0$$

and likewise, $v(P)exp(ip.x)$ satisfies the Dirac equation iff $(-p^0 + (\alpha, P) - \beta m)v(P) = 0$ Thus, $u(P)$ is an eigenvector of the matrix $H_D(P) = (\alpha, P) + \beta m$ with eigenvalue p^0 and $v(-P)$ is an eigenvector of $H_D(P)$ with eigenvalue p^0. Now since $H_D(P)$ is a 4×4 Hermitian matrix, it has four real eigenvalues taking all multiplicities into account. These eigenvalues are $\pm E(P), E(P) = \sqrt{m^2 + P^2}$ with each one have a multiplicity of two. We denote the corresponding mutually orthogonal eigenvectors by $u(P,\sigma), v(-P,\sigma), \sigma = \pm 1/2$. On applying second quantization, the free Dirac Hamiltonian becomes

$$H_{DQ} = \int \psi(x)^*((\alpha, -i\nabla) + \beta m)\psi(x)d^3x$$

and it is easy to verify that the normalizations of $u(P,\sigma)$ and $v(P,\sigma)$ are chosen so that

$$H_{DQ} = \int E(P)(a(P,\sigma)^* a(P,\sigma) + b(P,\sigma)b(P,\sigma)^*)d^3P$$

and if we postulate the anticommutation relations

$$\{a(P,\sigma), a(P',\sigma')^*\} = \{b(P,\sigma), b(P',\sigma')^*\} = \delta_{\sigma,\sigma'}\delta^3(P - P')$$

then and only then we can ensure the canonical anticommutation relations (CAR)

$$\{\psi_l(t,r), \pi_m(t,r')\} = i\delta_{lm}\delta^3(r - r')$$

where π_m is the canonical momentum associated with the canonical position field ψ_m. From the free Dirac Lagrangian density

$$L_D = \psi(x)^*(i\partial_0 - (\alpha, -i\nabla) - \beta m)\psi(x)$$
$$= \psi(x)^*\gamma^0(i\gamma^\mu\partial_\mu - m)\psi(x)$$

we infer that

$$\pi_m(x) = \frac{\partial L_D}{\partial \psi_{l,0}} = i\psi_l(x)^*$$

so that the CAR gives

$$\{\psi_l(t,r), \psi_m(t,r')^*\} = \delta_{lm}\delta^3(r - r')$$

Thus in particular, we can subtract an infinite constant from the second quantized Dirac Hamiltonian to get

$$H_{DQ} = \int E(P)(a(P,\sigma)^* a(P,\sigma) - b(P,\sigma)^* b(P,\sigma)) - - - (1)$$

This equation has the following nice interpretation: $a(P,\sigma)^*$ creates an electron with momentum P and spin σ, $a(P,\sigma)$ annihilates an electron with momentum P and spin σ. $b(P,\sigma)^*$ creates positron with momentum P and spin σ while

$b(P,\sigma)$ annihilates a positron with momentum P and spin σ. $a(P,\sigma)^*a(P,\sigma)$ is the number operator density for electrons and $b(P,\sigma)^*b(P,\sigma)$ is the number operator for positrons. Since the presence of an additional electron increases the energy of the Dirac sea of electrons by $E(P)$ while the presence of an additional positron decreases the energy of the Dirac sea by $E(P)$, equn (1) has the correct physical interpretation for the energy of the second quantized Dirac field. Now suppose we have a collection of photons, electrons and positrons. The total Lagrangian density is then

$$L = L_{EM} + L_D + L_{int}$$

$$= (-1/4)F_{\mu\nu}F^{\mu\nu} + \psi^*\gamma^0(\gamma^\mu(i\partial_\mu + eA_\mu) - m)\psi$$

so that

$$L_{EM} = (-1/4)F_{\mu\nu}F^{\mu\nu}, L_D = \psi^*\gamma^0(i\gamma^\mu\partial_\mu - m)\psi,$$

$$L_{int} = -J^\mu A_\mu,$$

$$J^\mu = -e\psi^*\gamma^0\gamma^\mu\psi$$

J^μ is the Dirac four current density. It is easily verified to be conserved even when an electromagnetic field is present. In other words, we can verify using the Dirac equation

$$[\gamma^\mu(i\partial_\mu + eA_\mu) - m]\psi = 0$$

that

$$\partial_\mu J^\mu = 0$$

ie, the current is conserved. We can further show that the matrices

$$K^{\mu\nu} = (-1/4)[\gamma^\mu, \gamma^\nu]$$

satisfy the same commutation relations as do the standard skew-symmetric generators of the Lorentz group do. Hence these matrices furnish a representation of the Lie algebra of the Lorentz group. Let D denote the corresponding representation of the Lorentz group. D is called the Dirac spinor representation of the Lorentz group and if Λ is any Lorentz transformation, we write

$$D(\Lambda) = exp(\omega_{\mu\nu}K^{\mu\nu})$$

where

$$\Lambda = exp(\omega_{\mu\nu}L^{\mu\nu})$$

with ω a skew symmetric matrix and $L^{\mu\nu}$ the standard generators of the Lorentz group:

$$(L^{\mu\nu})^{\alpha\beta} = \eta^{\mu\alpha}\eta^{\nu\beta} - \eta^{\mu\beta}\eta^{\nu\alpha}$$

Further, we note the following:

$$D(\Lambda)\gamma^\mu D(\Lambda)^{-1} = \Lambda^\mu_\nu\gamma^\nu$$

and hence, the Dirac equation is invariant under Lorentz transformations ie if $x^\mu \to \Lambda^\mu_\nu x^\nu$ and $\psi(x) \to D(\Lambda)\psi(x)$, $A^\mu \to \Lambda^\mu_\nu A^\nu$, then the Dirac equation remains invariant. Further, the existence of the positron follows from the fact that if we start with the Dirac equation, conjugate it and multiply by the unitary matrix $i\gamma^2$, then we get

$$[(i\gamma^2)(\bar{\gamma}^\mu)(i\gamma^{2-1})(-i\partial_\mu + eA_\mu) - m]i\gamma^2\bar{\psi} = 0$$

It is easily verified that this equation is the same as

$$[\gamma^\mu(i\partial_\mu - eA_\mu) - m]\tilde{\psi} = 0$$

where

$$\tilde{\psi} = i\gamma^2]\bar{\psi}$$

In other words $\tilde{\psi}$ satisfies the Dirac equation in an electromagnetic field but with the charge e replaced by $-e$ or equivalently, $-e$ replaced by e. This observation led Dirac to conclude the existence of the positron, namely the antiparticle of the electron, having the same mass but opposite charge as that of the electron. The positron was discovered in an accelerator later by Anderson. Another property of the Dirac equation in an external electromagnetic field is obtained by considering

$$(\gamma^\mu(i\partial_\mu + eA_\mu) + m)(\gamma^\nu(i\partial_\nu + eA_\nu) - m)\psi = 0$$

If $A_\mu = 0$, this reduces to the free KG equation

$$[\partial_\mu \partial^\mu + m^2]\psi = 0$$

since

$$\{\gamma^\mu, \gamma^{|nu}\} = 2\eta^{\mu\nu}$$

In case however $A_\mu \neq 0$, we get

$$[\gamma^\mu \gamma^\nu (i\partial_\mu + eA_\mu)(i\partial_\nu + eA_\nu) - m^2]\psi = 0$$

or equivalently,

$$((1/2)\{\gamma^\mu, \gamma^\nu\} + (1/2)[\gamma^\mu, \gamma^\nu])[(i\partial_\mu + eA_\mu)(i\partial_\nu + eA_\nu) - m^2]\psi = 0$$

or since $\{\gamma^\mu, \gamma^\nu\} = 2\eta^{\mu\nu}$, we get

$$[(\partial_\mu - eA_\mu)(\partial^\mu - ieA^\mu) + m^2 + (1/4)[\gamma^\mu, \gamma^\nu][\partial_\mu - ieA_\mu, \partial_\nu - ieA_\nu]]\psi = 0$$

or since

$$[\partial_\mu - ieA_\mu, \partial_\nu - ieA_\nu] =$$
$$-i(A_{\nu,\mu} - A_{\mu,\nu}) = -iF_{\mu\nu}$$

we can write

$$[(\partial_\mu - eA_\mu)(\partial^\mu - ieA^\mu) + m^2 - (i/4)[\gamma^\mu, \gamma^\nu]F_{\mu\nu}]\psi = 0$$

This equation is the same as the KG equation in an external electromagnetic field obatined by replacing ∂_μ by $\partial_{mu} - ieA_\mu$ except for the last term which displays explicitly the interaction of the electromagnetic field $F_{\mu\nu}$ (whose non-zero components are the electric and magnetic fields) with the spin of the electron described by the antisymmetric "spin tensor" $(-i/4)[\gamma^\mu, \gamma^\nu]$.

Exercise: Write down explicitly in terms of the components of the electric and magnetic fields, the spin-field interaction component and display in particular, the spin magnetic dipole moment and the spin electric dipole moment.

The photon and electron propagator:

We make this calculation using first the Feynman path integral for fields and then leave as an exercise to demonstrate the same result using the operator expansion of the fields. First, note that the photon propagator is defined in space-time as

$$D_{\mu\nu}(x, y) =< 0|T\{A_\mu(x)A_\nu(y)\}|0 >$$

and the electron propagator as

$$S_{lm}(x, y) =< 0|T\{\psi_l(x)\psi_m(y)^*\}|0 >$$

Here we are using the Lorentz gauge in which case even A_0 is a component of the electromagnetic field potential, not a matter field. In the absence of matter, $A_\mu(x)$ satisfies the wave equation

$$\partial_\mu \partial^\mu A_\alpha = 0$$

and this equation has solutions

$$A^\mu(x) = \int [a(K, \sigma)e^\mu(K, \sigma)exp(-ik.x)/\sqrt{2|K|} + a(K, \sigma)^* \bar{e}^\mu(K, \sigma)exp(ik.x)/\sqrt{2|K|}]d^3K$$

where

$$k.x = k_\mu x^\mu = |K|t - K.r$$

The Lorentz gauge condition

$$\partial_\mu A^\mu = 0$$

implies

$$k_\mu e^\mu(K, \sigma) = 0$$

which means that e^μ has three degrees of freedom. Formally, we do not take the canonical momentum π_0 as zero, even though the Lagrangian density L_{EM} of the electromagnetic field does not depend on $A^0_{,0}$ and hence implies $\pi_0 = 0$. The way out is to introduce a small perturbing Lagrangian density to the Lagrangian density of the electromagnetic field involving $A^0_{,0}$ replace L_{EM} by this perturbed Lagrangian density and define the canonical momenta as

$$\pi_\mu = \frac{\partial L_{EM}}{\partial A^\mu_{,0}}$$

Then, we introduce the commutation relations

$$[A^\mu(t,r), A_\nu(t,r')] = \delta^\mu_\nu \delta^3(r-r')$$

This is satisfied provided

$$[a(K,\sigma), a(K',\sigma')^*] = \delta_{\sigma,\sigma'}\delta^3(K-K')$$

Then, we find that since $a(K,\sigma)|0>=0, <0|a(K,\sigma)^*=0$, we have

$$D_{\mu\nu}(x,x') = \int \theta(t-t') <0|a(K,\sigma)a(K',\sigma')^*|0> e_\mu(K,\sigma)\bar{e}_\nu(K',\sigma')(exp(-i(k.x-k'.x'))/2|K|)d^3Kd^3K'$$

$$+\int \theta(t'-t) <0|a(K',\sigma')a(K,\sigma)^*|0> e_\nu(K',\sigma')\bar{e}_\mu(K,\sigma)(exp(i(k.x-k'.x'))/2|K|)d^3Kd^3K'$$

$$= \theta(t-t')\int e_\mu(K,\sigma)\bar{e}_\nu(K',\sigma')\delta_{\sigma,\sigma'}\delta^3(K-K')/2|K|)(exp(-i(k.x-k'.x'))/2|K|)d^3Kd^3K'$$

$$+\theta(t'-t)\int e_\nu(K',\sigma')\bar{e}_\mu(K,\sigma)\delta_{\sigma,\sigma'}\delta^3(K-K')(exp(i(k.x-k'.x'))/2|K|)d^3Kd^3K'$$

$$= \int [e_\mu(K,\sigma)\bar{e}_\nu(K,\sigma)\theta(t-t')exp(-i|K|(t-t')+iK.(r-r'))+e_\nu(K,\sigma)\bar{e}_\mu(K,\sigma)\theta(t'-t)exp(i|K|(t-t')-iK.(r-r'))]d^3K/2|K|$$

The normalization condition

$$e_r(K,\sigma)\bar{e}_s(K,\sigma) = \delta_{rs}, r,s=1,2,3$$

(summation over $\sigma = 1,2,3$) had to be imposed (look at the treatment in the Coulomb gauge) in order to guarantee the correct commutation relations for the creation and annihilation operators following from the canonical commutation relations (CCR) for the position and momentum fields. From this we get

$$D_{rs}(x,x') = \delta_{rs}\int (2|K|)^{-1}[\theta(t-t')exp(-ik.(x-x')) + \theta(t'-t)exp(ik.(x-x'))]d^3K$$

where $k^0 = |K|$. This is a function of only $x-x'$ and so, we can denote it by $D_{rs}(x-x')$. Now, consider k^0 to be a variable and consider the identity

$$\int_\Gamma dk^0/(k^{02}-|K|^2) = \pi i/|K|$$

where Γ is a contour along the real axis from $-\infty$ to $+\infty$ making a small encirclement of the pole at $|K|$ above the real axis but excluding the pole at $-|K|$ and then completed into a big infinite semicircle below the real axis (so that on this semicircle, k^0 has a negative imaginary part). Then it is clear that for $t > t'$, the contour integral

$$\int_\Gamma exp(-ik.(x-x'))dk^0/(k^{02}-|K|^2) = i\pi exp(-i|K|(t-t')+iK.(r-r'))/|K|$$

and likewise for the other term $t' > t$. Thus, it follows from the above formula that

$$\int D_{rs}(x).exp(-ik.x)d^4x = \frac{-\delta_{rs}}{k^{02}-|K|^2+i\epsilon}$$

and to preserve Lorentz invariance, we may assume

$$\hat{D}_{\mu\nu}(k) = \int D_{\mu\nu}(x).exp(-ik.x)d^4x = \frac{\eta_{\mu\nu}}{k^{02}-|K|^2+i\epsilon} = \frac{\eta_{\mu\nu}}{k^2+i\epsilon}$$

where

$$k^2 = k^{02}-|K|^2 = k_\mu k^\mu$$

We can repeat this calculation for the electron propagator and show that

$$\hat{S}_{lm}(p) = \int S_{lm}(x)exp(-ip.x)d^4x = (\gamma^0\gamma^\mu p_\mu - m)^{-1}$$

These results can be derived directly from the FPI:

$$D_{\alpha\beta}(y,z) = \int exp((-i/4)\int F_{\mu\nu}(x)F^{\mu\nu}(x)d^4x)A_\alpha(y)A_\beta(z)DA_\mu$$

$$F_{\mu\nu}(x)F^{\mu\nu}(x) = (A_{\nu,\mu} - A_{\mu,\nu})(A^{\nu,\mu} - A^{\mu,\nu})$$
$$= 2A_{\nu,\mu}A^{\nu,\mu} - 2A_{\nu\,mu}A^{\mu,\nu}$$
$$= -2A_\nu\partial_\mu\partial^\mu A^\nu + 2A_\nu\partial^\nu\partial_\mu A^\mu$$

on ignoring perfect divergences which do not contribute to the action integral. Thus,

$$S[A] = (-1/4)\int F_{\mu\nu}(x)F^{\mu\nu}(x)d^4x =$$

$$\int K^{\mu\nu}(x-y)A_\mu(x)A_\nu(y)d^4xd^4y$$

or equivalently in the Fourier domain

$$S[A] = \int \hat{K}^{\mu\nu}(k)\hat{A}_\mu(k)\hat{A}_\nu(k)^*d^4k$$

where

$$\hat{K}^{\mu\nu}(k) = k^2\eta^{\mu\nu} - k^\mu k^\nu$$

The matrix $\hat{K}(k)$ is singular and hence its inverse cannot be evaluated. However, we evaluate its pseudo-inverse and use it as the propagator, or since $\hat{K}^{\mu\nu}(k)k_\nu = 0$, we use as the photon propagator, the solution to the equation

$$\hat{K}(k)\hat{D}(k) = P(k)$$

where $P(k)$ is the orthogonal projecton onto the spatial variable subspace, ie,

$$P_{\mu\nu}(k) = I - k_\mu k_\nu/k^2$$

or equivalently,

$$P^{\mu\nu}(k) = I - k^\mu k^\nu/k^2$$

We are using the fact that the second moment matrix of a Gaussian distribution is its covariance matrix, ie, in this case, the inverse/pseudo-inverse of the matrix $\hat{K}(k)$. For the electron propagator, we find via the operator formalism that taking

$$\psi_l(x) = \int (u_l(p,\sigma)a(p,\sigma)exp(-ip.x) + v_l(p,\sigma)b(p,\sigma)^*exp(ip.x))d^3P$$

with

$$p^0 = E(P) = \sqrt{P^2 + m^2}$$

and $u(P,.), v(-P,.)$ eigenfunctions of the free Dirac Hamiltonian $(\alpha, P) + \beta m$, $\alpha^r = \gamma^0\gamma^r, \beta = \gamma^0$, normalized in such a way that the CAR for $\psi_l, \pi_l = i\psi_l$ imply

$$\{a(P,\sigma), a(P',\sigma')^*\} = \delta^3(P-P')\delta_{\sigma,\sigma'},$$
$$\{b(P,\sigma), b(P',\sigma')^*\} = \delta^3(P-P')\delta_{\sigma,\sigma'},$$
$$\{a(P,\sigma), b(P',\sigma')\} = 0,$$
$$\{a(P,\sigma), b(P',\sigma')^*\} = 0$$

the electron propagator is given by

$$S_{lm}(x,y) =< 0|T(\psi_l(x)\psi_m(y)^*)|0 >=$$

$$\int [\sum_l (u_l(P,\sigma)u_m(P,\sigma)^*exp(-ip.(x-y)) + v_l(P,\sigma)v_m(P,\sigma)^*)exp(ip.(x-y))]d^3P$$

$$= \int \hat{S}_{lm}(p)exp(ip.(x-y))d^4p$$

where

$$\hat{S}(p) = (\gamma^0\gamma^\mu p_\mu - m + i\epsilon)^{-1}$$

We leave this derivation as an exercise to the reader. Alternatively using the FPI,

$$S_{lm}(x,y) = \int exp(i\int \psi(x)^*(i\gamma^0\gamma^\mu\partial_\mu - m)\psi(x)d^4x)\psi_l(y)\psi_m(z)^*D\psi D\psi^*$$

gives directly the answer using the standard formula for the second moment of a Gaussian distribution. Now we discuss interactions. Suppose the initial state of the field is

$$|i>=|p_{im},\sigma_{im},p'_{il},\sigma'_{il},k_{in},s_{in},m=1,2,...,N_{i1},l=1,2,...,N_{i2},n=1,2,...,N_{i3}>$$

where p_{im},σ_{im} are the four momenta and spins of the mth electron, p'_{il},σ'_{il} are the four momenta and spins of the lth positron and k_{in},s_{in} are the four momenta and helicities of the nth photon. Then, we can write

$$|i>=\Pi_{m,l,n}a(p_{im},\sigma_{im})^*b(p'_{il},\sigma'_{il})^*c(k_{in},s_{in})^*|0>$$

where we are using the notation $c(k,s)$ for the photon annihilation operators. Likewise for the final state. Since transitions in the state occur only because of interactions, we work in the interaction representation in which the Hamiltonian of the electron-positron-photon field is given by

$$H_I(t)=-e\int A_\mu(x)\psi(x)^*(i\gamma^0\gamma^\mu\partial_\mu\psi(x)d^3x$$

Substituting the operator expressions for the quantum fields A_μ,ψ,ψ^*, it follows that H_I can be expressed as a trilinear functional of $(c(k,s),c(k,s)^*),(a(P,\sigma),b(P,\sigma)^*),(a(P,\sigma)^*,b(P,\sigma))$. The transition probability amplitude from $|i>$ at time $t\to-\infty$ to $|f>$ at time $t\to+\infty$ is given by

$$<f|T\{exp(-i\int_{-\infty}^{\infty}H_I(t)dt)\}|i>$$

$$=\delta(f-i)+\sum_{n=1}^{\infty}(-i)^n\int_{-\infty<t_n<t_{n-1}<...<t_1<\infty}<f|H_I(t_1)...H_I(t_n)|i>dt_1...dt_n$$

The computation of the various perturbation terms

$$<f|H_I(t_1)...H_I(t_n)|i>$$

involves in view of the above discussions, computation of the average value of products of the various creation and annihilation operators at different momenta and spins/helicities in the vacuum state. This is achieved by using the CCR for photonic operators and the CAR for electron-positron operators and involves pushing all the annihilation operators to the right and all the creation operators to the left in the vacuum expectation expression. The result of this average value of products is a product of delta functoins in the momentum domain which when integrated with the polarization functions $e^r(K,s)/\sqrt{2|K|},u(P,\sigma),v(P,\sigma)$ gives the final transition amplitude in terms of these polarization functions evaluated at the initial and final momenta and spins/helicities. The calculation of these integrals is a very tedious process. Feynman simplified this calculation by associating a diagram with each perturbtation term and rules for computing each perturbation term in the transition amplitude from the diagram. Feynman also proposed a path integral approach to this computation wherein the creation and annihilation operators in the momentum domain are expressed as Fourier integrals of the position space wave fields and then the problem reduces to computing the path integrals

$$\int exp(i\int(L_{EM}+L_D+L_{int})d^4x)F(\psi,\psi*,A_\mu)D\psi D\psi^*DA_\mu$$

This gives the vacuum expectation of

$$T(exp(i\int H_I(t)dt))F(\psi,\psi^*,A_\mu))$$

where the term $F(\psi,\psi^*,A_\mu)$ is obtained by expressing the creation and annihilation operators of the electron-positron-photon field appearing in $|i>$ and $|f>$ as operators acting on the vacuum $|0>$ as Fourier integrals of the position space field operators. To proceed further, we observe that $\int(L_{EM}+L_D)d^4x$ is a quadratic functional of the position fields and L_{int} is small. So perturbation theory gives for the above FPI,

$$\int exp(i\int(L_{EM}+L_D)d^4x)(1+i(\int L_{int}d^4x)+(i^2/2!)(\int L_{int}d^4x)^2+...)F(\psi,\psi^*,A_\mu)D\psi D\psi^*DA_\mu$$

where

$$L_{int}=e\int\psi^*(i\gamma^0\gamma^\mu\partial_\mu-m)\psi A_\mu d^4x$$

$$=-\int J^\mu A_\mu d^4x$$

Then the various terms in the intergral are evaluated using the standard formulae for the moments of a Gaussian distribution on noting that $exp(i\int(L_{EM}+L_D)d^4x)$ is a Gaussian density functional, it being a quadratic functional of ψ, ψ^*, A_μ. The propagators then enter naturally into the picture when we express the higher moments of even order as products of second moments, ie propagators.

Renormalization, an example: Consider the Hamiltonian

$$H = H_0 - g|e><e|$$

Let $|k>, k = 1, 2, ...$ denote the energy eigenstates of H_0 with

$$H_0|k>= E_k|k>$$

Let $|\psi>$ denote an energy eigenstate of H with energy eigenvalue E. Then,

$$H|\psi>= E|\psi>$$

gives

$$H_0|\psi> -g|e><e|\psi>= E|\psi>$$

or

$$|\psi>= g(H_0 - E)^{-1}|e><e|\psi>$$

$$= g\sum_k |k> (E_k - E)^{-1} <k|e><e|\psi>$$

The normalization condition $<\psi|\psi>= 1$ then implies

$$g^2\sum_k (E_k - E)^{-2}|<k|e>|^2|<e|\psi>|^2 = 1$$

This implies that

$$|<e|\psi>|^2 = (g^2\sum_k (E_k - E)^2|<k|e>|^2)^{-1}$$

If the above sum is divergent, then we would get $<e|\psi>= 0$ which may nor be the case. To avoid such divergences, we make an ultraviolet cutoff meaning thereby that the sum over k is truncated to k such that $E_k \leq \Lambda$ where Λ is a finite positive constant. For a given ultraviolet cutoff Λ, we may thus define a renormalized coupling constant $g = g(\Lambda)$ so that

$$g^2(\Lambda) = (\sum_{k:E_k\leq\Lambda} (E_k - E)^2 <k|e>|^2)^{-1}$$

and get with this cutoff imposed that

$$|<e|\psi>|^2 = 1$$

Thus, divergence problems are avoided by redefining the coupling constant. In quantum field theory, when we calculate the transition probability amplitudes like vacuum polarization, self energy of the electron or anomalous magnetic moment of the electron, the integrals obtained using the Feynman diagrams for the various Dyson series terms diverge. So to ge meaningful answers, we renormalize the fields and coupling constants like charge and mass, by scaling these with constant factors and then split the resulting Lagrangian density into a term not involving the coupling constants Z and an interaction term involving the Lagrangian terms scaled by $Z-1$. The latter terms are regarded as perturbations and we expand the resulting exponential in powers of $Z-1$ and calculate the modified matrix elements or propagators. Finally, Z may be made to tend to infinity in such a way so as to cancel out the infinities arising in the matrix elements or propagators computed without the Z. This method was first demonstrated by Dyson to lead to the experimentally correct values for the above phenomena.

[33] Dirac's wave equation in a gravitational field.
[34] Canonical quantization of the gravitational field.
Let $\Lambda(x)$ be a local Lorentz transformation and let $\Lambda \to D(\Lambda)$ be the Dirac spinor representation of the Lorentz group. Let $g_{\mu\nu}(x)$ be the metric of curved space-time and η_{ab} the Minkowski metric of flat space-time. Let $V^a_\mu(x)$ be the associated tetrad, ie,

$$\eta_{ab}V^a_\mu(x)V^b_\nu(x) = g_{\mu\nu}(x)$$

$((V^a_\mu(x)))$ can be regarded as a locally inertial frame, ie, $\xi^\mu \to V^a_\mu \xi^\mu = \xi^a$ transforms a vector field $\xi^\mu(x)$ in curved space-times to a Minkowski vector, ie each component ξ^a is a scalar. Let $((\gamma^a))$ be the Dirac Gamma matrices. They determine the Dirac equation in flat space-time:

$$(i\gamma^\mu \partial_\mu - m)\psi = 0$$

orin the presence of an electromagnetic field,

$$(\gamma^\mu(i\partial_\mu + eA_\mu(x)) - m)\psi = 0$$

This equation is invariant under global Lorentz transformations ,ie, if Λ is a constant 4×4 Lorentz transformation matrix, then

$$D(\Lambda)(\gamma^\mu(i\partial_\mu + eA_\mu) - m)\psi = 0$$

implies

$$[D(\Lambda)\gamma^\mu D(\Lambda)^{-1}(i\partial_\mu + eA_\mu) - m]D(\Lambda)\psi = 0$$

which implies

$$[\Lambda^\mu_\nu \gamma^\nu(i\partial_\mu + eA_\mu) - m]D(\Lambda)\psi = 0$$

or equivalently,

$$[\gamma^\nu(i\partial'_\nu + eA'_\nu(x')) - m]\psi'(x') = 0$$

where

$$x'^\mu = \Lambda^\mu_\nu x^\nu,$$

so that

$$i\partial'_\nu = \Lambda^\mu_\nu \partial_\mu$$

and

$$A'_\nu(x') = \Lambda^\mu_\nu A_\mu(x), \psi'(x') = D(\Lambda)\psi(x)$$

ie, $\psi'(x') = D(\Lambda)\psi(x)$ satisfies Dirac's equation in the Lorentz transformed system (x'^μ) with the electromagnetic field also transformed in accordance with the Lorentz transformation Λ. Note that the Lorentz generators of the Dirac representation are given by $J^{\mu\nu} = (i/4)[\gamma^\mu, \gamma^\nu]$. We wish to define a Dirac equation in curved space-time that is invariant under local Lorentz transformations in accordance with the equivalence principle of general relativity. To do this, we must use the tetrad which will transform the non-inertial metric to the inertial metric. So we assume our curved space-time Dirac equation to be given by

$$[\gamma^a V^\mu_a(i\partial_\mu + i\Gamma_\mu(x) + eA_\mu(x)) - m]\psi(x) = 0$$

where $\Gamma_\mu(x)$ is a 4×4 matrix interpreted as the gravitational connection of space-time in the Dirac spinor representation. Applying a local Lorentz transformation $D(\Lambda(x))$ gives

$$[D(\Lambda(x))\gamma^a D(\Lambda(x))^{-1} V^\mu_a(x)(iD(\Lambda(x))(\partial_\mu + \Gamma_\mu(x))D(\Lambda(x))^{-1}$$

$$+eA_\mu(x) - m]D(\Lambda(x))\psi(x) = 0$$

or

$$[\Lambda^a_b(x)\gamma^b V^\mu_a(x)(i(\partial_\mu + D(\Lambda(x))\Gamma_\mu(x)D(\Lambda(x))^{-1}) + iD(\Lambda(x))(\partial_\mu D(\Lambda(x))^{-1})$$

$$+eA_\mu(x)) - m]D(\Lambda(x))\psi(x) = 0$$

If we define

$$V^\mu_a(x) \to \Lambda^b_a(x)V^\mu_b(x) = V^{\mu'}_b(x)$$

as the transformation of the tetrad under the local Lorentz transformation $\Lambda(x)$, then we can express the above equation as

$$[\gamma^b V^{\mu'}_b(x)(i(\partial_\mu + \Gamma_{\mu'}(x)) + eA_\mu(x)) - m]D(\Lambda(x))\psi(x) = 0$$

where $\Gamma_{\mu'}(x)$ is the transformed gravitational connection under the local Lorentz transformation $\Lambda(x)$:

$$\Gamma'_\mu(x) = D(\Lambda(x))\Gamma_\mu(x)D(\Lambda(x))^{-1} + D(\Lambda(x))(\partial_\mu D(\Lambda(x))^{-1})$$

Equivalently, if $\omega(x) = ((\omega_{ab}(x)))$ is an infinitesimal local Lorentz transformation so that

$$\Lambda(x) = I + dD(\omega(x)) = I + \omega_{ab}(x)J^{ab}$$

and with neglect of $O(\| \omega(x) \|^2)$ terms,

$$D(\Lambda(x))^{-1} = I - \omega_{ab}(x)J^{ab}$$

then under the infinitesimal local Lorentz transformation $\Lambda(x)$, the gravitational connection $\Gamma_\mu(x)$ transforms to

$$\Gamma'_\mu(x) = (I + \omega_{ab}(x)J^{ab})\Gamma_\mu(x)(I - \omega_{cd}(x)J^{cd})$$

$$-(I + \omega_{ab}(x)J^{ab})\omega_{cd,\mu}(x)J^{cd}$$

$$= \Gamma_\mu(x) + \omega_{ab}(x)[J^{ab}, \Gamma_\mu(x)] - \omega_{ab,\mu}(x)J^{ab}$$

We need to look for a Dirac gravitational connection $\Gamma_\mu(x)$ that satisfies such a transormation law under an infinitesimal local Lorentz transformation $I + \omega(x)$. It is easily seen that

$$\Gamma_\mu(x) = J^{ab}V_a^\nu(x)V_{\nu b,\mu}(x)$$

does the job. Indeed, under $I + \omega(x)$, this transforms to

$$J^{ab}(V_a^\nu + \omega_{ac}V^{\nu c})(V_{\nu b} + \omega_{bd}V_\nu^d)_{,\mu}$$

$$= J^{ab}V_a^\nu V_{\nu b,\mu} + J^{ab}(V_a^\nu V_{\nu,\mu}^d \omega_{bd} + V^{\nu c}V_{\nu b,\mu}\omega_{ac})$$

$$= \Gamma_\mu + J^{db}\omega_{bd,\mu} + J^{ab}(V_a^\nu V_{\nu,\mu}^d \omega_{bd} + V^{\nu c}V_{\nu b,\mu}\omega_{ac})$$

This is seen to coincide with

$$\Gamma_\mu - \omega_{ab,\mu}J^{ab} + \omega_{ab}[J^{ab}, \Gamma_\mu]$$

on noting that

$$[J^{ab}, \Gamma_\mu] = [J^{ab}, J^{cd}]V_c^\nu V_{\nu d,\mu}$$

and using the Lie algebra commutation rules for the Lorentz group generators $((J^{ab}))$. Finally, we should replace ordinary partial derivatives of the tetrad field by covariant derivatives, ie,

$$\Gamma_\mu(x) = J^{ab}V_a^\nu(x)V_{\nu b:\mu}(x)$$

[35] The scattering matrix for the interaction between photons, electrons, positrons and gravitons. Calculating the scattering matrix using the Feynman path integral and also using the operator formalism with the Feynman diagrammatic rules.

[36] Atom interacting with a Laser; The general theory based on quantum electrodynamics. Representing the quantum electromagnetic field using finite sets of creation and annihilation operators. Representing any density operator for the quantum electromagnetic field via the diagonal Glauber-Sudarshan representation. Representing any state of the laser interacting with the spin of an atom using the Galuber-Sudarshan representation having matrix coefficients. Expressing the evolution equation for the density operator of the laser-atom system using the Glauber-Sudarshan representation.

The Hamiltonian of the field is given by

$$H_F = \sum_{k=1}^{p} \omega_k a_k^* a_k, a = (a_1, ..., a_p), a^* = (a_1^*, ..., a_p^*),$$

$$[a_k, a_m^*] = \delta_{km}$$

and all the other commutators vanish. The Hamiltonian of the atom is H_A an $N \times N$ Hermitian matrix and finally, the interaction Hamiltonian between the atom and the field is given by

$$H_I(t) = \sum_{k=1}^{p} (A_k(t)F_k(a, a^*) + A_k(t)^* F_k(a, a^*)^*)$$

where $A_k(t)$ is a time varying $N \times N$ matrix and $F_k's$ are ordinary complex valued functions which become field operators when their complex arguments are replaced by the field operators a, a^*. The evolution of the joint state $\rho(t)$ of the atom and field follows the Schrodinger equation

$$i\rho'(t) = [H(t), \rho(t)], H(t) = H_F + H_A + H_I(t)$$

Note that H_A and H_F commute. For obtaining the interaction representation, we define

$$\tilde{\rho}(t) = U_0(t)^* \rho(t) U(t)$$

where

$$U_0(t) = U_{0A}(t) U_{0F}(t),$$
$$U_{0A}(t) = exp(-itH_A), U_{0F}(t) = exp(-itH_F)$$

We note that

$$U_{0F}(t)^* a_k U_{0F}(t) = exp(-i\omega_k t) a_k = a_k(t), U_{0F}(t)^* a_k^* U_{0F}(t) = exp(i\omega_k t) a_k^* = a_k(t)^*$$

Thus,

$$U_{0F}(t)^* H_I(t) U_{0F}(t) = \sum_k (A_k(t) F_k(a(t), a(t)^*) + A_k(t)^* F_k(a(t), a(t)^*))$$

and defining the $N \times N$ complex matrices

$$B_k(t) = U_{0A}(t)^* A_k(t) U_{0A}(t)$$

we get the interaction picture Schrodinger equation for the atom and field

$$i\tilde{\rho}'(t) = [\tilde{H}_I(t), \tilde{\rho}(t)]$$

where

$$\tilde{H}_I(t) = U_0(t)^* H_I(t) U_0(t) = \sum_k B_k(t) F_k(a(t), a(t)^*) + B_k(t)^* F_k(a(t), a(t)^*)^*$$

We write

$$F_k(a(t), a(t)^*) = F(\{a_k exp(-i\omega_k t)\}, \{a_k^* exp(i\omega_k t)\}) = F_k(t, a, a^*)$$

so

$$\tilde{H}_I(t) = \sum_k (B_k(t) F_k(t, a, a^*) + B_k(t)^* F_k(t, a, a^*)^*)$$

and the Schrodinger equation in the interaction picture becomes

$$i\tilde{\rho}'(t) = [\tilde{H}_I(t), \tilde{\rho}(t)]$$

To solve this differential equation, we adopt the Galuber-Sudarshan diagonal represention: Let $a(z) = \sum_k \bar{z}_k a_k, a(z)^* = \sum_k z_k a_k^*$ where $z = (z_k) \in \mathbb{C}^p$. Then write

$$|e(z)> = \sum_n z^n a^{*n} |0> / n! = exp(a(z)^*)|0>$$

with the obvious p-tuple notation. The normalized energy eigenstates of the field are

$$|n> = a^{*n}|0> / \sqrt{n!}$$

and hence

$$|e(z)> = \sum_n z^n |n> / \sqrt{n!}$$

Thus,

$$< e(u)|e(z) > = exp(< u|z >)$$

We can normalize $|e(z)>$ by multiplying it by a function $\phi(z)$ of z so that

$$I = \int |e(z) > \phi(z) < e(z)| d^{2n} z$$

We have

$$a(u)|e(z)> = < u|z > |e(z)>, a_k|e(z)> = z_k|e(z)>$$

Also,

$$a_k^*|e(z)> = \sum_n z^n a_k^* a * n|0> / n! = \frac{\partial}{\partial z}|e(z)>$$

We can evaluate

$$a_k|e(z)><e(z)| = \bar{z}_k|e(z)><e(z)|,$$

$$a_k^*|e(z)><e(z)| = \frac{\partial}{\partial z_k}|e(z)><e(z)|$$

Thus,

$$F(t,a,a^*)|e(z)><e(z)| = F(t,z,\frac{\partial}{\partial z})|e(z)><e(z)|$$

assuming that in the expression $F(t,a,a^*)$, all the $a's$ appear to the left of all the $a^{*'}s$. Such a representation is possible in view of the commutation relations between the $a's$ and the $a^{*'}s$. Likewise, we have

$$|e(z)><e(z)|F(t,a,a^*)$$

$$= |e(z)> (F(t,a,a^*)^*|e(z)>)^*$$

and

$$F(t,a,a^*)^*|e(z)>= \bar{F}(t,a^*,a)|e(z)>$$

where now in the expression for $\bar{F}(t,a^*,a)$ all the a^*s will appear to the left of all the $a's$. We thus get

$$|e(z)><e(z)|F(t,a,a^*) = \bar{F}(t,\frac{\partial}{\partial \bar{z}},\bar{z})|e(z)><e(z)|$$

Here, $\bar{F}(t,u,v)$ is the function obtained by conjugating all the coefficients in the Taylor series expansion of $F(t,.)$. Now, with this understanding, we note that

$$\tilde{H}_I(t) = \sum_k (B_k(t)F_k(t,a,a^*) + B_k(t)^*F_k(t,a,a^*)^*)$$

can be written as

$$\tilde{H}_I(t) = \sum_k C_k(t)G_k(t,a,a^*)$$

where in the $G_k's$ all the $a's$ appear to the left of all the $a^{*'}s$. Then, by the above observation,

$$[G_k(t,a,a^*),|e(z)><e(z)|] = [G_k(t,z,\frac{\partial}{\partial z}) - \bar{G}_k(t,\bar{z},\frac{\partial}{\partial \bar{z}})]|e(z)><e(z)|$$

It should be noted that $\frac{\partial}{\partial z}$ acts only on the factor $|e(z)>$ while $\frac{\partial}{\partial \bar{z}}$ acts only on the factor $<e(z)|$ in the term $|e(z)><e(z)|$. This is because $z \to |e(z)>$ is an analytic Hilbert space valued function of the complex variable z and $\bar{z} \to <e(z)|$ is therefore an analytic function of the complex variable $\bar{z}$. Using

$$<e(u),e(z)>= exp(<u,z>)$$

it is easy to show that

$$\phi(z) = \pi^{-p}exp(\| z \|^2)$$

Indeed, then a simple Gaussian integral evaluation gives

$$<e(u)|\int \phi(z)|e(z)><e(z)|d^{2p}z|e(v)>= \int \phi(z)exp(<u|z>+<z|v>)d^{2p}z = 1$$

We can express the joint state of the atom and the field in the interaction representation as a Glauber-Sudarshan integral:

$$\tilde{\rho}(t) = \int \psi(t,z,\bar{z}) \otimes |e(z)><e(z)|d^{2p}z$$

where

$$\tilde{\psi}(t,z,\bar{z}) \in \mathbb{C}^{N\times N}$$

We then get

$$\tilde{\rho}'(t) = \int \frac{\partial\psi(t,z,\bar{z})}{\partial t} \otimes |e(z)><e(z)|d^{2p}z$$

and further,

$$[\tilde{H}_I(t), \tilde{\rho}(t)] =$$

$$\sum_k \int [C_k(t)\psi(t,z,\bar{z}) \otimes G_k(t,a,a^*)|e(z)><e(z)| - \psi(t,z,\bar{z})C_k(t) \otimes |e(z)><e(z)|G_k(t,z,z^*)]d^{2p}z$$

$$\sum_k \int [C_k(t)\psi(t,z,\bar{z}) \otimes G_k(t,z,\partial/\partial z)|e(z)><e(z)| - \psi(t,z,\bar{z})C_k(t) \otimes \bar{G}_k(t,\bar{z}.\partial/\partial\bar{z})|e(z)><e(z)|]d^{2p}z$$

$$= \sum_k \int C_k(t)(G_k(t,z,\partial/\partial z)^T - \bar{G}_k(t,\bar{z},\partial/\partial\bar{z})^T)\psi(t,z,\bar{z}) \otimes |e(z)><e(z)|d^{2p}z$$

where we have used integration by parts. Here for example,

$$(z_1^{m_1} z_2^{m_2} ... z_p^{m_p} \frac{\partial^{n_1+...+n_p}}{\partial z_1^{n_1} ... \partial z_p^{n_p}})^T$$

$$= (-1)^{n_1+...+n_p} \frac{\partial^{n_1+...+n_p}}{\partial z_1^{n_1} ... \partial z_p^{n_p}} z_1^{m_1} ... z_p^{m_p}$$

It follows that $\psi(t,z,\bar{z})$ satisfies the pde

$$i\frac{\partial\psi(t,z,\bar{z})}{\partial t} =$$

$$\sum_k C_k(t)(G_k(t,z,\partial/\partial z)^T - \bar{G}_k(t,\bar{z},\partial/\partial\bar{z})^T)\psi(t,z,\bar{z})$$

$$= \sum_k (G_k(t,z,\partial/\partial z)^T - \bar{G}_k(t,\bar{z},\partial/\partial\bar{z})^T)C_k(t)\psi(t,z,\bar{z})$$

Remark: The following notation has been used here: If $G(t,z,u)$ is a polynomial in variables z,u which may even be noncommuting operators, then by $\bar{G}(t,z,u)$, we mean the polnomial obatined from $G(t,z,u)$ by replacing its complex coefficients by their respective conjugates without making any change in the variables z,u.

[37] The classical and quantum Boltzmann equations.
Quantum Boltzmann equation: Let $\mathcal{H}_i, i = 1,2,...,N$ be Hilbert spaces and let

$$\mathcal{H} = \bigotimes_{k=1}^N \mathcal{H}_k$$

Let $\rho(t)$ be a density operator in $\mathcal{H}$. Let the Hamiltonian according to which ρ evolves have the form

$$H = \sum_{k=1}^N H_k + \sum_{1 \le k < j \le N} V_{kj}$$

where H_k acts in $\mathcal{H}_k$ and V_{kj} acts in $\mathcal{H}_k \otimes \mathcal{H}_j$. Assume that the Hilbert spaces $\mathcal{H}_k, k = 1,2,...,N$ are isomorphic and that the operators $H_k, k = 1,2,...,N$ are identical copies of H_1 and the operators $V_{kj}, j < k$ are identical copies of a $V = V_{12}$. Schrodinger's evolution equation for $\rho(t)$ is

$$i\rho'(t) = \sum_k [H_k, \rho(t)] + \sum_{k<j} [V_{kj}, \rho(t)] - - - (1)$$

We assume that all the k-marginals of $\rho(t)$ are the same for $k = 1,2,...,N$. Then taking the partial trace of (1) over $\bigotimes_{k>1} \mathcal{H}_k$ gives

$$i\rho_1'(t) = [H_1, \rho_1(t)] + (N-1)Tr_2([V_{12}, \rho_{12}(t)] - - - (2)$$

Taking the partial trace of (1) over $\mathcal{H}_k, k = 3,4,...,N$ gives

$$i\rho_{12}'(t) = [H_1 + H_2, \rho_{12}(t)] + [V_{12}, \rho_{12}(t)] + (N-2)Tr_3[V_{13} + V_{23}, \rho_{123}]$$

$$= [H_1 + H_2, \rho_{12}(t)] + [V_{12}, \rho_{12}(t)] + 2(N-2)Tr_3[V_{13}, \rho_{123}]$$

Neglecting the third marginal term, we get approximately,

$$i\rho_{12}'(t) = [H_1 + H_2, \rho_{12}(t)] + [V_{12}, \rho_{12}(t)] - - - (3)$$

Now, if the initial state is separable, then at time $t > 0$, the second marginal state $\rho_{12}(t)$will be a small perturbation of $\rho_1(t) \otimes \rho_2(t)$ (Note that, we are assuming that $\rho_2(t)$ is an identical copy of $\rho_1(t)$). Thus, if the interaction potential V_{12} is small, then in (3), we can replace $[V_{12}, \rho_{12}(t)]$ by $[V_{12}, \rho_1(t) \otimes \rho_2(t)]$. Then, the approximate solution to (3) is given by

$$\rho_{12}(t) = exp(-itad(H_1 + H_2))(\rho_{12}(0)) + \int_0^t exp(-i(t-s)ad(H_1 + H_2))ad(V_{12})(\rho_1(s) \otimes \rho_1(s))ds - - - (4)$$

and this can be substituted into (2) to get the following nonlinear integro-differential equation for the first marginal $\rho_1(t)$:

$$i\rho_1'(t) = [H_1, \rho_1(t)] + (N-1)Tr_2[ad(V_{12})T_{12}(t)(\rho_{12}(0))] + (N-1)Tr_2[\int_0^t ad(V_{12})T_{12}(t-s)ad(V_{12})(\rho_1(s) \otimes \rho_1(s))ds] - - - (5)$$

where

$$T_{12}(t) = exp(-itad(H_1 + H_2)) = exp(-itad(H_1)).exp(-itad(H_2)) = exp(-itad(H_2)).exp(-itad(H_1))$$

since H_1 and H_2 commute. (5) may be termed as the quantum Boltzmann equation. Other versions of this equation exist like we can solve (3) to get

$$\rho_{12}(t) = exp(-itad(H_{12}))(\rho_{12}(0))$$

where

$$H_{12} = H_1 + H_2 + V_{12}$$

(2) then gives

$$i\rho_1'(t) = ad(H_1)(\rho_1(t)) + (N-1)Tr_2 ad(V_{12})exp(-itad(H_{12})(\rho_{12}(0)) - - - (6)$$

The third version is to note that the last term in (4) already contains a multiplicative factor $ad(V_{12})$ and hence since we are interested in terms only upto linear orders in V_{12}, we can replace $\rho_1(s)$ in (4) by

$$exp(-is.ad(H_1))(\rho_1(0))$$

(4)then becomes

$$\rho_{12}(t) = exp(-itad(H_1 + H_2))(\rho_{12}(0)) + \int_0^t exp(-i(t-s)ad(H_1 + H_2))ad(V_{12})exp(-isad(H_1 + H_2))(\rho_1(0) \otimes \rho_1(0))ds$$

Thus, (2) can be approximated by

$$i\rho_1'(t) = [H_1, \rho_1(t)] + (N-1)Tr_2[ad(V_{12}exp(-it.ad(H_1+H_2))(\rho_{12}(0))] + (N-1)Tr_2[ad(V_{12}) \int_0^t exp(-i(t-s)ad(H_1+H_2))ad(V_{12})ex$$

[38] Bands in a semiconductor: Derivation using the Bloch wave functions in a 3-D periodic lattice.

$V : \mathbb{R}^3 \to \mathbb{R}$ is the potential of the periodic lattice produced by nuclei located at different sites of the crystal. The Lattice vectors are a_1, a_2, a_3 and the periodicity gives

$$V(r + n_1a_1 + n_2a_2 + n_3a_3) = V(r + n.a) = V(r), n_1, n_2, n_3 \in \mathbb{Z}$$

$V(r)$ can be expressed as a Fourier series

$$V(r) = \sum_{m \in \mathbb{Z}^3} c(m)exp(2\pi i m^T M r)$$

where the reciprocal lattice matrix M is calculated as

$$M[a_1, a_2, a_3] = I$$

ie

$$M = A^{-1}, A = [a_1, a_2, a_3]$$

It follows that

$$m^T M(n_1a_1 + n_2a_2 + n_3a_3) = m^T MAn = m^T n \in \mathbb{Z}, n \in \mathbb{Z}^3$$

This ensures that the above Fourier series for $V(r)$ is periodic along the a_1, a_2, a_3 directions. The wave function $\psi(r)$ satisfies

$$[-\nabla^2/2m + V(r)]\psi(r) = E\psi(r)$$

Changing r to $r + An$ $(An = n_1a_1 + n_2a_2 + n_3a_3)$ and using $V(r + An) = V(r)$ gives

$$H\psi(r + An) = E\psi(r + An)$$

so by uniqueness (assuming non-degeneracy),

$$\psi(r + An) = C(n)\psi(r), C(n) \in \mathbb{C}, |C(n)| = 1$$

We write

$$C_k = C(a_k), k = 1, 2, 3$$

Then if N_k nuclei are along the a_k direction, $k = 1, 2, 3$, then by imposing periodicity relations for the wave function at the crystal boundaries, we get

$$C_k^{N_k} = 1, k = 1, 2, 3$$

and hence

$$C_k = exp(2\pi i l_k/N_k), k = 1, 2, 3$$

for some $l_k = 0, 1, ..., N_k - 1$. We define the Bloch wave function corresponding to $l = (l_1, l_2, l_3)$ by

$$\psi(r) = u_l(r)exp(2\pi i l^T Kr) =$$

where K is some 3×3 matrix. Then, for u_l to be periodic with periods a_1, a_2, a_3, we require that

$$\psi(r + a_k)exp(-2\pi i l^T Ka_k) = \psi(r), k = 1, 2, 3$$

or equivalently,

$$C_k = exp(2\pi i l^T Ka_k)$$

Thus we require that

$$exp(2\pi i l_k/N_k) = ex[(2\pi i l^T Ka_k)$$

and so we can take

$$K = N^{-1}M = N^{-1}A^{-1}, N = diag[N_1, N_2, N_3]$$

We note that

$$l^T N^{-1}A^{-1}a_k = l^T N^{-1}e_k = l_k/N_k, k = 1, 2, 3$$

We write

$$b = b(l) = 2\pi K^T l$$

Then,

$$\psi(r) = u(r)exp(ib^T r), u = u_l$$

We substitute this into the Schrodinger equation and derive the pde satisfied by $u(r)$. After that since u is periodic with period A, we can expand it in a Fourier series as we did for $V(r)$:

$$u(r) = u_l(r) = \sum_{m \in \mathbb{Z}^3} d(m)exp(2\pi i m^T Mr)$$

and then derive a difference equation for the Fourier coefficients $d(m)$ of $u(r)$. For each $l = (l_1, l_2, l_3) \in \times_{k=1}^3\{0, 1, ..., N_k - 1\}$ we thus obtain a sequence of solutions $u_{lk}(r), k = 1, 2, ...$ with energy eigenvalues $E = E(l, k), k = 1, 2, ...$ We say that each l defines an energy band.

[39] The Hartree-Fock apporoximate method for computing the wave functions of a many electron atom.

The Hamiltonian of the system comprising N particles has the form

$$H = \sum_{k=1}^{N} H_k + \sum_{1 \le k < j \le N} V_{kj}$$

where H_k acts in the Hilbert space $\mathcal{H}_k$ of the kth particle and V_{kj} acts in $\mathcal{H}_k \otimes \mathcal{H}_j$. The wave function is assumed to be of the separable form

$$|\psi>= \otimes_{k=1}^N |\psi_k>$$

where $|\psi_k> \in \mathcal{H}_k$. We substitute this into the expression $<\psi|H|\psi>$ for the average energy and extremize this w.r.t. the component wave functions $|\psi_k>, k = 1, 2, ..., N$ subject to the constraints $<\psi_k|\psi_j>= \delta_{kj}$. Incorporating these constraints using Lagrange multiplier $\lambda(k, j)$ gives us the functional to be extremized as

$$S[\{\psi_k, \lambda(k, j)\}] =<\psi|H|\psi> - \sum_{1\le k\le j\le N} \lambda(k, j)(<\psi_k|\psi_j> -\delta_{kj})$$

We observe that

$$<\psi|H|\psi>= \sum_k <\psi_k|H_k|psi_k> + \sum_{1\le k<j\le N} <\psi_k \otimes \psi_j|V_{kj}|\psi_k \otimes \psi_j>$$

$$\sum_{1\le k\le j\le N} \lambda(k, j)(<\psi_k|\psi_j> -\delta_{kj})$$

The variational equations

$$\delta S/\delta\psi_k^* = 0$$

gives

$$H_k|\psi_k> + \sum_{j:j>k} <I_k \otimes \psi_j|V_{kj}|\psi_k \otimes \psi_j> + \sum_{j:j<k} <\psi_j \otimes I_k|V_{kj}|\psi_k \otimes \psi_j>$$

$$-\lambda(k, k)|\psi_k> - \sum_{j:k<j} \lambda(k, j)|\psi_j>= 0$$

For example, if $\mathcal{H}_k = L^2(\mathbb{R}^3)$ and

$$H_k = -\nabla_k^2/2m + U(r_k), 1 \le k \le N$$

and

$$V_{kj} = e^2/|r_k - r_j| = V(|r_k - r_j|), k < j$$

then the above equations give in the position representation

$$-\nabla_k^2\psi_k(r_k)/2m + U(r_k)\psi_k(r_k) + (\sum_{j:j\ne k} V(|r_k - r_j|)|\psi_j(r_j)|^2 d^3r_j)\psi_k(r_k)$$

$$-E_k\psi_k(r_k) - \sum_{j:k<j} \lambda(k, j)\psi_j(r_j) = 0$$

where

$$E_k = \lambda(k, k)$$

Exercise: Since Fermions have antisymmetric wave functions, ideally, our trial wave function should be of the form

$$|\psi>= (n!)^{-1/2} \sum_{\sigma\in S_n} sgn(\sigma)|\psi_{\sigma 1}> \otimes ...|psi_{\sigma n}>$$

Carry out for this trial wave function the above extremization of $<\psi|H|\psi>$ subject to the constraints $<\psi_k|psi_j>= \delta_{kj}$ and specialize to the position representation. Next, take into account the spin of each electron so that the component wave function $|\psi_k>$ depends on both the position r_k and spin variable $s_k = \pm 1/2$. The trial wave function is then

$$\psi(r_1, s_1, ..., r_n, s_n) = (n!)^{-1/2} \sum_{\sigma\in S_n} \psi_{\sigma 1, s_{\sigma 1}}(r_1) \otimes ... \otimes \psi_{\sigma n, s_{\sigma n}}(r_n)$$

Note that the constraints are

$$<\psi_{k,s}|\psi_{k',s'}>= \delta_{kk'}\delta_{ss'}$$

[40] The Born-Oppenheimer approximate method for computing the wave functions of electrons and nuclei in a lattice.

Nucleon positions are $R_k, k = 1, 2, ..., N$. electron positions associated with the kth nucleus are $r_{kl}, l = 1, 2, ..., Z$. Nucleon mass is M, electron mass is m. Total Hamiltonian of the system is

$$H = T_N + T_e + V_{ee} + V_{NN} + V_{eN}$$

where

$$T_N = T_N(R) = -\sum_k \nabla^2_{R_k}/2M$$

is the total kinetic energy operator of all the nucleons.

$$T_e = T_e(r) = -\sum_{k,l} \nabla^2_{r_{kl}}/2m$$

is the total kinetic energy of all the electrons.

$$V_{ee} = \sum_{(k,l)\neq(m,j)} e^2/2|r_{kl} - r_{mj}|$$

is the total electron-electron interaction potential energy.

$$V_{NN} = \sum_{k\neq m} Z^2e^2/|R_k - R_m|$$

is the total nucleon-nucleon interaction potential energy.

$$V_{eN} = -\sum_{k,m,l} Ze^2/|R_k - r_{ml}|$$

is the total electron-nucleon interaction potential energy. We first solve

$$(T_e + V_{ee} + V_{eN})\Phi(r, R) = E_e(R)\Phi(r, R)$$

ie, the eigenfunctions for the electrons with fixed values of the nuclear positions. The energy levels of the electrons then depend on the nucleon positions R. We then Assume that the total wave function of the electrons and nucleons is

$$\Psi(r, R) = \Phi(r, R)\chi(R)$$

Substituting this into the complete electron-nucleon eigenvalue equation gives

$$(T_e + T_N + V_{ee} + V_{NN} + V_{eN})(\Phi(r, R)\chi(R)) = E\Phi(r, R)\chi(R)$$

or using the above electron eigenvalue equation,

$$(T_N + V_{NN} + E_e(R))(\Phi(r, R)\chi(R)) = E(\Phi(r, R)\chi(R))$$

or equivalently,

$$\Phi(r, R)^{-1}T_N(\Phi(r, R)\chi(R)) + (V_{NN} + E_e(R))\chi(R) = E\chi(R)$$

Now,

$$T_N(\Phi(r, R)\chi(R)) =$$

$$(-\sum_k \nabla^2_{R_k}/2M)(\Phi(r, R)\chi(R))$$

$$= -(2M)^{-1}\sum_k[\chi(R)(\nabla^2_{R_k}\Phi(r, R)) + \Phi(r, R)\nabla^2_{R_k}\chi(R) + 2(\nabla_{R_k}\Phi(r, R), \nabla_{R_k}\chi(R))]$$

Chapter 5
Quantum Gates, Scattering Theory, Noise in Schrodinger and Dirac's Equations

[41] The performance of quantum gates in the presence of classical and quantum noise.

Suppose we design a quantum gate by perturbing the Hamiltonian H_0 of a quantum system to $H_0+\delta\sum_k f_k(t)V_k$ and running the unitary evolution for a duration of T seconds. Upto $O(\delta^2)$, the evolution gate at time T is given by

$$U(T) = U_0(T)W(T), U_0(T) = exp(-iTH_0),$$

$$W(T) = I - i\delta\sum_k\int_0^T f_k(t)\tilde{V}_k(t)dt - \delta^2\sum_{k,m}\int_{0<s<t<T} f_k(t)f_k(s)\tilde{V}_k(t)\tilde{V}_m(s)dtds$$

where

$$\tilde{V}_k(t) = U_0(-t)V_kU_0(t)$$

We choose the real valued functions $f_k(t)$ so that

$$E[f] =\| U(T) - U_g \|^2=\| W(T) - W_g \|^2$$

is a minimum where if U_g is the given unitary gate to be designed, then $W_g = U_0(-T)U_g$. We expand $E[f]$ upto $O(\delta^2)$ (note that the Frobenius norm is being used for computational simplicity as the above optimization leads to linear integral equations for the functions f_k. Other norms like the spectral norm or matrix norms induced by L^p norms lead to highly nonlinear integral equations). Now after designing the functions f_k, suppose noise creeps into these functions in the sense that the optimal $f_k(t)$ gets replaced by $f_k(t) + w_k(t)$ where the $w_k(t)'s$ are zero mean random processes. Then, we can calculate the effect of this noise on the error energy using as perform index the nsr (noise to signal ratio):

$$nsr = \frac{\mathbb{E}E[f+w]}{\| U_g \|^2}$$

To calculate the above expectation, we approximate $E[f + w]$ by replacing $W(T)$ with

$$W(T) = I - i\delta\sum_k\int_0^T (f_k(t) + w_k(t))\tilde{V}_k(t)dt - \delta^2\sum_{k,m}\int_{0<s<t<T}(f_k(t) + w_k(t))(f_k(s) + w_k(s))\tilde{V}_k(t)\tilde{V}_m(s)dtds$$

$$= W_0(T) - i\delta\sum_k\int_0^T w_k(t)\tilde{V}_k(t)dt - \delta^2\sum_{k,m}\int_{0<s<t<T}(f_k(t)w_m(s) + w_k(t)f_m(s))\tilde{V}_k(t)\tilde{V}_m(s)dtds$$

$$-\delta^2\int_{0<s<t<T} w_k(t)w_m(s)\tilde{V}_k(t)\tilde{V}_m(s)dtds$$

with

$$W_0(t) = W(T) = I - i\delta\sum_k\int_0^T f_k(t)\tilde{V}_k(t)dt - \delta^2\sum_{k,m}\int_{0<s<t<T} f_k(t)f_k(s)\tilde{V}_k(t)\tilde{V}_m(s)dtds$$

We can express this equation as

$$W(T) = W_0(T) + \sum_k\int w_k(t)S_k(t)dt + \sum_{k,m}\int_0^T\int_0^T w_k(t)w_m(s)Q_{km}(t,s)dtds$$

where $S_k(t), Q_{km}(t,s)$ are operators depending on the $\tilde{V}_k(t)'s$ and the non-random optimal functions $f_k(t)$. We then have

$$\mathbb{E}E[f+w] = \mathbb{E}\| W_0(T) - W_g + \sum_k\int w_k(t)S_k(t)dt + \sum_{k,m}\int w_k(t)w_m(s)Q_{km}(t,s)dtds \|^2$$

$$= E[f] + \int_0^T\int_0^T a_{km}(t,s)\mathbb{E}(w_k(t)w_m(s))dtds$$

where $a_{km}(t,s)$ are real valued functions of $W_0 - W_g, S_k(t), Q_{km}(t,s)$ obtained in terms of the trace of products of such operators or their adjoints. It should be noted that we could in principle, get a better design, ie, a lower nsr by first averaging over the noise distribution and then minimizing $\mathbb{E}E[f+w]$ w.r.t the nonrandom $f_k's$. We leave this as an exercise to the reader.

[42] Design of quantum gates by applying a time varying electromagnetic field on atoms and oscillators.

Suppose an electromagnetic field described by a time varying but spatially invariant electric and magnetic fields $E(t), B(t) \in \mathbb{R}^3$ is incident upon an atom/oscillator with unperturbed Hamiltonian

$$H_0 = p^2/2m + V(r)$$

Taking spin effects into account, the perturbed Hamiltonian is then given by

$$H(t) = -(\nabla + ieA(t,r))^2/2m + V(r) - e(E(t), r) + g(\sigma, B(t))/2m$$

where

$$A(t,r) = B(t) \times r/2$$

We can thus write

$$H(t) = H_0 - (ie/2m)(divA(t,r) + 2(A(t,r), \nabla)) - e^2A^2(t,r)/2m - e(E(t),r) + g(\sigma, B)/2m$$

$$= H_0 + (e/2m)(B(t), L) - e^2(B(t) \times r)^2/2m - e(E(t),r) + g(\sigma, B(t))/2m$$

$$= H_0 + (e/2m)(B(t), L + g\sigma) - e^2(B(t) \times r)^2/2m - e(E(t),r)$$

where

$$L = -ir \times \nabla$$

is the angular momentum vector operator. We are given a unitary gate U_g in $L^2(\mathbb{R}) \otimes \mathbb{C}^2$ and wish to design this gate by choosing the $\mathbb{R}^6$ valued function of time $\xi(t) = (E(t)^T, B(t)^T)^T$ so that

$$\mathcal{E}(\xi) = \| U(T) - U_g \|^2$$

is a minimum. We approximate $U(T)$ by retaining terms only upto quadratic orders in ξ and then set up the optimal equations. Specifically, $U(T)$ is approximated by

$$U(T) = U_0(T)W(T), U_0(T) = exp(-iTH_0),$$

$$W(T) \approx I - i\int_0^T [(e/2m)(B(t), L + g\sigma) - e(E(t),r) - e^2(B(t) \times r)^2/2m]dt$$

$$-\int_{0<s<t<T} ((e/2m)(B(t), L + g\sigma) - e(E(t),r))((e/2m)(B(s), L + g\sigma) - e(E(s),r))dtds$$

When the unperturbed system is a $3-D$ harmonic oscillator, we can write

$$H_0 = \sum_{k=1}^{3} \omega_k a_k^* a_k$$

and

$$L_1 = x_2p_3 - x_3p_2, L_2 = x_3p_1 - x_1p_3, L_3 = x_1p_2 - x_2p_1$$

where

$$a_k = (p_k - im\omega_k x_k)/\sqrt{2}, a_k^* = (p_k + im\omega_k x_k)/\sqrt{2}$$

Thus, L is a quadratic function of $a_k, a_k^*, k = 1,2,3$. Thus, we can express the matrix elements of $W(T)$ with respect to the eigenbasis $|n_1, n_2, n_3, s>$ of H_0 where

$$H_0|n_1,n_2,n_3,s> = (n_1 + n_2 + n_3 + 3/2)|n_1,n_2,n_3,s>, s = \pm 1/2, n_1,n_2,n_3 = 0,1,2,...,$$

$$a_k|n,s> = \sqrt{n_k}|n - e_k, s>, a_k^*|n,s> = \sqrt{n_k+1}|n + e_k, s>$$

where

$$n = (n_1,n_2,n_3), e_k = (\delta_{k1}, \delta_{k2}, \delta_{k3})$$

$|s>$ is the eigenstate of σ_3 with eigenvalue $2s, s = \pm 1/2$. It is then easy to calculate the Frobenius norm

$$\sum_{n,s,n',s'} | <n,s|W(T)|n',s'> - <n,s|W_g|n',s'> |^2$$

as a linear-quadratic functional of $\xi(t), 0 \le t \le T$ and perform the minimization.

[43] Solution of Dirac's equation in the Coulomb potential: Relativistic correction to the levels of the hydrogen atom.
[44] Dirac's equation in general radial potentials.
The stationary Dirac equation in a potential $V(r)$ is given by

$$[(\alpha, p) + \beta m + V(r)]\psi = E\psi$$

equivalently, writing

$$\psi = [f^T, g^T]^{\prime} f(r,\hat{r}), g(r,\hat{r}) \in \mathbb{C}^2$$

we have

$$(\sigma,p)g+(V-E+m)f=0,$$
$$-(\sigma,p)f+(E-V+m)g=0$$
$$J=L+\Sigma/2$$

is conserved, ie if

$$H=(\alpha,p)+\beta m+V(r)$$

then

$$[H,J]=0$$

Proof:

$$J_z=L_z+\Sigma_z/2$$

Note that

$$\Sigma=\begin{pmatrix}\sigma & 0\\ 0 & \sigma\end{pmatrix}$$

$$[H,L_z]=[(\alpha,p),L_z]=\alpha_x[p_x,L_z]+\alpha_y[p_y,L_z]$$
$$=\alpha_x[p_x,xp_y-yp_x]+\alpha_y[p_y,xp_y-yp_x]$$
$$=-i\alpha_x p_y+i\alpha_y p_x=-i(\alpha\times p)_z$$

Thus

$$[H,L]=-i\alpha\times p=\begin{pmatrix}0 & -i\sigma\times p\\ -i\sigma\times p & 0\end{pmatrix}$$

Also,

$$[H,\Sigma_z]=[(\alpha,p),\Sigma_z]$$

since $[\beta,\Sigma_z]=0$. Now,

$$[\alpha_x,\Sigma_z]=\begin{pmatrix}0 & -2i\sigma_y\\ -2i\sigma_y & 0\end{pmatrix}$$

Likewise, for α_y and α_z. We thus get

$$[H,\Sigma_z]=\begin{pmatrix}0 & 2i\sigma\times p\\ 2i\sigma\times p & 0\end{pmatrix}$$

Thus,

$$[H,L_z+\Sigma_z/2]=0$$

ie

$$[H,J_z]=0$$

Likewise $[H,J_x]=[H,J_y]=0$. Thus,

$$[H,J]=0$$

Also

$$[H,L^2]=0$$

since $[L^2,V(r)]=0,[L^2,p]=0$. Thus, we can write

$$f(r,\hat{r})=\begin{pmatrix}c_1Y_{l,j_z-1/2}(\hat{r})\\ c_2Y_{l,j_z+1/2}(\hat{r})\end{pmatrix}u(r)$$

$$g(r,\hat{r})=\begin{pmatrix}d_1Y_{l,j_z-1/2}(\hat{r})\\ d_2Y_{l,j_z+1/2}(\hat{r})\end{pmatrix}v(r)$$

where l takes the values $j-1/2,j+1/2$. The physical interpretation of these spinor expressions is as follows: In the up spin state, $s_z=\sigma_z/2$ takes the value $1/2$ and according to the above expression for f L_z takes the same value $j_z-1/2$ and since $j_z-1/2+1/2=j_z$, it follows that in the up spin state, J_z takes the value j_z. Likewise in the down spin state in the expression for f, L_z takes the value $j_z+1/2$ while $s_z=\sigma_z/2$ takes the value $-1/2$ so $J_z=j_z$ once again. The same l is used in all expressions because L^2 commutes with H and hence the energy eigenstates can be parametrized by definite values of l ($l(l+1)$ is the eigenvalue of L^2 in the state Y_{lm}). Note that since J_z also commutes with H and with L^2, our energy eigenstates can be labeled by $|E,l,j_z>$. Further by the law of composition of angular momenta, the eigenvalues of J^2 are $j(j+1)$ where j assumes values $|l-1/2|$ and $|l+1/2|$ for fixed l. Equivalently, with j fixed, l assumes the values $j-1/2, j+1/2$. j assumes values $1/2, 3/2, 5/2, ...$ ie, half integers. Now,

$$(\sigma, r)(\sigma, p) = (r, p) + i(\sigma, L) = rp_r + i(\sigma, L)$$

The equation

$$(\sigma, p)g + (V - E + m)f = 0$$

therefore gives

$$rp_r g + i(\sigma, L)g + (V - E + m)(\sigma, r)f = 0$$

and the equation

$$-(\sigma, p)f + (E - V + m)g = 0$$

gives

$$-rp_r f - i(\sigma, L)f + (E - V + m)(\sigma, r)g = 0$$

We also note the following relations:

$$g = (E - V + m)^{-1}(\sigma, p)f$$

and so

$$(\sigma, p)(E - V + m)^{-1}(\sigma, p)f + (V - E + m)f = 0$$

Premultiplying this equation by (σ, r) gives

$$(rp_r + i(\sigma, L))(E - V + m)^{-1}(\sigma, p)f + (V - E + m)(\sigma, r)f = 0$$

We note that since V is radial, it commutes with L and hence with (σ, L). Defining

$$\phi = (\sigma, r)f$$

we get

$$(\sigma, r)\phi = r^2 f$$

and hence, the above equation can be expressed as

$$(rp_r + i(\sigma, L))(E - V + m)^{-1}r^{-2}(\sigma, p)(\sigma, r)\phi + (V - E + m)\phi = 0$$

or using

$$(\sigma, p)(\sigma, r) = (p, r) + i(\sigma, L)$$

$$= (r, p) - 3i + i(\sigma, L) = rp_r - 3i + i(\sigma.L)$$

we get

$$(rp_r + i(\sigma, L))(E - V + m)^{-1}r^{-2}(rp_r - 3i + i(\sigma, L))\phi + (V - E + m)\phi = 0$$

We note that (σ, L) commutes with rp_r and further,

$$(\sigma, L) = (L + \sigma/2)^2 - L^2 - \sigma^2/4 = J^2 - L^2 - 3/4$$

$$= j(j+1) - l(l+1) - 3/4 = \lambda$$

say, when acting on f or g. We note that

$$(\sigma, L)\phi = [(\sigma, L), (\sigma, r)]f + (\sigma, r)(\sigma, L)f$$

$$= \lambda\phi + [(\sigma, L), (\sigma, r)]f$$

Now,

$$[(\sigma, L), (\sigma, r)] = i(\sigma, L \times r - r \times L)$$

and hence ϕ is not an eigenvector of (σ, L).

Remark: We note that

$$(\sigma, r)(\sigma, L) = (r, L) + i(\sigma, r \times L)$$

$$= i(\sigma, r \times L)$$

and

$$r \times L = r \times (r \times p) = r(r,p) - r^2 p$$
$$= r^2 \hat{r} p_r - r^2 p$$

We also note the following: f and g are eigenvectors of J_x, J_y, J_z. We can determine the corresponding eigenvalues of these operators using the Ladder operators J_+, J_- defined as

$$J_+ = J_x + iJ_y, J_- = J_x - iJ_y$$

Writing $f = f_{j,j_z}$, we have

$$J_+ f_{j,j_z} = a(j,j_z) f_{j,j_z+1}, J_- f_{j,j_z} = b(j,j_z) f_{j,j_z-1}$$

and likewise for $g = g_{j,j_z}$. It is also possible to discuss the action of (σ, r) on f and g.

An easier analysis of the Dirac equation in a radial potential: We define

$$\psi = [f^T, g^T]^T$$

where

$$f = \begin{pmatrix} c_1 Y_{l,j_z-1/2}(\hat{r}) \\ c_2 Y_{l,j_z+1/2}(\hat{r}) \end{pmatrix} u(r)$$

$$g(r,\hat{r}) = (\sigma, \hat{r}) \begin{pmatrix} d_1 Y_{l,j_z-1/2}(\hat{r}) \\ d_2 Y_{l,j_z+1/2}(\hat{r}) \end{pmatrix} v(r)$$

Now the Dirac equation gives

$$(\sigma, p)g + (m + V - E)f = 0,$$
$$(\sigma, p)f + (V - m - E)g = 0$$

Now,

$$(\sigma, p)g = (\sigma, p)(\sigma, \hat{r}) \begin{pmatrix} d_1 Y_{l,j_z-1/2}(\hat{r}) \\ d_2 Y_{l,j_z+1/2}(\hat{r}) \end{pmatrix} v(r)$$

$$= ((p, \hat{r}) + i(\sigma, p \times \hat{r})) \begin{pmatrix} d_1 Y_{l,j_z-1/2}(\hat{r}) \\ d_2 Y_{l,j_z+1/2}(\hat{r}) \end{pmatrix} v(r)$$

Now,

$$(p, \hat{r}) = p_x(x/r) + p_y(y/r) + p_z(z/r) = [p_x, x/r] + [p_y, y/r] + [p_z, z/r] + p_r$$

$(p_r = (\hat{r}, p) = -i\partial/\partial r)$

$$= -i(\partial_x(x/r) + \partial_y(y/r) + \partial_z(z/r)) + p_r$$
$$= -2i/r + p_r$$

Also

$$(p \times \hat{r})_x = p_y(z/r) - p_z(y/r) = r^{-1}(zp_y - yp_z) + i(yz/r^2 - yz/r^2) = -r^{-1} L_x$$

and likewise for the other components. Thus,

$$p \times \hat{r} = -r^{-1} L$$

Thus,

$$(\sigma, p)(\sigma, \hat{r}) = p_r - 2i/r - ir^{-1}(\sigma, L)$$

Now, we let

$$\sigma_+ = \sigma_1 + i\sigma_2, \sigma_- = \sigma_1 - i\sigma_2,$$
$$L_+ = L_1 + iL_2, L_- = L_1 - iL_2$$

Thus,

$$(\sigma, L) = (\sigma_+ + \sigma_-)(L_+ + L_-)/4 - (\sigma_+ - \sigma_-)(L_+ - L_-)/4 + \sigma_3 L_3$$
$$= (\sigma_+ L_- + \sigma_- L_+)/2 + \sigma_3 L_3$$

Note that

$$\sigma_+ = \begin{pmatrix} 0 & 2 \\ 0 & 0 \end{pmatrix},$$

$$\sigma_- = \begin{pmatrix} 0 & 0 \\ 2 & 0 \end{pmatrix}$$

Thus,

$$(\sigma, L)\begin{pmatrix} d_1 Y_{l,j_z-1/2}(\hat{r}) \\ d_2 Y_{l,j_z+1/2}(\hat{r}) \end{pmatrix}$$

$$= \begin{pmatrix} d_2 L_- Y_{l,j_z+1/2} + (j_z - 1/2) d_1 Y_{l,j_z-1/2} \\ d_1 L_+ Y_{l,j_z-1/2} + (j_z + 1/2) d_2 Y_{l,j_z+1/2} \end{pmatrix}$$

$$= \begin{pmatrix} (d_2 d(l, j_z + 1/2) - d_1(j_z - 1/2)) Y_{l,j_z-1/2} \\ (d_1 c(l, j_z - 1/2) + j_z + 1/2) d_2) Y_{l,j_z-1/2}) \end{pmatrix}$$

We choose d_1, d_2 so that

$$d_2 d(l, j_z + 1/2) - d_1(j_z - 1/2) = \lambda d_1,$$

$$d_1 c(l, j_z - 1/2) + d_2(j_z + 1/2) = \lambda d_2$$

We will get two solutions for λ. We select one solution. Then, we get

$$(\sigma, p) g = ((-2i/r + p_r) v(r)) \begin{pmatrix} d_1 Y_{l,j_z-1/2}(\hat{r}) \\ d_2 Y_{l,j_z+1/2}(\hat{r}) \end{pmatrix}$$

$$-(i\lambda/r) \begin{pmatrix} d_1 Y_{l,j_z-1/2}(\hat{r}) \\ d_2 Y_{l,j_z+1/2}(\hat{r}) \end{pmatrix} v(r)$$

Thus, the first spinor component of Dirac's equation in a radial potential $V(r)$ is given by provided we take $c_1 = d_1, c_2 = d_2$,

$$(-2i/r + p_r) v(r) - (i\lambda/r0v(r) + (m + V(r) - E) u(r) = 0$$

The second spinor component is the same as (on premultiplying that equation by $(\sigma, \hat{r})$,

$$(\sigma, \hat{r})(\sigma, p) f + (V(r) - m + E) \begin{pmatrix} d_1 Y_{l,j_z-1/2}(\hat{r}) \\ d_2 Y_{l,j_z+1/2}(\hat{r}) \end{pmatrix} v(r) = 0$$

Now,

$$(\sigma, \hat{r})(\sigma, p) = -(\hat{r}, p) + i(\sigma, \hat{r} \times p)$$

$$= p_r - (i/r)(\sigma, L)$$

So as before, we get

$$(p_r - i/r) u(r) - (i\lambda/r) u(r) + (V(r) - m - E) v(r) = 0$$

Thus, $\{u(r), v(r)\}$ satisfy two coupled linear first order differential eigen-equations from which the possible values of the energy eigenvalue E can be determined by a Taylor series expansion.

[45] The Schrodinger equation in an electromagnetic field described as a quantum stochastic process. Calculation of transition probabilities for the atom and bath.

[46] Dirac's equation in an electromagnetic field described as a quantum stochastic process:

$$((\alpha, -i\nabla + eA_t) dt + \beta m dt - eV_t) \psi_t = i d\psi_t$$

Here, $A_t = A_t(r)$ is the magnetic vector potential and $V_t = V_t(r)$ is the electric potential. In the Lorentz gauge, we have

$$V_t = -\int_0^t div A_s ds$$

Now, let $A_t[k], A_t^*[k]$ be quantum creation and annihilation processes (not the magnetic vector potential) and let $\Lambda_t[k, m]$ be the conservation processes, the three satisfying quantum Ito's formulae. We assume that the vector potential has the form

$$\sum_k (c[k](dA_t[k]/dt)\phi_k(r) + \bar{c}[k](dA_t^*[k]/dt)\phi_k(r)^* + \sum_{k,m} d[k, m](d\Lambda_t[k, m]/dt)\chi_{km}(r)$$

where ϕ_k are ordinary linearly independent functions of position and so are χ_{km}. Then by the above Lorentz gauge condition,

$$V_t = -\sum_k (c[k] A_t[k] div\phi_k(r) + \bar{c}[k] A_t^*[k] div\phi_k(r)^*) - \sum_{k,m} d[k, m]\Lambda_t[k, m] div\chi_{km}(r)$$

It is more convenient to use the Hudson-Parthasarathy notation for these noise operators, namely $\Lambda_b^a(t)$. These satisfy the quantum Ito formula

$$d\Lambda_b^a(t).d\Lambda_d^c(t) = \epsilon_d^a d\Lambda_d^b(t), a, b, c, d \geq 0$$

where

$$\Lambda_0^0(t) = t, \Lambda_0^k(t) = A_t[k], \Lambda_k^0(t) = A_t[k]^*, k \geq 1,$$

$$\Lambda_m^k(t) = \lambda(\chi_{[0,t]}|m >< k|), k, m \geq 1$$

ϵ_b^a is defined to be zero if either a or b are zero and δ_b^a otherwise. Thus, the above quantum Ito formulae read

$$dA_t[k]dA_t[m]^* = \delta[k-m]dt, dA_t[k]d\Lambda[r,s] = d\Lambda_0^k.d\Lambda_s^r = \delta[k-s]d\Lambda_0^r$$

or equivalently,

$$dA_t[k]d\Lambda[r,s](t) = \delta[k-s]dA_t[r]$$

$$d\Lambda[r,s](t).dA_t[k]^* = d\Lambda_s^r(t)d\Lambda_k^0(t) = \delta[r-k]d\Lambda_s^0(t) = \delta[r-k]dA_t[s]^*$$

Our noisy Dirac equation in this new notation is obtained by expressing the magnetic vector potential as

$$A_t(r) = \phi_a^b(r)d\Lambda_b^a(t)/dt$$

and hence by the Lorentz gauge condition, the electric scalar potential is

$$V_t(r) = -div(\phi_a^b(r))\Lambda_b^a(t)$$

where the Einstein summation convention has been adopted, ie summation over $a, b \geq 0$. Thus, the qsde describing the evolution of the unitary kernel $U_t(r, r')$ in the Dirac equation is

$$idU_t(r,r') = [(\alpha, -i\nabla_r) + \beta m]U_t(r,r')dt + e[(\alpha, \phi_a^b(r))d\Lambda_b^a(t) - \chi_a^b(r)\Lambda_b^a(t)]U_t(r,r')dt$$

Formally, using operator theoretic notation, this equation can be expressed as

$$idU_t = [(\alpha, p) + \beta m - e\chi_a^b\Lambda_b^a(t)]U_t dt + e[(\alpha, \phi_a^b)d\Lambda_b^a(t)]$$

where

$$\chi_b^a = div\phi_b^a$$

Note that

$$\phi_b^a : \mathbb{R}^3 \to \mathbb{C}^3, \chi_b^a : \mathbb{R}^3 \to \mathbb{C}$$

and to ensure that the vector and scalar potentials are real valued functions, we require that

$$\phi_b^a = \bar{\phi}_a^b$$

The functions phi_b^a must be chosen to maintain unitarity of the evolution operator U_t. We leave it as an exercise to determine these conditions by application of the quantum Ito's formula to

$$0 = d(U^*U) = dU^*.U + U^*dU + dU^*dU$$

[47] General Scattering theory, the Moller and wave operators, the scattering matrix, the Lippman-Schwinger equation for the scattering matrix, Born scattering.

H_0 is the unperturbed/free projectile Hamiltonian ($= P^/2m$), $H = H_0 + V$ is the perturbed Hamiltonian where V is the interaction Hamiltonian between the scatterer and the projectile. The in state is $|\phi_i>$. This is an eigenstate of the free projectile at time $t = -\infty$ projected to time $t = 0$ via the free projectile evolution $exp(-itH_0)$. More precisely, $|\phi_i>$ belongs to the continuous spectrum state of H_0, ie $E_{ac}(H_0)|\phi_i> = |\phi_i>$ where $E_{ac}(H_0)$ is the integral of the spectral measure of H_0 over its continuous spectrum which in the case of $H_0 = P^2/2m$ is the entire real line. The in scattered state projected to time $t = 0$ is $|\psi_i>$. It evolves according to $exp(-itH)$. It follows that

$$lim_{t \to -\infty} exp(-itH)|\psi_i> - exp(-itH_0)|\phi_i> = 0$$

so that

$$|\psi_i> = \Omega_-|\phi_i>, \Omega_- = s.lim_{t \to \infty} exp(itH).exp(-itH_0)$$

Likewise $|\phi_o>$ is the out state ie, the free projectile state at $t=+\infty$ projected to time $t=0$. It also evolves according to the free particle Hamiltonian H_0. The out scattered state $|\psi_o>$ from which it evolves, evolves according to $exp(-itH)$. Thus,

$$lim_{t\to\infty}exp(-itH)|\psi_o> - exp(-itH_0)|phi_o>=0$$

or equivalently,

$$|\psi_o>=\Omega_+|\phi_o>, \Omega_+ = s.lim_{t\to\infty}exp(itH).exp(-itH_0)$$

We define the scattering operator by

$$S=\Omega_+^*\Omega_-$$

It follows that

$$<\phi_o|S|\phi_i>=<\psi_o|\psi_i>$$

ie, the probability of scattering from an in scattered state $|\psi_i>$ to an out scattered state $|\psi_o>$ equals $|<\psi_o|\psi_i>|^2 = |<\phi_o|S|\phi_i>|^2$ where $|\phi_i>$ is the input free state of the projectile and $|\phi_o>$ is the output free state of the projectile. The domain of Ω_+ is not the whole of $\mathcal{H}$ but rather contained in the absolutely continuous spectrum of H_0. We denote the spectral measure of H_0 by E_0 and that of H by E. We then have

$$\Omega_+ = I + \int_0^\infty \frac{d}{dt}(exp(itH).exp(-itH_0)dt$$

$$= I + i\int_0^\infty exp(itH)V.exp(-itH_0)dt$$

and hence

$$\Omega_+^* = I - i\int_0^\infty exp(itH_0)V.exp(-itH)dt$$

Likewise, we have

$$\Omega_- = I - \int_{-\infty}^0 \frac{d}{dt}(exp(itH).exp(-itH_0))dt$$

$$= I - i\int_{-\infty}^0 exp(itH)V.exp(-itH_0)dt$$

$$= I - i\int_0^\infty exp(-itH)V.exp(itH_0)dt$$

and hence

$$\Omega_-^* = I + i\int_0^\infty exp(-itH_0)V.exp(itH)dt$$

It is clear that Ω_+ and Ω_- are isometries, ie,

$$\Omega_+^*\Omega_+ = I, \Omega_-^*\Omega_- = I$$

defined respectively on the domains of Ω_+ and Ω_-. We have

$$R = S - I = \Omega_+^*\Omega_- - \Omega - -^*\Omega_-$$

Now,

$$S = \Omega_+^*\Omega_- =$$

$$(I - i\int_0^\infty exp(itH_0)V.exp(-itH)dt).\Omega_-$$

$$= \Omega_- - i\int_0^\infty exp(itH_0)V.\Omega_-.exp(-itH_0)dt$$

where we have used the identity

$$exp(-itH)\Omega_- = lim_{s\to\infty}exp(-itH).exp(isH).exp(-isH_0)$$

$$= lim_{s\to\infty}exp(i(s-t)H).exp(-i(s-t)H_0).exp(-itH_0)$$

$$\Omega_-.exp(-itH_0)$$

Note that likewise, we have

$$exp(-itH)\Omega_+ = \Omega_+ exp(-itH_0)$$

We now also have

$$I = \Omega_-^*\Omega_- = (I + i\int_0^\infty exp(-itH_0)V.exp(itH)dt)\Omega_-$$

$$= \Omega_- + i\int_0^\infty exp(-itH_0)V.\Omega_-.exp(itH_0)dt$$

Hence,

$$R = S - I =$$

$$-i[\int_0^\infty exp(itH_0)V.\Omega_-.exp(-itH_0)dt + \int_0^\infty exp(-itH_0)V.\Omega_- exp(itH_0)dt]$$

$$= -i\int_{-\infty}^\infty exp(itH_0)V.\Omega_-.exp(-itH_0)dt$$

$$= -i\int_{\mathbb{R}^3} E_0(d\lambda)V\Omega_- E_0(d\mu).exp(it(\lambda-\mu))dt$$

$$= -i\int_{\mathbb{R}^3} E_0(d\lambda)V(I - i\int_0^\infty exp(-isH)V.exp(isH_0)ds)E_0(d\mu).exp(it(\lambda-\mu))$$

$$= -i\int_{\mathbb{R}^3} E_0(d\lambda)V.E_0(d\mu)exp(it(\lambda-\mu))$$

$$-\int_{\mathbb{R}^3\times\mathbb{R}_+} E_0(d\lambda)Vexp(-isH)Vexp(isH_0)E_0(d\mu).exp(it(\lambda-\mu))dtds$$

$$= -2\pi i\int E_0(d\lambda)V.E_0(d\mu)\delta(\lambda-\mu)$$

$$-2\pi\int_{\mathbb{R}^3\times\mathbb{R}_+} E_0(d\lambda)V.exp(-is(H-x))VE_0(dx)E_0(d\mu)\delta(\lambda-\mu)dxds$$

$$= -2\pi i\int E_0(d\lambda)V.E_0(d\mu)\delta(\lambda-\mu)$$

$$+2\pi i\int_{\mathbb{R}^3} E_0(d\lambda)V.(H-x-i0)^{-1}VE_0(dx)E_0(d\mu)\delta(\lambda-\mu)dxds$$

$$= -2\pi i\int_{\mathbb{R}^2} E_0(d\lambda)(V - V(H-\lambda-i0)^{-1}V)E_0(d\mu)\delta(\lambda-\mu)$$

From this it follows that the initial projectile energy μ is conserved after the scattering process, ie, $\lambda = \mu$ and if $|\phi_i>$ is a state of the incident projectile having energy in the range $\lambda + d\lambda$ and likewise for $|\phi_o>$, then

$$<\phi_o|S-I|\phi_i> \approx -2\pi i <\phi_o|V - V(H-\lambda-i0)^{-1}V|\phi_i>$$

In other words, if $R(\lambda) = (S-I)(\lambda)$ denotes the scattering matrix elements for the energy λ between two free particle states having momenta along the directions ω and ω' ($\in S^2$) respectively, then this matrix element is given by $-2\pi i$ times the diagonal matrix D_V minus the product of the matrices D_V, $(K + D_V - \lambda I)^{-1}$ and D_V where D_V is the diagonal matrix having elements $V(\sqrt{2m\lambda'}\omega')\delta(\sqrt{2m\lambda'}\omega' - \sqrt{2m\lambda''}\omega'')$. K is the non-diagonal matrix $(-1/2m)\nabla^2\delta(\sqrt{2m\lambda'}\omega' - \sqrt{2m\lambda''}\omega'')$ The final matrix $D_V - D_V(K + D_V - \lambda I)^{-1}D_V$ is to be evaluated at the row index (λ, ω') and column index (λ, ω) with λ being fixed.

[48] Design of quantum gates using time dependent scattering theory.

$$H(t) = H_0 + V(t)$$

$$\Omega(+) = lim_{t\to\infty}T\{exp(-i\int_0^t H(s)ds)\}^{-1}.exp(-itH_0)$$

$$\Omega(-) = lim_{t\to\infty}T\{exp(-i\int_{-t}^0 H(s)ds)\}.exp(itH_0)$$

$$S = \Omega(+)^*\Omega(-) = lim_{t\to\infty} exp(itH_0).T\{exp(-i\int_0^t H(s)ds)\}.T\{exp(-i\int_{-t}^0 H(s)ds)\}.exp(itH_0)$$

$$= lim_{t\to\infty} exp(itH_0).T\{exp(-i\int_{-t}^t H(s)ds)\}.exp(itH_0)$$

For large t, the approximate value of the transition probability from an initial state $|i>$ to a final state $|f>$ is given by (assuming $V(t)$ is an operator valued stochastic process),

$$P_{2t}(|i>\to|f>) =$$

$$\mathbb{E}[|<f|exp(itH_0).T\{exp(-i\int_{-t}^t H(s)ds)\}.exp(itH_0)|i>|^2]$$

We have upto linear orders in V, the truncated Dyson series

$$T\{exp(-i\int_{-t}^t H(s)ds)\} = exp(-itH_0)(I - i\int_{-t}^t \tilde{V}(s)ds).exp(-itH_0)$$

where

$$\tilde{V}(s) = exp(isH_0)V(s).exp(-isH_0)$$

Thus in this approximation, the scattering operator is approximately given by

$$S = I - i\int_{-t}^t \tilde{V}(s)ds$$

and the transition probability from the state $|i>$ to the state $|f>$ is given by

$$\mathbb{E}|<f|\int_{-t}^t \tilde{V}(s)ds|i>|^2$$

The exact expression for the matrix elements of the scattering operator is given by

$$\mathbb{E}|<f|\sum_{n=1}^{\infty}(-i)^n\int_{-\infty<s_n<...<s_1<\infty}\tilde{V}(s_1)...\tilde{V}(s_n)ds_1...ds_n|i>|^2$$

Chapter **6**

Quantum Stochastics, Filtering, Control and Coding, Hypothesis Testing and Non-Abelian Gauge Fields

[49] Evans-Hudson flows and its application to the quantization of the fluid dynamical equations in noise. Consider the following classical stochastic differential equation:

$$dx_i(t) = f_i(t, x(t))dt + \sum_j g_{ij}(t, x(t))dB_j(t), i = 1, 2, ..., n$$

For a function $\phi(x)$ on $\mathbb{R}^n$, we have

$$d\phi(x(t)) = L_t\phi(x(t))dt + sum_{i,j}\phi_{,i}(x(t))g_{ij}(t, x(t))dB_j(t)$$

where L_t is the generator of the diffusion $x(t)$, ie,

$$L_t\phi(x) = \sum_i f_i(t, x)\partial_i\phi(x) + \frac{1}{2}\sum_{i,j} a_{ij}(t, x)\partial_i\partial_j\phi(x)$$

with

$$a_{ij}(t, x) = (gg^T)_{ij} = \sum_k g_{ik}(t, x)g_{jk}(t, x)$$

Define

$$j_t(\phi) = \phi(x(t))$$

Then, j_t is a homomorphism from the algebra of all twice differentiable functions on $\mathbb{R}^n$ into the algebra of all random variables on the given probability space. We can write

$$dj_t(\phi) = j_t(L_t\phi)dt + \sum_j j_t(\theta_{tj}(\phi))dB_j(t)$$

where

$$\theta_{tj}(\phi) = \sum_i g_{ij}\phi_{,i}$$

Based on this idea, we can generalize the notion of a stochastic differential to the non-commutative quantum setting: Let j_t be a homomorphism from the algebra $\mathcal{B}(\mathfrak{h})$ of bounded operators in the system Hilbert space $\mathfrak{h}$ into the algebra of operators in $\mathfrak{h}\otimes\Gamma_s(\mathcal{H}_{t]})$ where $\mathcal{H}_{t]} = L^2([0, t])\otimes\mathbb{C}^d$. Then, if $\Lambda_b^a(t), 0 \leq a, b \leq d$ are the standard noise operators, we can define a quantum stochastic differnetial equation by the rule

$$dj_t(X) = j_t(\theta_b^a(X))d\Lambda_a^b(t)$$

where $\theta_b^a : \mathcal{B}(\mathfrak{h})rightarrow\mathcal{B}(\mathfrak{h})$ are linear operators satisfying certain structure equations. These structure equations are derived by applying the quantum Ito formula to the relation expressing the fact that j_t must be a homomorphism:

$$dj_t(XY) = d(j_t(X)j_t(Y)) = dj_t(X).j_t(Y) + j_t(X).dj_t(Y) + dj_t(X).dj_t(Y)$$

$$= j_t(\theta_b^a(X))j_t(Y)d\Lambda_a^b + j_t(X)j_t(\theta_b^a(Y))d\Lambda_a^b + j_t(\theta_b^a(X))j_t(\theta_d^c(Y))d\Lambda_a^b d\Lambda_c^d$$

$$= j_t(\theta_b^a(X)Y + X\theta_b^a(Y)d\Lambda_a^b + j_t(\theta_b^a(X)\theta_d^c(Y))\epsilon_c^b d\Lambda_a^d$$

$$= j_t(\theta_b^a(X)Y + X\theta_b^a(Y) + \epsilon_c^d\theta_d^a(X)\theta_b^c(Y))d\Lambda_a^b$$

Comparing this with the expression

$$dj_t(XY) = j_t(\theta_b^a(XY))d\Lambda_a^b$$

gives us the structure equations

$$\theta_b^a(XY) = \theta_b^a(X)Y + X\theta_b^a(Y) + \epsilon_c^d\theta_d^a(X)\theta_b^c(Y), X, Y \in \mathcal{B}(\mathfrak{h})$$

[50] Classical non-linear filtering: The Kushner, Stratanovich, Kallianpur-Striebel formulas.
The state model that $X(t), t \geq 0$ is a Markov process with state space $\mathbb{R}^d$ with generator K_t, ie,

$$\mathbb{E}(\phi(X(t+h))|X(t)=x) = \phi(x) + h.K_t\phi(x) + o(h)$$

where

$$K_t\phi(x) = \int K_t(x,y)\phi(y)dy$$

The measurement model is

$$dz(t) = h_t(X(t))dt + \sigma_v dV(t)$$

where V is a vector valued Brownian motion process. Let

$$Z_t = \sigma(z(s) : s \leq t)$$

then the aim of filtering is to recursively compute the conditional density $p(X(t)|Z_t)$ or more precisely $p(t,x|Z_t)$, the conditional density of $X(t)$ given measurements upto time t. Using Bayes rule, and the independence of the processes $X(.)$ and $V(.)$, we have

$$p(X(t+dt)|Z_{t+dt}) = p(X(t+dt), Z(t), dz(t))/p(Z(t), dz(t))$$

$$= \int p(dz(t)|X(t+dt), X(t))p(X(t+dt)|X(t))p(X(t)|Z_t)dX(t)/\int (numerator)dX(t+dt)$$

Now, conditioned on $X(t)$, we have that $dz(t)$ has its random part coming only from $dV(t)$ which is independent of $X(t+dt)$. Hence, the above can also be expressed as

$$p(X(t+dt)|Z(t+dt)) = p(dz(t)|X(t))p(X(t+dt)|X(t))p(X(t)|Z_t)/\int numeratordX(t)$$

For any function ϕ on $\mathbb{R}^d$, we define

$$\pi_t(\phi) = \mathbb{E}(\phi(X(t))|Z_t)$$

and then the above equation gives

$$\pi_{t+dt}(\phi) = \frac{\sigma_{t+dt}(\phi)}{\sigma_{t+dt}(1)}$$

where

$$\sigma_{t+dt}(\phi) = \int p(dz(t)|X(t))(\phi(X(t)) + K_t\phi(X(t))dt)p(X(t)|Z_t)dX(t)$$

$$= C\int exp(-(dz(t) - h_t(X(t))dt)^T(dz(t) - h_t(X(t))dt)/2dt\sigma_v^2)(\phi(X(t)) + K_t\phi(X(t))dt)p(X(t)|Z_t)dX(t)$$

Note that repeated application of this equation implies that

$$p(X(T)|Z_T) =$$

$$\frac{\int exp(\int_0^T (h_t(X(t))^T dz(t) - h_t(X(t))^T h_t(X(t))dt))p(X(t), 0 \leq t \leq T)\pi_{0 \leq t < T} dX(t)}{\int exp(\int_0^T (h_t(X(t))^T dz(t) - h_t(X(t))^T h_t(X(t))dt))\Pi_{0 \leq t \leq T} dX(t)}$$

[51] Derivation of the extended Kalman filter (EKF) as an approximation to the Kushner filter.
State model: $X(t)$ is a Markov process with generator K_t. Measurment model:

$$dz(t) = h_t(X(t)) + dV(t)$$

where V is a Levy process, ie, a process having independent increments and independent of $X(.)$. $\phi(X(t))$ is an observable to be estimated based on past measurement data $Z_t = (z(s) : s \leq t)$.

$$\pi_t(\phi) = \mathbb{E}(\phi(X(t))|Z_t)$$

We may assume that

$$d\pi_t(\phi) = F_t(\phi)dt + \sum_{k=1}^{\infty} G_{tk}(\phi)(dz(t))^k$$

where $F_t(\phi), G_{tk}(\phi)$ are Z_t measurable. Let

$$dC(t) = C(t)\sum_{k\geq 1} f_k(t)(dz(t))^k, C(0) = 1$$

Then $C(t)$ is Z_t measurable. $f_k's$ are arbitrary non-random functions of time. We have

$$\mathbb{E}[(\phi(X(t)) - \pi_t(\phi))C(t)] = 0$$

Applying Ito's formula to the differential of this equation and using the arbitrariness of the functions f_k gives

$$\mathbb{E}[d\phi(X(t)) - d\pi_t(\phi)|Z_t] = 0,$$

$$\mathbb{E}[(\phi(X(t)) - \pi_t(\phi))(dz(t))^k|Z_t] + \mathbb{E}[(d\phi(X(t)) - d\pi_t(\phi))(dz(t))^k|Z_t] = 0, k = 1, 2, ...$$

This equation is even valid if $X(t)$ is not independent of $V(t)$. In that case $d\phi(X(t)).dz(t)$ need not be zero, however, if $X(.)$ is independent of $z(.)$, then $d\phi(X(t)).dz(t) = 0$.

[52] Belavkin's theory of non-demolition measurements and quantum filtering in coherent states based on the Hudson-Parthasarathy Boson Fock space theory of quantum noise; The quantum Kallianpur-Striebel formula.

Quantum Kallianpur-Striebel formula: Let η_t be the family of output non-demolition measurement filtration algebras. Then if η_t^i is the family of input measurement algebras, we have

$$\eta_t = U(t)^* \eta_t^i U(t)$$

where $U(t)$ is the unitary evolution described by the Hudson-Parthasarathy noisy Schrodinger equation, ie,

$$dU(t) = L_\beta^\alpha U(t) d\Lambda_\alpha^\beta(t)$$

where the system operators L_β^α may be time dependent and are constrained to make $U(t)$ unitary. We have that if $\mathfrak{h}$ is the system Hilbert space and $\Gamma_s(L^2(\mathbb{R}_+) \otimes \mathbb{C}^d)$ is the Boson Fock space for the noise processes (input), then every operator in $\mathfrak{h}$ commutes with every operator in $\Gamma_s(L^2(\mathbb{R}_+) \otimes \mathbb{C}^d)$. Thus, the algebra η_t^i commutes with the algebra $L(\mathfrak{h})$. It follows that the algebra $U_t^* \eta_t^i U(t)$ commutes with the algebra $U(t)^* L(\mathfrak{h}) U(t)$. But since $U(t)$ is unitary, it is easy to see from the commutativity of $L(\mathfrak{h})$ and η_t^i, that $U(T)^* \eta_t^i U(T) = U(t)^* \eta_t^i U(t) = \eta_t$ and hence it follows further that η_t commutes with $j_T(X) = U(T)^* X U(T)$ for all $T \geq t$ for any operator X in $L(\mathfrak{h})$. Now for any system observable X, we define

$$\mathbb{E}_t(X) = \mathbb{E}(U(t)^* X U(t)) = \mathbb{E}(j_t(X))$$

where $\mathbb{E}$ is defined w.r.t the state $|f\phi(u)>$ where $|f> \in \mathfrak{h}$ and $|\phi(u)> = exp(- \| u \|^2 /2)|e(u)>$ where $|e(u)>$ is the exponential vector in the Boson Fock space. Then we claim that

$$\mathbb{E}(j_t(X)|\eta_t) = U(t)^* \mathbb{E}_t(X|\eta_t^i) U(t)$$

Note that η_t^i commutes with X and hence η_t commutes with $j_t(X)$. To prove the above formula, we observe that if if $Z \in \eta_t^i$, then $U(t)^* Z U(t) \in \eta_t$ and hence

$$\mathbb{E}[(j_t(X) - U(t)^* \mathbb{E}_t(X|\eta_t^i) U(t)) U(t)^* Z U(t)]$$

$$= \mathbb{E}[U(t)^* X Z U(t) - U(t)^* \mathbb{E}_t(X|\eta_t^i) Z U(t)]$$

$$= \mathbb{E}[U(t)^*(XZ - \mathbb{E}_t(X|\eta_t^i) Z) U(t)] = \mathbb{E}_t(XZ - \mathbb{E}_t(XZ|\eta_t^i)) = 0$$

by the definition of conditional expectation. This proves the claim. Now suppose $F(t)$ is a system operator, ie, in $L(\mathfrak{h})$ such that

$$\mathbb{E}_t(X) = \mathbb{E}(U(t)^* X U(t)) = \mathbb{E}(F(t)^* X F(t)) = \mathbb{E}_{F(t)}(X)$$

Note that $[F(t), \eta_t^i] = 0$. It follows clearly that

$$\mathbb{E}(F(t)^* F(t)) = \mathbb{E}(U(t)^* U(t)) = 1$$

Then, we claim that

$$\mathbb{E}_{F(t)}(X|\eta_t^i) = \mathbb{E}(F(t)^* X F(t)|\eta_t^i) / \mathbb{E}(F(t)^* F(t)|\eta_t^i)$$

To prove this, we observe that the lhs and the numerator and denominator of the rhs all commute with each other since η_t^i is an Abelian algebra. Further, we have for any operator $Z \in \eta_t^i$ that

$$\mathbb{E}_{F(t)}[(X - (\mathbb{E}(F(t)^* X F(t)|\eta_t^i) / BbbE(F(t)^* F(t)|\eta_t^i)) Z]$$

$$= \mathbb{E}[F(t)^* X Z F(t) - F(t)^* (\mathbb{E}(F(t)^* X F(t) Z|\eta_t^i) / \mathbb{E}(F(t)^* F(t)|\eta_t^i)) F(t)]$$

$$= \mathbb{E}[F(t)^* X Z F(t)] - \mathbb{E}[\mathbb{E}(F(t)^* X Z F(t)|\eta_t^i) F(t)^* F(t) / \mathbb{E}(F(t)^* F(t)|\eta_t^i)]$$

since $F(t)$ commutes with η_t^i and $Z \in \eta_t^i$. Further, conditioning the second term above on η_t^i and then taking the expectation gives

$$\mathbb{E}[\mathbb{E}(F(t)^* XZF(t)|\eta_t^i)F(t)^* F(t)/\mathbb{E}(F(t)^* F(t)|\eta_t^i)]$$

$$= \mathbb{E}[\mathbb{E}(F(t)^* XZF(t)|\eta_t^i)\mathbb{E}(F(t)^* F(t)|\eta_t^i)/\mathbb{E}(F(t)^* F(t)|\eta_t^i)]$$

$$= \mathbb{E}(F(t)^* XZF(t))$$

We have thus proved that

$$\mathbb{E}_{F(t)}[(X - (\mathbb{E}(F(t)^* XF(t)|\eta_t^i)/BbbE(F(t)^* F(t)|\eta_t^i))Z] = 0$$

and hence the claim.

The Hudson-Parthasarathy noisy Schrodinger evolution is described by the qsde

$$dU(t) = (L_\beta^\alpha d\Lambda_\alpha^\beta(t))U(t)$$

where summation over the repeated indices $\alpha, \beta \geq 0$ is understood and the basic processes Λ_β^α satisfy quantum Ito's formula

$$d\Lambda_\beta^\alpha . d\Lambda_\nu^\mu = \epsilon_\nu^\alpha d\Lambda_\beta^\mu$$

where ϵ_ν^μ is zero if either $\mu = 0$ or $\nu = 0$ or $\mu \neq \nu$ and is one otherwise. The system operators L_β^α satisfy the following conditions for $U(t)$ to describe a unitary evolution:

$$0 = d(U^* U) = dU^* .U + U^* .dU + dU^* .dU =$$

$$U^*(L_\alpha^{\beta *} d\Lambda_\alpha^\beta + L_\beta^\alpha d\Lambda_\alpha^\beta + L_\alpha^{\beta *} L_\nu^\mu d\Lambda_\alpha^\beta d\Lambda_\mu^\nu)U$$

so that

$$(L_\alpha^{\beta *} d\Lambda_\alpha^\beta + L_\beta^\alpha d\Lambda_\alpha^\beta + L_\alpha^{\beta *} L_\nu^\mu d\Lambda_\alpha^\beta d\Lambda_\mu^\nu)U$$

which gives

$$L_\alpha^{\beta *} + L_\beta^\alpha + L_\alpha^{\nu *} L_\beta^\mu \epsilon_\mu^\nu = 0$$

The Evans-Hudson flow corresponding to this HP equation is obtained by taking a self-adjoint operator X in the system Hilbert space $\mathfrak{h}$ and setting

$$j_t(X) = U(t)^* XU(t)$$

Then application of quantum Ito's formula gives

$$dj_t(X) = dU^* XU + U^* XdU + dU^* XdU =$$

$$U^*(L_\alpha^{\beta *} X + XL_\beta^\alpha + \epsilon_\mu^\nu L_\alpha^{\nu *} XL_\beta^\mu)U.d\Lambda_\alpha^\beta$$

$$= j_t(\theta_\beta^\alpha(X))d\Lambda_\alpha^\beta$$

where the structure maps θ_β^α are given by

$$\theta_\beta^\alpha(X) =$$

$$L_\alpha^{\beta *} X + XL_\beta^\alpha + \epsilon_\mu^\nu L_\alpha^{\nu *} XL_\beta^\mu$$

We note that they satisfy the structure equations since j_t as defined is a $*$ unital algebra homomorphism. The Belavkin input measurement processes are taken as

$$Y_{in,k}(t) = c_\beta^\alpha[k]\Lambda_\alpha^\beta, k = 1, 2, ..., r, t \geq 0$$

where the $\alpha = \beta = 0$ term is omitted. These processes jointly generate an Abelian family of Von-Neumann algebras.The corresponding output processes are

$$Y_{out,k}(t) = Y_k(t) = U(t)^* Y_{in,k}(t)U(t)$$

and since $U(t)$ is unitary, and $Y_{in,k}(t)$ commute with the system operators, it follows that

$$Y_k(t) = U(T)^* Y_{in,k}(t)U(T), T \geq t$$

and hence

$$[Y_k(t), j_s(X)] = 0, s \geq t$$

ie the output measurements are non-demolition processes. Let $\eta_t = \sigma(Y_k(s) : s \leq t, k = 1, 2, ..., r)$. Then η_t is an Abelian Von-Neumann algebra and belongs to the commutant of the algebra generated by $j_t(X)$ as X ranges over the system operators. We write

$$\pi_t(X) = \mathbb{E}(j_t(X)|\eta_t)$$

where the expectation is taken in the state $|f\phi(u)>$ with $|f>\in \mathfrak{h}, <f,f>=1$ and $\phi(u) = exp(-\parallel u \parallel^2/2)|e(u)>$ where $u \in L^2(\mathbb{R}_+)\otimes \mathbb{C}^d$ and $|e(u)>$ is the corresponding exponential vector in $\Gamma_s(L^2(\mathbb{R}_+)\otimes \mathbb{C}^d)$. Now we can write

$$d\pi_t(X) = F_t(X)dt + G_{kmt}(X)(dY_m(t))^k$$

where the summation in the last term is over $m \geq 1, k \geq 1$ and $F_t(X), G_{kmt}(X)$ are all η_t measurable operators, ie, they can be regarded as commutative stochastic processes. We can write

$$dY_m(t) = dY_{in,m}(t) + dU(t)^*.dY_{in,m}(t).U(t) + U(t)^*dY_{in,m}(t).dU(t)^* =$$

$$+j_t(S^\alpha_\beta[m])d\Lambda^\beta_\alpha$$

and hence for $k \geq 1$

$$(dY_m(t))^k = j_t(S^\alpha_\beta[m,k])d\Lambda^\beta_\alpha(t)$$

where $S^\alpha_\beta[k]$ are system operators, ie, operators in $\mathfrak{h}$ and $S^0_0[k] = 0, k \geq 2$. The equations

$$\mathbb{E}[(dj_t(X) - d\pi_t(X))(dY_m(t))^k|\eta_t] + \mathbb{E}[(j_t(X) - \pi_t(X))(dY_m(t))^k] = 0, k \geq 1$$

and

$$\mathbb{E}[(dj_t(X) - d\pi_t(X))|\eta_t] = 0$$

follow by taking the differential of the expression

$$\mathbb{E}[(j_t(X) - \pi_t(X))C(t)] = 0$$

where $C(t)$ satisfies the qsde

$$dC(t) = \sum_{m,k\geq 1} f_{m,k}(t)C(t)(dY_m(t))^k, C(0) = 1$$

and using the arbitrariness of the complex valued functions $f_{m,k}(t)$. We have

$$\mathbb{E}[dj_t(X)(dY_m(t))^k|\eta_t] =$$

$$\mathbb{E}[j_t(\theta^\alpha_\beta(X))j_t(S^\mu_\nu[m,k])d\Lambda^\beta_\alpha(t).d\Lambda^\nu_\mu(t)|\eta_t]$$

$$= \epsilon^\beta_\mu \pi_t(\theta^\alpha_\beta(X).S^\mu_\nu[m,k])u_\nu(t)\bar{u}_\alpha(t)dt$$

$$\mathbb{E}[d\pi_t(X)(dY_m(t))^k|\eta_t] =$$

$$G_{rst}(X)\mathbb{E}[(dY_s(t))^r(dY_m(t))^k|\eta_t]$$

$$= G_{rst}(X)j_t(S^\alpha_\beta[s,r]S^\mu_\nu[m,k])\mathbb{E}[d\Lambda^\beta_\alpha(t).d\Lambda^\nu_\mu(t)|\eta_t]$$

$$= G_{rst}(X)\pi_t(S^\alpha_\beta[s,r]S^\mu_\nu[m,k])\epsilon^\beta_\mu u_\nu(t)\bar{u}_\alpha(t)dt$$

[53] Classical control of a stochastic dynamical system by error feedback based on a state observer derived from the EKF.

[54] Quantum control using error feedback based on Belavkin quantum filters for the quantum state observer.

Let $Y(t)$ be a measurement noise process and $V(t)$ a system operator process say like $A(t) + A(t)^*$. Consider a qsde

$$dU_c(t) = (-iV(t)dY(t) - V(t)^2dt/2)U_c(t)$$

We are assuming that $V(t)$ is Hermitian. If $V(t) = V$ does not vary with time, we can write the solution to the above qsde as

$$U_c(t) = exp(-iVY(t))$$

$U_c(t)$ is a unitary matrix and its application to the state evolved from a qsde can remove the effect of noise if $Y(t)$ is present as a noise in the qsde. For example, consider the following qsde

$$dU(t) = (-(iH + V^2/2)dt - iVdY(t))U(t)$$

We can write this approximately as

$$U(t + dt) = (I - dt(iH + V^2/2) - iVdY(t))U(t)$$

We now apply the control unitary

$$(I - iV(t)dY(t) - V^2(t)dt/2)^{-1} = I + iV(t)dY(t) - V^2(t)dt/2$$

to $U(t + dt)$ to get

$$(I + iV(t)dY(t) - V^2(t)dt/2)(I - iHdt - iV(t)dY(t) - V^2(t)dt/2)U(t)$$

$$= (I - iHdt)U(t)$$

ie, the effect of the noise is cancelled out. We note that the output measurement is given by $Y_o(t) = U(t)^*Y(t)U(t)$ and it is this process which follows the non-demolition property. We can measure only Y_o without disturbing the dynamics generated by $U(t)$ on the system Hilbert space $\mathfrak{h}$. We cannot measure the input measurement Y without disturbing the system dynamics.

Now Belavkin's quantum filtering equation can be expressed as

$$d\pi_t(X) = F_t(X)dt + G_t(X)dY_t$$

where

$$j_t(X) = U(t)^*XU(t), \pi_t(X) = \mathbb{E}[j_t(X)|\eta_t]$$

$$\eta_t = \sigma\{Y_s : s \leq t\}$$

is the output measurement Abelian algebra at time t. $U(t)$ is generated by the Hudson-Parthasarathy noisy Schrodinger equation and expectations are calculated in the pure state $|f\phi(u) >$ where f is a normalized system vector and $\phi(u)$ is a normalized coherent vector (See the paper by John Gough and Kostler). $F_t(X), G_t(X)$ are linear functions of the system observable and belong to the measurement algebra η_t. More precisely, we can express the above filtering equation as

$$d\pi_t(X) = \pi_t(\theta_0(X))dt + [\pi_t(L_1X + XL_1^*) - \pi(L_2)\pi(X)](dY_t - \pi_t(L_3X + XL_3^*)dt)$$

where L_2 is a Hermitian matrix. For quadrature measurements, $Z_t = Y_t - \int_0^t \pi_s(L_3X + XL_3^*)ds$ is a Brownian motion process. θ_0 is the Gorini-Kossakowski-Sudarshan-Lindblad (GKSL) generator on the system operator space. More generally, for general non-demolition measurements, like photon counting and a combination of quadrature and photon counting measurements, the Belavkin equation has the form

$$d\pi_t(X) = \pi_t(\theta_0(X))dt + [\sum_{k\geq 1} f_k(\pi_t(M_k))\pi_t(\theta_k(X))](dY_t - \sum_{k\geq 1} g_k(\pi_t(N_k))\pi_t(\phi_k(X))dt)$$

where $\theta_k, \phi_k, k \geq 1$ are linear operators on the Banach space of system operators. If we write ρ_t for the density operator on the system space conditioned on the measurements upto time t, we have

$$\pi_t(X) = Tr(\rho_t X)$$

ρ_t should be viewed as a random density matrix on the system operator algebra where the randomness comes from conditioning on the measurements upto time t. Thus, the above Belavkin equation reads

$$d\rho_t = \theta_0^*(\rho_t)dt + [\sum_{k\geq 1} f_k(Tr(\rho_t M_k))\theta_k^*(\rho_t)][dY_t - \sum_{k\geq 1} g_k(Tr(\rho_t N_k))\phi_k^*(\rho_t)dt]$$

It is to be noted that the process

$$M_t = Y_t - \int_0^t \sum_{k\geq 1} g_k(Tr(\rho_s N_k))\phi_k^*(\rho_s)ds$$

is a Martingale (Ph.d thesis of Luc Bouten). The above equation for ρ_t is called a Stochastic Schrodinger Equation. Now, we wish to remove the noise from the above Belavkin equation by an appropriate control so that the evolution equation of the density has just the first term $\theta_0^*(\rho_t)$, ie, we wish to recover the $GKSL$ equation. Consider now the control unitary $U_c(t) = exp(iWY_t)$ where W is a system observable and Y_t the above output measurement. Assume for simplicity, that we take quadrature measurements, ie, M_t is a Wiener process. We apply $U_c(t)$ to ρ_t to get

$$\rho_{c,t} = U_c(t)\rho_t U_c(t)^*$$

Then, by Quantum Ito's formula,

$$d\rho_{c,t} = dU_c(t)\rho_t U_c(t)^* + U_c(t)d\rho_t U_c(t)^* + U_c(t)\rho_t dU_c(t)^* + dU_c(t)d\rho_t U_c(t)^* + U_c(t)d\rho_t dU_c(t)^*$$

[55] Lyapunov's stability theory with application to classical and quantum dynamical systems.

[56] Imprimitivity systems as a description of covariant observables under a group action. Construction of imprimitivity systems, Wigner's theorem on the automorphisms of the orthogonal projection lattice.

$(\Omega, \mathcal{F}, p\mu)$ is a measure space. P is a spectral measure on this space, ie, for each $E \in \mathcal{F}$, $P(E)$ is an orthogonal projection operator in a Hilbert space $\mathcal{H}$. The set of all orthogonal projections on $\mathcal{H}$ is denoted by $\mathcal{P}(\mathcal{H})$. It is also called the projection lattice. Let τ be an automorphism of $\mathcal{P}(\mathcal{H})$, ie, if $\tau : \mathcal{P}(\mathcal{H}) \rightarrow \mathcal{P}(\mathcal{H})$ is such that if $P_1, P_2, ...$ are mutually orthogonal, then $\tau(\sum_j P_j) = \sum_j \tau(P_j)$. More generally, we require that $\tau(max(P_1, P_2)) = max(\tau(P_1), \tau(P_2))$ and $\tau(min(P_1, P_2)) = min(\tau(P_1), \tau(P_2))$ for any two $P_1, P_2 \in \mathcal{P}(\mathcal{H})$. Then, Wigner proved that there exists a unitary or antiunitary operator U in $\mathcal{H}$ such that

$$\tau(P) = UPU^*, P \in \mathcal{P}(\mathcal{H})$$

It follows that if G group that acts on the measure space $(\Omega, \mathcal{F}, P)$ and $g \in G \to \tau_g$ is a homomorphism from G into $aut(\mathcal{P}(\mathcal{H}))$ such that $\tau_g(P(E)) = P(g.E)$, then there exists a projective unitary-antiunitary representation $g \to U_g$ of G into $UA(\mathcal{H})$ such that

$$P(gE) = U_g P(E) U_g^*, g \in G, E \in \mathcal{F}$$

We note that

$$\tau_{g_1}\tau_{g_2}(P(E)) = P(g_1 g_2 E) = \tau_{g_1 g_2}(P(E))$$

so the requirement that $g \to \tau_g$ be a homomorphism is natural from the viewpoint of covariant transformation of observables. Let now H be a subgroup of G. Consider the homogeneous space $X = G/H$. G acts transitively on X. Let μ be a quasi-invariant measure on X, ie, for any $g \in G$, the measures $\mu.g^{-1}$ and μ are absolutely continuous with respect to each other. Consider $f \in L^2(X, \mu)$. Consider for $f \in L^2(X, \mu)$,

$$U_g f(x) = (d\mu.g^{-1}/d\mu)^{1/2} f(g^{-1}x), x \in X$$

Then,

$$\int \| U_g f(x) \|^2 \, d\mu(x) == \int_X |f(g^{-1}x)|^2 d\mu.g^{-1}(x) = \int_X |f(x)|^2 d\mu(x)$$

This proves that U_g is a unitary operator. It is easy to see that U is also a representation:

$$U_{g_1 g_2} f(x) = (d\mu.(g_1 g_2)^{-1}(x)/d\mu) f(g_2^{-1} g_1^{-1} x)$$

On the other hand,

$$U_{g_1}(U_{g_2} f)(x) = (d\mu.g_1^{-1}(x)/d\mu)^{1/2} (U_{g_2} f)(g_1^{-1} x)$$

$$= (d\mu.g_1^{-1}(x)/d\mu)^{1/2} (d\mu.g_2^{-1} g_1^{-1}(x)/d\mu.g_1^{-1} x)^{1/2} f(g_2^{-1} g_1^{-1} x)$$

$$= (d\mu.g_2^{-1} g_1^{-1}(x)/d\mu) f(g_2^{-1} g_1^{-1} x)$$

Thus,

$$U_{g_1 g_2} = U_{g_1}.U_{g_2}, g_1, g_2 \in G$$

Let $A(g, x)$ be a map from $G \times X$ into the algebra of linear operators in a Hilbert space (such that

$$A(g_1 g_2, x) = A(g_1, g_2 x) A(g_2, x)$$

Then consider the operator U_g defined on $L^2(\mu, \mathfrak{h})$

$$U_g f(x) = (d\mu.g^{-1}(x)/d\mu)^{1/2} A(g, g^{-1}x) f(g^{-1}x)$$

We see that

$$U_{g_1}(U_{g_2} f)(x) = (d\mu.g_1^{-1}(x)/d\mu)^{1/2} A(g_1, g_1^{-1}x)(U_{g_2} f)(g_1^{-1}x)$$

$$= (d\mu.g_1^{-1}(x)/d\mu)^{1/2} A(g_1, g_1^{-1}x)(d\mu.g_2^{-1.g_1^{-1}}(x)/d\mu.g_1^{-1})^{1/2} A(g_2, g_2^{-1} g_1^{-1} x) f(g_2^{-1} g_1^{-1} x)$$

$$= (d\mu.g_2^{-1} g_1^{-1}(x)/d\mu)^{1/2} A(g_1 g_2, g_2^{-1} g_1^{-1} x) f((g_1 g_2)^{-1} x)$$

$$= U_{g_1 g_2} f(x)$$

Thus, $g \to U_g$ is a representation of G in $L^2(X, \mathfrak{h})$.

[57] Schwinger's analysis of the interaction between the electron and a quantum electromagnetic field. Let $A(t, r)$ denote the vector potential corresponding to a quantum electromagnetic field. The dynamical variables q, p of the electron bound to the nucleus commute with A. Let $\Phi(t, r)$ denote the quantum scalar potential of the quantum electromagnetic field. Note that if we adopt the Lorentz gauge, $\Phi = -c^2 \int_0^t divA dt$ while if we adopt the Coulomb gauge, then $divA = 0, \Phi = 0$ where for the latter, we are assuming that there is no externally charged matter to generate the scalar potential. The Hamiltonian of the atom interacting with the quantum electromagnetic field is given by

$$H(t) = (p + eA)^2/2m + V(q) - e\Phi + H_{em}$$

where the field Hamiltonian H_{em} has also been added. $H(t)$ is thus the total Hamiltonian of the atom interacting with the quantum em field. We have

$$H(t) = p^2/2m + V(q) + ((p, A) + (A, p))/2m - e\Phi + H_{em}$$

We note that

$$(p, A) + (A, p) = 2(A, p) - i.div(A) = -2i(A, \nabla) - idiv(A)$$

Schrodinger's equation for the wave function $\psi(t)$ of the atom and field is given by

$$i\psi'(t) = H(t)\psi(t)$$

Making the transformation

$$\psi(t) = exp(-itH_{em}))\phi_1(t)$$

and assuming the Coulomb gauge gives

$$i\phi_1'(t) = (p^2/2m + V(q) + (\tilde{A}, p)/m)\phi_1(t)$$

where

$$\tilde{A} = exp(itH_{em}).A.exp(-itH_{em})$$

Making another transformation

$$\phi_1(t) = exp(-iS(t))\phi_2(t)$$

where $S(t)$ is linear in the creation and annihilation operators of the em field, we get

$$\phi_1'(t) = exp(-iS(t))(-iS'(t) + [S(t), S'(t)]/2)\phi_2(t) + exp(-iS(t))\phi_2'(t)$$

since $[S(t), S'(t)]$ is a c-number function of time owing to the commutation relations between the creation and annihilation operators of the em field. Taking $S(t) = (p, \int_0^t \tilde{A}dt)/m = (p, Z(t))$ where $Z(t) = \int_0^t \tilde{A}dt/m$ gives

$$i\phi_2'(t) = exp(iS(t))(p^2/2m - S'(t) - i[S(t), S'(t)]/2 + V(q) + S'(t))exp(-iS(t))\phi_2(t)$$

$$= (p^2/2m + V(q + Z(t)))\phi_2(t)$$

where the c-number function $-i[S(t), S'(t)]$ has been neglected as it only gives an additional phase factor to the wave function.

[58] Quantum control: The system observable X evolves to $j_t^{(1)}(X)$ after time t. The desired system observable X_d evolves to $j_t^{(2)}(X_d)$. Here,

$$j_t^{(k)} = (H_k(t), L_k), k = 1, 2$$

which means that

$$j_t^{(k)}(Z) = U_k(t)^* Z U_k(t), k = 1, 2, Z = X, X_d$$

where $U_k(t)$ satisfies the HP equation

$$dU_k(t) = (-(iH_k(t) + Q_k)dt + L_k dA(t) - L_k^* dA(t)^*)U_k(t), Q_k = L_k L_k^*$$

Define

$$\tilde{X} = \begin{pmatrix} X & 0 \\ 0 & X_d \end{pmatrix},$$

$$j_t(\tilde{X}) = \begin{pmatrix} j_t^{(1)}(X) & 0 \\ 0 & j_t^{(2)}(X_d) \end{pmatrix}$$

Let $\mathfrak{h}$ denote the system Hilbert space and $\Gamma_s(L^2(\mathbb{R}_+))$ the Boson Fock space. j_t is a $*$ unital homomorhism from the algebra $\mathcal{B}(\mathfrak{h}) \oplus \mathcal{B}(\mathfrak{h})$ into the algebra $(\mathcal{B}(\mathfrak{h} \otimes \Gamma_s(L^2(\mathbb{R}_+))) \oplus \mathcal{B}(\mathfrak{h} \times \Gamma_s(L^2(\mathbb{R}_+)))$. Define the operators

$$H(t) = diag[H_1(t), H_2(t)], L = diag[L_1, L_2], U(t) = diag[U_1(t), U_2(t)], Q = diag[Q_1, Q_2]$$

Then, the HP equations for $U_k, k = 1, 2$ can be expressed as a single qsde:

$$dU(t) = [-(iH(t) + Q)dt + LdA(t) - L^* dA(t)^*]U(t)$$

The corresponding observables $\tilde{X}$ in $\mathcal{B}(\mathfrak{h}) \oplus \mathcal{B}(\mathfrak{h})$ evolve to $U(t)^* \tilde{X} U(t) = j_t(\tilde{X})$. A non demolition measurement for the process $j_t(\tilde{X})$ in the sense of Belavkin is given by

$$Y^o(t) = U(t)^* Y^i(t) U(t), Y^i(t) = A(t) + A(t)^*$$

It satisfies the sde

$$dY^o(t) = dY^i(t) + dU(t)^* dY^i(t) U(t) + U(t)^* dY^i(t).dU(t)$$

$$= dY^i(t) - U(t)^*(L + L^*)U(t)dt = dY^i(t) + j_t(S)dt = dA(t) + dA(t)^* + j_t(S)dt$$

where

$$S = -(L + L^*)$$

Let $\eta_t = \sigma(Y_s^o : s \leq t)$. Then η_t is an Abelian Von-Neumann algebra and we define the conditional expectation

$$\mathbb{E}j_t(\tilde{X})|\eta_t) = \pi_t(\tilde{X})$$

We can assume that

$$d\pi_t(Z) = F_t(Z)dt + G_t(Z)dY^o(t)$$

where $F_t(Z)$ and $G_t(Z)$ are in η_t. Then, its evident [See the paper by Gough and Koestler] that

$$\mathbb{E}[(j_t(Z) - \pi_t(Z))dY^o(t)|\eta_t] + \mathbb{E}[(dj_t(Z) - d\pi_t(Z))dY^o(t)|\eta_t] = 0 - - - (1)$$

$$\mathbb{E}[(dj_t(Z) - d\pi_t(Z))|\eta_t] = 0 - - - (2)$$

We can write

$$dj_t(Z) = j_t(\theta_0(Z))dt + j_t(\theta_1(Z))dA(t) + j_t(\theta_2(Z))dA(t)^*$$

where for $Z = diag[Z_1, Z_2]$, we have

$$j_t(Z) = diag[j_t^{(1)}(Z_1), j_t^{(2)}(Z_2)]$$

and

$$dj_t^{(1)}(Z_1) = j_t^{(1)}(\theta_{10}(Z_1))dt + j_t^{(1)}(\theta_{11}(Z_1))dA(t) + j_t^{(1)}(\theta_{12}(Z_1))dA(t)^*$$

$$dj_t^{(2)}(Z_2) = j_t^{(2)}(\theta_{20}(Z_2))dt + j_t^{(2)}(\theta_{21}(Z_1))dA(t) + j_t^{(2)}(\theta_{22}(Z_2))dA(t)^*$$

Thus,

$$\theta_k(Z) = diag[\theta_{1k}(Z_1), \theta_{2k}(Z_2)], k = 0, 1, 2$$

[59] Quantum error correcting codes: Let $\mathcal{H} = \mathbb{C}^N$ be the Hilbert space of the quantum system and let $\mathcal{C}$ be a subspace of $\mathcal{H}$. $\mathcal{C}$ is called the code subspace. If ρ is a density operator in $\mathcal{H}$, we say that $\rho \in \mathcal{C}$ if $Range(\rho) \subset \mathcal{C}$. If $\mathcal{N}$ is a subspace of the space $\mathbb{C}^{N\times N}$ of all linear operators in $\mathcal{H}$, then we say $\mathcal{C}$ is an $\mathcal{N}$ error correcting code, if there exist operators $R_0, ..., R_r \in \mathbb{C}^{N\times N}$ such that $\sum_k R_k^* R_k = I$ and whenever $E_1, ..., E_r \in \mathcal{N}$ are such that $\sum_k E_k E_k^* = I$, and $\rho \in \mathcal{C}$, then

$$\sum_k R_k E(\rho) R_k^* = c\rho$$

where $c \in \mathbb{C}$ and

$$E(\rho) = \sum_k E_k \rho E_k^*$$

is the output state of the noisy channel $\{E_k\}$. $\{R_k\}$ are called recovery operators for the noise subspace $\mathcal{N}$ and the code subspace $\mathcal{C}$. If there exist such recovery operators, then we say that $\mathcal{C}$ is an $\mathcal{N}$ error correcting code.

Theorem (Knill-Laflamme): $\mathcal{C}$ is an $\mathcal{N}$ error correcting code iff $PN_2^*N_1P = \lambda(N_2 * N_1)P$ for all $N_1, N_2 \in \mathcal{N}$ and some complex scalar $\lambda(N_2^*N_1)$ (dependent on $N_2^*N_1$). Here, P is the orthogonal projection onto $\mathcal{C}$.

Proof: Suppose first that $\mathcal{C}$ is an $\mathcal{N}$ error correcting code with recovery operators $R_0, ..., R_r$. Let $E_k, k = 1, 2, ..., m$ be a quantum channel in $\mathcal{N}$. Then we have for all $|\psi> \in \mathcal{C}$ the relation

$$\sum_{k,s} R_k E_s |\psi><\psi| E_s^* R_k^* = \lambda(\psi)|\psi><\psi|$$

where $\lambda(\psi)$ is a complex scalar possibly dependent on $|\psi>$. It follows that for all $|\psi> \perp |\psi>$, ie, $<\phi|\psi>= 0$, we have

$$\sum_{k,s} |<\phi|R_k E_s|\psi>|^2 = \lambda|<\phi|\psi>|^2 = 0$$

Thus,

$$R_k E_s|\psi> \perp |\phi> \forall |\phi> \perp |\psi$$

ie,

$$R_k E_s|\psi> = \beta_{ks}(\psi)|\psi>, \forall |\psi> \in \mathcal{C}$$

and some complex numbers $\beta_{ks}(\psi)$. By linearity of the operators, it is clear that β_{ks} cannot depend on $|\psi>$. We get therefore,

$$R_k E_s P = \beta_{ks} P, \forall k, s$$

Thus,

$$PE_q^* R_k^* R_k E_s P = \beta_{ks}\bar{\beta}_{kq} P$$

and summing this equation over k and using $\sum_k R_k^* R_k = I$, we get

$$PE_q^* E_s P = a_{qs} P, a_{qs} \in \mathbb{C}$$

We have thus proved that

$$PN_2^* N_1 P = \propto P, \forall N_1, N_2 \in \mathcal{N}$$

Conversely, suppose that this relation holds. Then, let

$$\mathcal{N}_0 = \{N \in \mathcal{N} : \lambda(N^*N) = 0\}$$

It is clear that $\mathcal{N}_0$ is a subspace of $\mathcal{N}$. Indeed, suppose $N_1, N_2 \in \mathcal{N}_0$. Then,

$$PN_k^* N_k P = \lambda(N_k^* N_k) P = 0, k = 1, 2$$

and hence

$$N_k P = 0, k = 1, 2$$

Thus,

$$PN_j^* N_k P = 0, j, k = 1, 2$$

and hence,

$$P(c_1 N_1 + c_2 N_2)^*(c_1 N_1 + c_2 N_2) P = 0, .c_1, c_2 \in \mathbb{C}$$

Now, define for $N \in \mathcal{N}$,

$$[N] = N + \mathcal{N}_0 \in \mathcal{N}/\mathcal{N}_0$$

We define

$$< [N_1], [N_2] >= \lambda(N_1^* N_2), N_1, N_2 \in \mathcal{N}$$

This definition is valid since for $N_0 \in \mathcal{N}_0$ and $N \in \mathcal{N}$ we have

$$\lambda(N^* N_0) = \lambda(N_0^* N) = 0$$

This is because,

$$PN_0^* N_0 P = \lambda(N_0^* N_0) P = 0$$

implies

$$PN_0^* = N_0 P = 0$$

and hence,

$$0 = PN_0^* NP = \lambda(N_0^* N) P,$$

$$0 = PN^* N_0 P = \lambda(N * N_0) P$$

It is clear therefore that $< .,. >$ defines an inner product on $\mathcal{N}/\mathcal{N}_0$. So, we can choose an onb $\{[N_k], k = 1, 2, ..., r\}$ for $\mathcal{N}/\mathcal{N}_0$ w.r.t. this inner product. Now, define

$$R_k = PN_k^*, P_k = N_k P N_k^*, k = 1, 2, ..., r, R_0 = I - \sum_{k=1}^{r} P_k$$

Then,

$$P_k P_j = N_k P N_k^* N_j P N_j^* = \lambda(N_k^* N_j) N_j P N_j^* = \delta_{kj} P_j$$

proving that $P_j, j = 1, 2, ..., r$ for a set of mutually orthogonal orthogonal projections in $\mathcal{H}$. Thus R_0 is also an orthogonal projection. Now, suppose $\rho \in \mathcal{C}$. Then $\rho = P\rho = \rho P = P\rho P$. Then, for $N \in \mathcal{N}$,

$$\sum_k R_k N \rho N^* R_k^* = \sum_{k \geq 1} PN_k^* NP\rho PN^* N_k P + R_0 N \rho N^* R_0$$

$$= \sum_{k\geq 1} |\lambda(N_k^* N)|^2 \rho + R_0 N \rho N^* R_0$$

$$= \lambda(N^* N)\rho + R_0 N \rho N^* R_0$$

Here, we have used the Bessel equality/Parseval relation for orthonormal bases for a Hilbert space. Further,

$$R_0 NP = NP - \sum_k P_k NP = NP - \sum_k N_k P N_k^* NP$$

$$= NP - \sum_k \lambda(N_k^* N) N_k P$$

$$= NP - NP = 0$$

by the generalized Fourier series in a Hilbert space and the relation $N_0 P = 0$ for all $N_0 \in \mathcal{N}$. Thus, we finally get

$$\sum_{k=0}^{r} R_k^* N \rho N^* R_k^* = \lambda(N^* N)\rho$$

and further,

$$\sum_{k=0}^{r} R_k^* R_k = \sum_{k\geq 1} N_k P N_k^* + R_0 = \sum_{k\geq 1} P_k + I - \sum_{k\geq 1} P_k = I$$

This completes the proof of the Knill-Laflamme theorem.

Construction of quantum error correcting codes using imprimitivity systems: Let A be a finite Abelian group and define the onb $|x>, x \in A$ for $L^2(A)$ so that

$$|x>= [\delta_{x,0}, ..., \delta_{x,N-1}]^T$$

where we are assuming

$$A = \{0, 1, ..., N-1\}$$

with addition modulo N. with each $n \in A$, define the character

$$\chi_n(x) = exp(2\pi i n x/N), x \in A$$

We may identify this character with $n \in A$. In this way, we get an isomorphism of $\hat{A}$ with A and we write

$$< n, x >=< x, n >= \chi_n(x)$$

Define the unitary operators U_x, V_x on $L^2(A)$ by

$$U_a|x >= |x + a >, V_a|x >=< a, x > |x >$$

Then define the Weyl operator

$$W(x, y) = U_x V_y, x, y \in A$$

on $L^2(A)$. These are N^2 unitary operators and in fact, as we shall soon see, they form an orthogonal basis for $L^2(A \times A)$, for the space of all linear operators in $L^2(A)$. We derive the Weyl commutation relations:

$$W(x, y)|a >= U_x V_y|a >= U_x < y, a > |a >=< y, a > |a + x >$$

$$V_y U_x|a >= V_y|a + x >=< y, a + x > |a + x >=< y, a >< y, x > |a + x >$$

Thus,

$$V_y U_x =< y, x > U_x V_y =< x, y > U_x V_y$$

Also,

$$W(x, y)W(u, v) = U_x V_y U_u V_v =< y, u > U_{u+x} V_{v+y} =< y, u > W(u + x, v + y)$$

So $(x, y) \rightarrow W(x, y)$ is a projective unitary representation of the Abelian group $G = A \times A$ into the space of operators in $L^2(A)$. Now,

$$W(x, y)|a >=< y, a > |a + x >$$

Hence,

$$Tr(W(x,y)) = \sum_a < a|W(x,y)|a >= \sum_a < y,a > \delta_{a+x,a}$$

It follows that $Tr(W(x,y)) = 0$ if $x \neq 0$ and if $x = 0$, then $Tr(W(x,y)) = Tr(W(0,y)) = \sum_a < y,a >= N\delta_{y,0}$ Thus, for all $x, y \in A$,

$$Tr(W(x,y)) = N\delta_{x,0}\delta_{y,0}$$

It follows that

$$W(x,y)^*W(u,v) = V_{-y}U_{-x}U_uV_v = V_{-y}U_{u-x}V_v =< y, x-u > U_{u-x}V_{v-y} =< y, x-u > W(u-x, v-y)$$

Thus,

$$Tr(W(x,y)^*W(u,v)) =< y, x-u > Tr(W(u-x, v-y)) = 0, (u,v) \neq (x,y)$$

and further,

$$Tr(W(x,y)^*W(x,y)) = Tr(I) = N$$

Thus, $\{W(x,y)/\sqrt{N} : x, y \in A\}$ forms an orthonormal basis for $L^2(A \times A)$ and in particular, this is a set of N^2 linearly independent operators. We can express this orthonormality relations as

$$Tr(W(x,y)^*W(u,v)) = N\delta[x-u]\delta[y-v]$$

Now let H be a Gottesman subgroup of $G = A \times A$. This means that for each pair $(u,v), (x,y) \in H$, the identity

$$< u, y >^* < v, x >= 1$$

holds. For arbitrary $(u,v) \in G$, we define the map $\omega_H(u,v) : H \to \mathbb{C}$ such that

$$\omega_H(u,v)(x,y) =< u, y >^* < v, x >$$

Then we have

$$\omega_H(u,v) = 1, (u,v) \in H$$

In other words, $H \subset Ker(\omega_H) = K$ say. Now we have that for $(x,y), (u,v) \in H$,

$$W(x,y)W(u,v) =< y, u > W(x+u, y+v), W(u,v)W(x,y) =< v, x > W(u+x, v+y)$$

and since

$$< y, u > / < v, x >=< y, u >< v, x >^* = 1$$

it follows that

$$[W(x,y), W(u,v)] = 0 \forall (x,y), (u,v) \in H$$

Thus, there exists an onb for $L^2(A)$ such that relative to this basis, $W(x,y)$ is diagonal for all $(x,y) \in H$.

$$W(x,y) = diag[\alpha_k(x,y), k = 1, 2, ..., N], (x,y) \in H$$

Define $\tilde{W}(x,y), (x,y) \in G$ by the equation

$$W(x,y) = \alpha_1(x,y)\tilde{W}(x,y)$$

Note that $|\alpha_k(x,y)| = 1 \forall k$ and also for $(x,y), (u,v) \in H$ we have

$$\tilde{W}(x,y)\tilde{W}(u,v) = (\alpha_1(x,y)\alpha_1(u,v))^{-1}W(x,y)W(u,v) =$$

$$(\alpha_1(x,y)\alpha_1(u,v))^{-1} < y, u > W(x+u, y+v)$$

$$= \frac{< y, u > \alpha_1(x+u, y+v)}{\alpha_1(x,y)\alpha_1(u,v)}\tilde{W}(x+u, y+v)$$

Now the relation

$$W(x,y)W(u,v) =< y, u > W(x+u, y+v)$$

implies using the diagonal representation when restricted to H that

$$\alpha_1(x,y)\alpha_1(u,v) =< y, u > \alpha_1(x+u, y+v)$$

and hence

$$\tilde{W}(x,y)\tilde{W}(u,v) = \tilde{W}(x+u, y+v), (x,y), (u,v) \in H$$

ie, the projective unitary representation W of G reduces to an ordinary unitary representation when restricted to H.

An example: Let $\mathbb{F}$ be a finite field and consider the vector spaces $V = \mathbb{F}^p$. Let $L : V \to V$ and $M : V \to V$ be two linear transformations such that $L^T M : V \to V$ is symmetric. Define $N : V \to V$ by $L^T M = N + N^T$. For $x, y \in V$, we define the Weyl operator $W(x,y) : L^2(V) \to L^2(V)$ in the usual way. We note that V is a finite vector space over $\mathbb{F}$ and if $\mathbb{F}$ consists of a elements, then V will consist of a^p elements and $dimL^2(V) = a^p$. Now, let χ_0 be a character of the field $\mathbb{F}$ viewed as an Abelian group under addition. Consider for a fixed $a \in V$

$$\tilde{W}(u) = \chi_0(a^T u + u^T N u)W(Lu, Mu), u \in V$$

Now

$$\tilde{W}(u+v) = \chi_0(a^T(u+v) + u^T Nu + v^T Nv)W(Lu, Mu)W(Lv, Mv)$$

$$\chi_0(a^T(u+v) + u^T Nu + v^T Nv)\chi_0(u^T M^T Lv)W(L(u+v), M(u+v))$$

$$= \chi_0(a^T(u+v) + u^T Nu + v^T Nv + u^T(N+N^T)v)W(L(u+v), M(u+v))$$

$$= \chi_0(a^T(u+v) + (u+v)^T N(u+v))W(L(u+v), M(u+v))$$

$$= \tilde{W}(u+v)$$

provided that we assume that the Weyl operator $W(x,y)$ has been defined so that in the expression

$$W(x,y)W(u,v) =< y,u > W(x+u, y+v),$$

$$< y,u >= \chi_0(y^T u)$$

or in other words, the character of V corresponding to any $y \in V$ when V is viewed as an Abelian group under addition, is given by $u \to \chi_0(y^T u)$. Thus $u \to \tilde{W}(u)$ is a unitary representation of V. We note that $H = \{(Lu, Mu) : u \in V\}$ is a Gottesman subgroup of $V \times V$. This is because, for $u, v \in V$,

$$< Lu, Mv >^* < Mu, Lv >= \bar{\chi}_0(u^T L^T Mv)\chi_0(u^T M^T Lv) = \chi_0(u^T(M^T L - L^T M)v) = 0$$

since

$$\bar{\chi}_0(x) = \chi_0(x)^{-1}$$

Construction of quantum error correcting codes using the Imprimitivity theorem of Mackey: Let G be a finite group acting on a finite set X. For each $x \in X$ let P_x be an orthogonal projection onto a Hilbert space $\mathcal{H}$ such that $P_x P_y = \delta_{x,y} I, \sum_{x \in X} P_x = I$. Assume that $g \to U_g$ is a unitary representation of G in $\mathcal{H}$ satisfying the Imprimitivity condition

$$U_g P_x U_g^* = P_{gx}, g \in G, x \in X$$

Let $E \subset X$ and then we have

$$U_g P(E) U_g^* = P(gE), g \in G$$

Now let $\mathcal{N}$ be a linear space of linear operators in $\mathcal{H}$ spanned by $\{U_g P(E) : g \in G\}$ where $E \subset X$ is fixed. We choose $x \in X$ and ask the question when does the quantum code P_x detect $\mathcal{N}$. This happens iff

$$P_x U_g P(E) P(E) U_g^* P_x$$

is proportional to P_x for all $g, h \in G$. This is the same as saying that

$$P_x P(gE) P_x = \lambda P_x$$

This happens iff either $x \notin gE$ in which case, the lhs is zero, or if $gE = x$ for all $g \in G$ which is impossible. Thus, P_x corrects $\mathcal{N}$ iff $g^{-1}x \notin E$. Now we take $F \subset X$ and derive the conditions for $P(F)$ to detect $\mathcal{N}$. This happens iff

$$P(F) U_g P(E) P(E) U_g^* P(F)$$

is proportional to $P(F)$, ie, iff $P(F \cap gE)$ is proportional to $P(F)$. This happens iff either $gE \cap F = \phi$ or $gE \subset F$. Now we consider the correction problem. $P(F)$ corrects $\mathcal{N}$ iff for all $g, h \in G$,

$$P(F) U_g P(E) P(E) U_h^* P(F) = \lambda P(F)$$

where λ may depend on g, h. This happens iff

$$P(F)P(gE)P(gh^{-1}F)U_{hg^{-1}} = \lambda P(F)$$

or equivalently, iff

$$P(F \cap gE \cap gh^{-1}F)U_{hg^{-1}} = \lambda P(F)$$

Post-multiplying both sides of this equation by their adjoints gives us

$$P(F \cap gE \cap gh^{-1}F) = |\lambda|^2 P(F)$$

This is a necessary condition for $P(F)$ to correct $\mathcal{N}$ and happens only if for each g, h either $F \cap gE \cap gh^{-1}F = \phi$ or $F \subset gE \cap gh^{-1}F$. As a special case, choosing $F = \{x\}$, we see that P_x corrects $\mathcal{N}$ only if for each g, h either $x \notin gE$ or or $x = gh^{-1}x$. We also derive the following conclusions from the above discussion. P_x corrects $\mathcal{N}$ iff for each g, h, either $x \notin gE$ or $x = gh^{-1}x$ and

$$U_{gh^{-1}}P_x = \lambda P_x$$

We also observe that if $U_a P_x = \lambda P_a$ for some $a \in G$, then,

$$U_a P_x U_a^* = |\lambda|^2 P_x$$

or equivalently,

$$P_{ax} = |\lambda|^2 P_x$$

which implies that $|\lambda| = 1$ and $ax = x$. Thus if we define

$$G_{eig}^x = \{g \in G : U_g P_x = \lambda P_x\}$$

and

$$G_{iso}^x = \{g \in G : gx = x\}$$

then we find that

$$G_{eig}^x \subset G_{iso}^x$$

and that P_x corrects $\mathcal{N} = span\{U_g P(E) : g \in G\}$ iff for each $g, h \in G$ (either $x \notin gE$ or $gh^{-1} \in G_{eig}^x\}$) In particular, P_x corrects $span\{U_g P_y : g \in G\}$ iff for each $g, h \in H$, (either $x \neq gy$ or $gh^{-1} \in G_{eig}^x$). As a special case, P_x corrects span $\{U_g P_x : g \in G\}$ if for each $g, h \in G$ either $g \notin G_{iso}^x$ or $gh^{-1} \in G_{eig}^x$.

Let C be a cross section of G/G_{iso}^x. Let $c_1, c_2 \in C, g, h \in G_{eig}^x$. Then

$$P_x U_{c_1 g}^* U_{c_2 h} P_x = P_x U_{g^{-1} c_1^{-1} c_2 h} P_x = \lambda P_x$$

since $G_{eig}^x \subset G_{iso}^x$. Hence, P_x corrects $span\{U_g : g \in CG_{eig}^x\}$ for each $x \in G$.

A linear algebraic example of a quantum error correcting code. Let the code subspace projection P be given by

$$P = \begin{pmatrix} I_r & 0 \\ 0 & 0 \end{pmatrix} \in \mathbb{C}^{n \times n}$$

Let U be a fixed $n \times n$ unitary matrix and let the noise subspace of operators $\mathcal{N}$ be the span of the adjoints of all $n \times n$ matrices N_k having the block structure

$$N_{kj} = \begin{pmatrix} \lambda_{kj} A_k & \lambda_{kj} B_k \\ C_{kj} & D_{kj} \end{pmatrix}, k = 0, 1, ..., n/k - 1$$

where $[A_k|B_k] \in \mathbb{C}^{r \times (n-r)}$ is obtained as the $kr+1$ to $(k+1)r$ rows of U arranged one below the other ($k = 0, 1, ..., (n/k) - 1$) and the λ_{kj} are arbitrary complex numbers. We find that since

$$A_k A_m^* + B_k B_m^* = \delta_{km} I_r$$

(obtained using $UU^* = I_n$), we have

$$N_{kj} N_{ml}^* = \begin{pmatrix} \lambda_{kj} \bar{\lambda}_{kl} \delta_{km} I_r & F_1 \\ F_2 & F_3 \end{pmatrix}$$

where F_1, F_2, F_3 are matrices of size $r \times (n-r)$, $(n-r) \times r$ and $(n-r) \times (n-r)$ respectively. Thus,

$$PN_{kj}N_{ml}^*P = \delta_{km}\lambda_{kj}\bar{\lambda}_{kl}P$$

and hence the quantum code P corrects $\mathcal{N}$.

[60] Quantum hypothesis testing: Let A, B be two density matrices and let $0 \leq T \leq I$. The probability of making a correct decision when the density is A and the measurement is T is given according to quantum mechanical rules by $Tr(AT)$. The probability of making a correct decision when the density is B is given by $Tr(B(I-T))$. Now we wish to choose T such that $Tr(AT)$ is a maximum subject to the constraint $Tr(BT) \leq \alpha$ where $\alpha \in [0,1)$ is given. Such a test corresponds to minimizing the error probability under A subject to the constraint that the error probability under B is smaller than a prescribed threshold. The function $c \to T(c) = \{A \geq cB\}$ is assumed to be continuous on $[0,\infty)$. We note that $T(0) = I, T(\infty) = 0$ (By $\{U \geq V\}$ for Hermitian operators U, V, we mean the orthogonal projection onto the space spanned by all vectors v for which $(U-V)v = \lambda v$ for some $\lambda \geq 0$.) Hence, $Tr(BT(c))$ takes all values in $[0,1)$ as c varies over $[0,\infty)$. Let $\alpha \in [0,1)$ and choose c such that $Tr(BT(c)) = \alpha$. Then, we claim that $T(c)$ is an optimal test. To see this, suppose $0 \leq T \leq I$ is any other test (ie measurement or positive operator) such that $Tr(BT) \leq \alpha$. Then

$$Tr(AT) = Tr((A-cB)T) + cTr(BT) \leq Tr((A-cB)T) + c\alpha$$

$$= Tr((A-cB)T(c)T) + Tr((A-cB)\{A < cB\}T) + c\alpha \leq Tr((A-cB)T(c)) + c\alpha$$

$$= Tr(AT(c))$$

Here, we make use of the fact that since $A - cB$ commutes with $T(c)$ and $T(c)^2 = T(c)$, we have

$$Tr((A-cB)T(c)T) = Tr(T(c)(A-cB)T(c)T) = Tr(T^{1/2}T(c)(A-cB)T(c)T^{1/2}) \leq Tr(T(c)(A-cB)T(c))$$

$$= Tr((A-cB)T(c))$$

since $0 \leq T^{1/2} \leq I$ and $T(c)(A-cB)T(c) \geq 0$. Now, we can derive bounds on the error probabilities. We have for $s \geq 0$,

$$P_1(c) = Tr(A(I-T(c)) = Tr(A\{A < cB\}) \leq c^s Tr(A^{1-s}B^s)$$

Thus,

$$log(P_1(c)) \leq s.log(c) + log(Tr(A^{1-s}B^s) = s(R + log(A^{1-s}B^s)/s)$$

where

$$R = log(c)$$

Consider now the function

$$f(s) = sR + log(Tr(A^{1-s}B^s), s \geq 0$$

We wish to select R so that $f(s)$ has a minimum at $s = 0$. For this, we require that $f'(0) = 0, f''(0) \geq 0$. Now,

$$f'(0) = R - D(A|B)$$

where

$$D(A|B) = Tr(A(logA - logB))$$

is the relative entropy between A and B. Thus, the first condition gives $R = D(A|B)$. The second condition $f''(0) \geq 0$ gives

$$\frac{d}{ds}[Tr(-A^{1-s}log(A)B^s + A^{1-s}B^s log(B))/Tr(A^{1-s}B^s)]|_{s=0} \geq 0$$

or equivalently,

$$Tr(A(log(A))^2 - 2Alog(A)log(B) + A(log(B)^2) + (Tr(Alog(A) - Alog(B)))^2 \geq 0$$

which is true since by the Schwarz inequality,

$$|Tr(Alog(A)log(B))|^2 \leq Tr(A(log(A))^2).Tr(A.(log(B))^2)$$

It follows that

$$P_1(e^R) = 0$$

if $R = D(A|B)$ and for the second kind of error probability, we get

$$P_2(c) = Tr(BT(c)) = Tr(B\{A < cB\}) = 1 - Tr(B\{A > cB\})$$

Also, for $s \leq 1$

$$Tr(B\{A > cB\}) \leq c^{s-1}Tr(A^{1-s}B^s)$$

so

$$log(1 - P_2(c)) \geq (s-1)log(c) + log(Tr(A^{1-s}B^s))$$

We now find on letting $s = 0$ that

$$log(1 - P_2(c)) \geq -log(c) + 1$$

so if we put $R = log(c)$, we get

$$log(1 - P_2(e^R)) \geq 1 - R$$

or equivalently,

$$P_2(e^R) \leq 1 - exp(1 - R)$$

We also note that

$$lim_{s \to 0} log(Tr(A^{1-s}B^s)/s = -D(A|B)$$

So, taking $log(c) = D(A|B) - \delta$ gives

$$log(1 - P_2(c)) \geq -log(c) + s(log(c) + log(Tr(A^{1-s}B^s))/s)$$

$$\geq -D(A|B) + \delta + s(D(A|B) - \delta - D(A|B)) = -D(A|B) + (1-s)\delta$$

for $s \leq 1$. We also recall that

$$log(P_1(c)) \leq s(log(c) + log(Tr(A^{1-s}B^s))/s)$$

$$= s(D(A|B) - \delta + D(A|B) + \epsilon(s)) = -s(\delta - \epsilon(s))$$

where

$$\epsilon(s) \to 0, s \to 0$$

Hence, if s is sufficiently small, say $s \leq s_0$, then $\epsilon(s) < \delta/2$ and for all such s, we have

$$log(P_1(c)) \leq -s\delta/2 \leq -s_0\delta/2$$

Note: Consider the function

$$g(s) = log(Tr(A^{1-s}B^s))/s$$

We have

$$g'(s) = Tr(-A^{1-s}log(A)B^s + A^{1-s}B^s log(B))/(s.Tr(A^{1-s}B^s)) - log(Tr(A^{1-s}B^s))/s^2$$

The limit of this as $s \to 0$ is the same as the limit of

$$-D(A|B)/s + D(A|B)/s$$

as $s \to 0$ which is zero. We also note that the limit of $g''(s)$ as $s \to 0$ is the same as the limit of

$$Tr(A(log(A))^2 + B(log(B))^2 - 2Alog(A)log(B))/s + D(A|B)/s^2 + (D(A|B)^2/s) + (D(A|B)/s^2) + (2/s^3)log(Tr(A^{1-s}B^s)$$

which is same as the limit of

$$Tr(A(log(A))^2 + B(log(B))^2 - 2Alog(A)log(B))/s + D(A|B)/s^2 + (D(A|B)^2/s) + (D(A|B)/s^2) - 2D(A|B)/s^2$$

The above equals

$$Tr(A(log(A))^2 + B(log(B))^2 - 2Alog(A)log(B))/s + (D(A|B)^2/s$$

which is positive. Hence, $g(s)$ for $s \geq 0$ attains its minimum value of $D(A|B)$ at $s = 0$.

Chapter 7
Master Equations, Non-Abelian Gauge Fields

[61] The Sudarshan-Lindblad equation for observables on the quantum system has the following general form:

$$X' = i[H, X] - \frac{1}{2}\sum_k (L_k^* L_k X + X L_k^* L_k - 2L_k^* X L_k)$$

Assume that

$$H = \frac{1}{2}p^T F_1 p + \frac{1}{2}q^T F_2 q$$

and

$$L_k = \lambda_k^T p + \mu_k^T q, k \geq 1$$

We have

$$p = ((p_n)), q = ((q_n)), [q_n, p_m] = i\delta_{n,m}$$

$$L_k^* L_k X + X L_k^* L_k - 2L_k^* X L_k = L_k^*[L_k, X] + [X, L_k^*]L_k = \theta_k(X)$$

say. We thus have

$$[L_k, q_n] = [\lambda_k^T p, q_n] = -i\lambda_k[n],$$

$$[q_n, L_k^*] = [q_n, \bar{\lambda}_k^T p] = i\bar{\lambda}_k[n]$$

so

$$\theta_k(q_n) = -i\lambda_k[n](\bar{\lambda}_k^T p + \bar{\mu}_k^T q) + i\bar{\lambda}_k[n](\lambda_k^T p + \mu_k^T q)$$

so

$$\sum_k \theta_k(q) = \sum_k [(-i\lambda_k \lambda_k^* + i\bar{\lambda}_k \lambda_k^T)p + (-i\lambda\mu_k^* + i\bar{\lambda}_k \mu_k^T)q]$$

$$= \sum_k [2Im(\lambda_k \lambda_k^*)p + 2Im(\lambda_k \mu_k^*)q]$$

Likewise,

$$[L_k, p_n] = [\mu_k^T q, p_n] = i\mu_k[n], [p_n, L_k^*] = [p_n, \bar{\mu}_k^T q] = -i\bar{\mu}_k[n]$$

so

$$\theta_k(p_n) = i\mu_k[n](\bar{\lambda}_k^T p + \bar{\mu}_k^T q) - i\bar{\mu}_k[n](\lambda_k^T p + \mu_k^T q)$$

so

$$\theta_k(p) = -2Im(\mu_k \lambda_k^*)p - 2Im(\mu_k \mu_k^*)q$$

We write

$$\theta = \sum_k \theta_k$$

Then,

$$[H, q_n] = [p^T F_1 p/2, q_n] = -i\sum_m F_1[n, m]p_m$$

ie,

$$[H, q] = -iF_2 p$$

Likewise,

$$[H, p] = iF_1 q$$

[62] The Yang-Mills field and its quantization using path integrals:

Let G be a finite dimensional Lie group and $\mathfrak{g}$ its Lie algebra. For simplicity, assume that G is a subgroup of $U(n, \mathbb{C})$. Then $\mathfrak{g}$ consists of $n \times n$ complex skew-Hermitian matrices. We can thus choose a basis $\{i\tau_a : a = 1, 2, ..., n\}$ for $\mathfrak{g}$ with the $\tau_a's$ Hermitian matrices. Thus, the commutation relations of these basis elements has the form

$$[i\tau_a, i\tau_b] = -C(abc)i\tau_c$$

where the $C(abc)$ are real constants and summation over the repeated index c is understood. Equivalently, these commutation relations can be expressed as

$$[\tau_a, \tau_b] = iC(abc)\tau_c$$

We let $A_\mu : \mathbb{R}^4 \to \mathfrak{g}$ be the gauge fields. Thus, the covariant derivatives in terms of these gauge fields are defined by

$$\nabla_\mu = \partial_\mu - eA_\mu$$

where e is a real constant. We can write

$$A_\mu(x) = iA_\mu^a(x)\tau_a$$

where the $A^a_\mu(x)'s$ are now real valued fields. There are in all $4n$ of such fields. We assume that the wave function $\psi(x) \in \mathbb{R}^{4n}$ satisfies an equation of the form

$$[\gamma^\mu i\partial_\mu \otimes I_n + eA^a_\mu(x)\gamma^\mu \otimes \tau_a - mI_{4n}]\psi(x) = 0$$

where γ^μ are the four Dirac γ matrices forming a basis for the Clifford algebra in $\mathbb{C}^{4\times 4}$. Formally, we can write the above wave equation as

$$\gamma^\mu(i\partial_\mu - ieA_\mu) - m)\psi = 0$$

or equivalently as

$$[\gamma^\mu(i\partial_\mu + eA^a_\mu\tau_a) - m]\psi = 0$$

This eqations can be derived from the variational principle

$$\delta_\psi S = 0$$

where

$$S[\psi] = \int \bar{\psi}^*\gamma^0(\gamma^\mu(i\partial_\mu + eA^a_\mu\tau_a) - m)\psi d^4x$$

$$= \int Ld^4x$$

Note that $\gamma^0\gamma^\mu$ are Hermitian matrices and so is $\gamma^0\gamma^\mu \otimes \tau_a$. Hence using integration by parts, it follows that $S[\psi]$ is real. In terms of the gauge covariant derivative defined above, we have

$$L = \psi^*\gamma^0(i\gamma^\mu\nabla_\mu - m)\psi$$

We consider the following transformation of the wave function

$$\psi(x) \to \psi'(x) = g(x)\psi(x)$$

where $g(x) \in G$ is to be intepreted as $I_4 \otimes g(x)$. This is called a local gauge transformation of the wave function. Then we consider a corresponding transformation of the gauge field $A_\mu(x)$:

$$A_\mu(x) \to A'_\mu(x)$$

so that if

$$\nabla'_\mu = \partial_\mu - eA'_\mu$$

then

$$\nabla'_\mu\psi' = g(x)\nabla_\mu\psi$$

If this happens, the the above Lagrangian density L will be invariant under a local gauge transformations of the wave function and the gauge fields. To get this, we must satisfy

$$g(x)(\partial_\mu - eA_\mu(x))\psi(x) = (\partial_\mu - eA'_\mu(x))\psi'(x)$$

$$= (\partial_\mu - eA'_\mu(x))g(x)\psi(x)$$

which is equivalent to

$$A'_\mu = gA_\mu g^{-1} + e^{-1}(\partial_\mu g)g^{-1}$$

Note that we get a gauge invariant Lagrangian by considering $\psi^*\gamma^0\gamma^\mu(\partial_\mu - eA_\mu)\psi$ which is the same as $\psi^* gamma^0\gamma^\mu\nabla_\mu\psi$ or more precisely,

$$\psi^*(\gamma^0\gamma^\mu \otimes \nabla_\mu)\psi$$

we note that the gauge transformation $g(x)$ is actually to be interpreted as $I_4 \otimes g(x)$ and it acts only on the second component in the tensor product of $\mathbb{C}^4 \otimes \mathbb{C}^n$.

[63] A general remark on path integral computations for gauge invariant actions.
Suppose that we have an action integral $I(\phi)$ of the fields $\phi(x)$ and we compute a path integral of the form

$$S = \int exp(iI(\phi))B[\phi]d\phi$$

Suppose that the action integral $I(\phi)$ is invariant under a Gauge transformation $\phi \rightarrow \phi_\Lambda$. Suppose also that the path integral measure $d\phi = \Pi_{x\in\mathbb{R}^4} d\phi(x)$ is invariant under the same Gauge transformation. Then, we can write

$$S = \int exp(iI(\phi_\Lambda)B[\phi_\Lambda]d\phi_\Lambda$$

$$= \int exp(iI(\phi))B[\phi_\Lambda]d\phi$$

If follows that for an function $C(\Lambda)$ on the gauge group ($\Lambda \in G$), we have

$$S\int C(\Lambda)d\Lambda = \int exp(iI(\phi))d\phi \int B[\phi_\Lambda]C(\Lambda)d\Lambda$$

Now let Λ, λ be two gauge group transformations. Then, we have

$$\int B[\phi_\Lambda]C(\Lambda)d\Lambda$$

$$= \int B[\phi_{\Lambda o\lambda}]C(\Lambda o\lambda)d\lambda$$

assuming that the measure $d\lambda$ on the gauge group is a left invariant Haar measure so that $d\Lambda o\lambda = d\lambda$. Define for any functional F of the fields ϕ

$$\mathcal{F}(\phi, x) = (\delta F(\phi_{\Lambda)})/\delta\Lambda)(x)|_{\Lambda=Id}$$

We have

$$(\delta F(\phi_{\Lambda o\lambda})/\delta\lambda)(x) =$$

$$(\delta F(\phi_{\Lambda o\lambda})/\delta(\Lambda o\lambda))(x)(\delta(\Lambda o\lambda)/\delta\lambda)(x)$$

It follows that on evaluating both sides at $\lambda = Id$, the identity gauge transformation, we get

$$\tilde{F}(\phi, \Lambda, x) = F(\phi, \Lambda, x)G(\Lambda)$$

where

$$\tilde{F}(\phi, \Lambda, x) = (\delta F(\phi_{\Lambda o\lambda})/\delta\lambda)(x)|_{\lambda=Id},$$

and

$$G(\Lambda) = (\delta(\Lambda o\lambda)/\delta\lambda)|_{\lambda=Id}$$

$$F(\phi, \Lambda, x) = (\delta F(\phi_\Lambda)/\delta\Lambda)(x)$$

Now consider

$$S = \int exp(iI(\phi))B[F[\phi]]F(\phi, x)\Pi d\phi(x)$$

$$= \int exp(iI(\phi))B[F[\phi_\Lambda]]\Pi_x F(\phi_\Lambda, x)d\phi(x)$$

provided that we assume invariance of the path measure $exp(iI(\phi))d\phi$ under gauge transformations Λ, ie,

$$exp(iI(\phi_\Lambda))d\phi_\Lambda = exp(iI(\phi))d\phi$$

It follows that

$$S\int C(\Lambda)d\Lambda =$$

$$\int exp(iI(\phi))B[F[\phi_\Lambda]]\Pi_x F(\phi_\Lambda, x)d\phi(x)C(\Lambda)d\Lambda$$

$$= \int exp(iI(\phi))B[F[\phi_\Lambda]](\delta F[\phi_\Lambda]/\delta\Lambda)G(\Lambda)^{-1}C(\Lambda)d\Lambda$$

So by choosing $C = G$, we get

$$S\int C(\Lambda)d\Lambda =$$

$$\int exp(iI(\phi))B[F[\phi]]dF[\phi] = \int exp(iI(\phi))B[\phi]d\phi$$

establishing the invariance of the scattering matrix S under the gauge fixing functional F.

[64] Calculation of the normalized spherical harmonics.

$$Y_{lm}(\theta, \phi) = P_{lm}(cos(\theta))exp(im\phi)$$

where $f(x) = P_{lm}(x)$ satisfies the modified Legendre equation

$$((1-x^2)f')' + (l(l+1) - m^2/(1-x^2))f = 0$$

or equivalently,

$$(1-x^2)^2 f'' - 2x(1-x^2)f' + (l(l+1)(1-x^2) - m^2)f = 0$$

or

$$(1-x^2)^2 f'' - 2x(1-x^2)f' + (l(l+1) - m^2 - l(l+1)x^2)f = 0$$

For simplicity of notation, let

$$\lambda = l(l+1)$$

Then, the above equation is the same as

$$(1 + x^4 - 2x^2)f'' - 2x(1-x^2)f' + (\lambda - m^2 - \lambda x^2)f = 0$$

Let

$$f(x) = \sum_{n \geq 0} c(n)x^n$$

Then, we get on substituting this into the above differential equation,

$$\sum_n c(n)n(n-1)(x^{n-2} + x^{n+2} - 2x^n) - 2\sum_n nc(n)(x^n - x^{n+2}) + (\lambda - m^2)\sum_n c(n)x^n - \lambda\sum_n c(n)x^{n+2} = 0$$

Equating coefficients of x^n gives

$$(n+2)(n+1)c(n+2) + (n-2)(n-3)c(n-2) - 2n(n-1)c(n) - 2nc(n) + 2(n-2)c(n-2) + (\lambda - m^2)c(n) - \lambda c(n-2) = 0$$

or

$$(n+2)(n+1)c(n+2) + ((n-1)(n-2) - \lambda)c(n-2) + (\lambda - m^2 - 2n^2)c(n) = 0$$

or

$$(n+4)(n+3)c(n+4) + (n(n+1) - l(l+1))c(n) + (l(l+1) - m^2 - 2(n+2)^2)c(n+2) = 0$$

If l is not a non-negative integer, then there will be nonzero $c(n)'s$ for arbitrarily large n and as $n \to \infty$, we would get

$$c(n+4) + c(n) - 2c(n+2) \approx 0, n \to \infty$$

or equivalently,

$$c(n+4) - c(n+2) - (c(n+2) - c(n)) \approx 0$$

which would imply that $c(n+2) - c(n)$ converges to a constant, say K. Then $c(2n)$ or $c(2n+1)$ for large n behaves as Kn and since the series $\sum_{n \geq 0} nx^{2n}$ behaves as $x/(1-x^2)^2$ which is not integrable over $x \in [0,1]$ since it has a singularity at $x = 1$ (ie, at $\theta = 0$), it follows that the series has to terminate at some finite N, ie, there must exist a finite integer N such that $c(n) = 0$ for all $n > N$ which is equivalent to saying that $f(x)$ must be a polynomial. This can happen only if we impose the condition $l = N$ and $c(N+2) = 0$. Assume first that $N = 2r$ is even. Then, we may assume that $c(2n+1) = 0$ for all n and

$$(2n+4)(2n+3)c(2n+4) + (2n(2n+1) - 2r(2r+1))c(2n) + (2r(2r+1) - m^2 - 4(n+1)^2)c(2n+2) = 0$$

for $n = 0, 1, ..., r-1$. Putting $n = r$ we then get $c(2r+2) = 0$ since we are assuming that $c(2r+4) = 0$. Thus, $c(2n) = 0$ for all $n > r$ and we get that

$$f(x) = \sum_{n=0}^{r} c(2n)x^{2n}$$

where $c(0)$ is arbitrary, $c(2)$ is obtained by putting $n = -1$ in the above difference equation and using $c(k) = 0, k < 0$:

$$c(2) = (m^2 - 2r(2r+1))c(0)/2$$

and

$$(2n+4)(2n+3)c(2n+4) = -[(2n(2n+1) - 2r(2r+1))c(2n) + (2r(2r+1) - m^2 - 4(n+1)^2)c(2n+2)]/(2n+4)(2n+3)$$

for $n = 0, 1, ..., r-2$. In a similar way, we can describe the polynomials $P_{lm}(x)$ for l odd, ie $l = 2r+1$. We assume that $c(2n) = 0$ for all n and get using the recursion

$$(n+4)(n+3)c(n+4) + (n(n+1) - (2r+1)(2r+3))c(n) + ((2r+1)(2r+3) - m^2 - 2(n+2)^2)c(n+2) = 0$$

at $n = -1$,

$$6c(3) + ((2r+1)(2r+3) - m^2 - 2)c(1) = 0$$

so that

$$c(3) = (m^2 + 2 - (2r+1)(2r+3))c(1)/6$$

and then replacing n by $2n+1$ in the above recursion,

$$c(2n+5) = -[(2n+5)(2n+4)]^{-1}[((2n+1)(2n+3) - (2r+1)(2r+3))c(2n+1) + ((2r+1)(2r+3) - m^2 - 2(2n+3)^2)c(2n+3)]$$

for $n = 0, 1, ..., r-2$ and we may assume that $c(2n+1) = 0$ for $n = r+1, r+2, ...$ The values of $c(0)$ and $c(1)$ for the two cases are determined by the normalization condition

$$2\pi \int_{-1}^{1} P_{lm}(x)^2 dx = 1$$

This guarantees that

$$\int_0^{\pi} \int_0^{2\pi} |Y_{lm}(\theta, \phi)|^2 sin(\theta) d\theta d\phi = 1$$

A technique for calculating the representation matrix $\pi_l(R)$ for $R \in SO(3)$. Here, $\pi_l(R)$ is defined by

$$Y_{lm}(R^{-1}\hat{n}) = \sum_{m'=-l}^{l} [\pi_l(R)]_{m'm} Y_{lm'}(\hat{n})$$

From this equation, it is clear that

$$[\pi_l(R)]_{m'm} = \int Y_{lm}(R^{-1}\hat{n}) \bar{Y}_{lm'}(\hat{n}) d\Omega(\hat{n})$$

For example, taking

$$R = R_x(\beta)$$

we have

$$R^{-1}[cos(\phi)sin(\theta), sin(\phi)sin(\theta), cos(\theta)]^T =$$

$$[cos(\phi)sin(\theta), sin(\phi)sin(\theta)cos(\beta) - cos(\theta)sin(\beta), sin(\phi)sin(\theta)sin(\beta) + cos(\theta)cos(\beta)]^T$$

Denoting, this new point by (θ', ϕ'), we get

$$\theta' = cos^{-1}(sin(\phi)sin(\theta)sin(\beta) + cos(\theta)cos(\beta)),$$

$$\phi' = tan^{-1}[(sin(\phi)sin(\theta)cos(\beta) - cos(\theta)sin(\beta))/cos(\phi)sin(\theta)]$$

We then have

$$[\pi_l(R_x(\beta))]_{m'm} = \int_0^{\pi} \int_0^{2\pi} \bar{Y}_{lm'}(\theta, \phi) Y_{lm}(\theta', \phi') sin(\theta) d\theta d\phi$$

A MATLAB programme for tabulating the matrix elements $[\pi_l(R_x(-\beta))]_{m'm}$ would then proceed along the following lines: We store $[\pi_l(R_x(-2\pi k/N))]_{m'm}$ with $k = 0, 1, ..., N-1, m', m = -l, -l+1, ..., l$ as a two dimensional array A with matrix elements $A[k+1, (2l+1)(m'+l) + m + 1]$ of size $N \times (2l+1)^2$. Thus,

for $k = 0 : N - 1$
for $m' = -l : l$
for $m = -l : l$
$sum = 0$;
for $r = 0 : N - 1$
for $s = 0 : N - 1$
$\theta = \pi r/N$
$\phi = 2\pi s/N$
$\beta = 2\pi k/N$

$$\theta' = cos^{-1}(sin(\phi)sin(\theta)sin(\beta) + cos(\theta)cos(\beta)),$$

$$\phi' = tan^{-1}[(sin(\phi)sin(\theta)cos(\beta) - cos(\theta)sin(\beta))/cos(\phi)sin(\theta)]$$

$sum = sum + conj(Y_{lm'}(\theta, \phi)) * Y_{lm}(\theta', \phi') * (\pi/N) * (2\pi/N)$;
end;
end;
$A[k + 1, (2l + 1)(m' + l) + m + 1] = sum$;
end;
end;
end;

[65] Volterra systems in quantum mechanics: The Hammiltonian has the form

$$H(t) = H_0 + f(t)V_0$$

Let $U(t)$ be the Schrodinger evolution operator:

$$U'(t) = -iH(t)U(t), U(0) = I$$

Then,

$$U(t) = U_0(t)W(t), U_0(t) = exp(-itH_0)$$

and $W(t)$ has the Dyson series

$$W(t) = I + \sum_{n\geq 1}(-i)^n \int_{0<t_n<...<t_1<t} f(t_1)...f(t_n)V(t_1)...V(t_n)dt_1...dt_n$$

where

$$V(t) = U_0(t)^* V_0 U_0(t)$$

The average value of an observable X at time t in the state ρ is given by

$$< X > (t) = Tr(\rho U(t)^* X U(t)) = Tr(\rho W(t)^* U_0(t)^* X U_0(t) W(t))$$

$$= Tr(\rho W(t)^* X_0(t) W(t))$$

$$= \sum_{n,m\geq 0} i^{n+m}(-1)^m \int_{0<t_n<...<t,0<s_m<...<s_1<t} f(t_1)...f(t_n)f(t_m)...f(t_m)Tr(\rho V(t_n)...V(t_1)X_0(t)V(s_1) ...V(s_m))dt_1...dt_n ds_1...ds_m$$

where

$$X_0(t) = U_0(t)^* X U_0(t)$$

This equation is expressible as

$$< X > (t) = Tr(\rho X) + \sum_{n\geq 1}\int_{[0,t]^n} h_n(t, t_1, ..., t_n)f(t_1)...f(t_n)dt_1...dt_n$$

Note that

$$V(t_n)...V(t_1)X_0(t)V(s_1)...V(s_m)$$

$$= U_0(-t_n)V_0U_0(t_n - t_{n-1})V_0U_0(t_{n-1} - t_{n-2})V_0...V_0U_0(t_1 - t)XU_0(t - s_1)V_0U_0(s_1 - s_2)...V_0U_0(s_m)$$

Suppose in particular, that $|k>, k = 0, 1, 2, ...$ are the orthonormal eigenstates of H_0 and

$$\rho = \sum_{k\geq 0} p[k]|k><k|$$

ie $[\rho, H_0] = 0$. It then follows that

$$Tr(\rho V(t_n)...V(t_1)X_0(t)V(s_1)...V(s_m))$$
$$= Tr(\rho U_0(s_m-t_n)V_0U_0(t_n-t_{n-1})V_0U_0(t_{n-1}-t_{n-2})V_0...V_0U_0(t_1-t)XU_0(t-s_1)V_0U_0(s_1-s_2$$

and it is immediately clear from this expression that this can be expressed as a function of $t - t_k, t - t - s_j, k = 1, 2, ..., n, j = 1, 2, ..., m$. Hence in this particular case, when the density matrix of the system commutes with the unperturbed Hamiltonian H_0 (for example, the Gibbs state $\rho = exp(-\beta H_0)/Tr(exp(-\beta H_0))$), it follows that the above Volterra system is time invariant, ie, it can be expressed as

$$<X>(t) = h_0 + \sum_{n\geq 1} \int_{0<t_n<...<t_1<t} h_n(t-t_1, ..., t-t_n)f(t_1)...f(t_n)dt_1...dt_n$$

The Volterra kernel l h_n depends on V_0, H_0, X, ρ. Thus identifying the Volterra kernel h_n from measurements of the input signal $f(t)$ and the output signal $<X>(t)$ will enable us to estimate the observables of the quantum system V_0, H_0, X, ρ. In the discrete domain, consider the second order Volterra approximation:

$$y[n] = \sum_{k=0}^{p} h[k]x[n-k] + \sum_{k=0}^{p} g[k,l]x[n-k]x[n-l] + w[n]$$

where we are assuming the time invariant case, ie $[\rho, H_0] = 0$. In that case, once we identify the kernels h, g we would get information about $V_0, H_0, X, \{p[k]\}$ where $\rho = \sum_k p[k]|k><k|$. Here, $w[n]$ is the noise process involved in the measurement. Now, we discuss the RLS lattice like algrorithm for estimating the kernels h, g using both order and time recursions. Write

$$x_n = [x[n], x[n-1], ..., x[0]]^T, x_{n,p} = [x[n], ..., x[n-p]]^T, y_n = [y[n], y[n-1], ..., y[0]]^T$$

Let z^{-1} denote the delay operator, ie for $r \geq 0$,

$$z^{-r}x_n = x_{n-r} = [x[n-r], x[n-r-1], ..., x[-r]]^T,$$

etc. Then the above Volterra model has the form

$$y_n = X_{n,p}h + Z_{n,p}g$$

where

$$X_{n,p} = [x_n, x_{n-1}, ..., x_{n-p}]$$
$$= [x_{n,p}, x_{n-1,p}, ..., x_{0,p}]^T \in \mathbb{R}^{n+1 \times p+1}$$
$$Z_{n,p} = [x_{n,p} \otimes x_{n,p}, ..., x_{0,p} \otimes x_{0,p}]^T \in \mathbb{R}^{n+1 \times (p+1)^2}$$

We define $h_{n,p}, g_{n,p}$ to be the optimal least squares estimate of h, g based on input-output data collected upto time n with filters of order p:

$$(h_{n,p}, g_{n,p}) = argmin_{h,g} \parallel y_n - X_{n,p}h - Z_{n,p}g \parallel^2$$

Thus,

$$[h_{n,p}^T, g_{n,p}^T]^T = [X_{n,p}|Z_{n,p}] \begin{pmatrix} X_{n,p}^T X_{n,p} & X_{n,p}^T Z_{n,p} \\ Z_{n,p}^T X_{n,p} & Z_{n,p}^Y Z_{n,p} \end{pmatrix} \begin{pmatrix} X_{n,p}^T \\ Z_{n,p}^T \end{pmatrix} y_n$$

Consider the matrix

$$U_{n,p} = [X_{n,p}|Z_{n,p}]$$

We have

$$U_{n+1,p} = \begin{pmatrix} x_{n+1,p}^T & x_{n+1,p}^T \otimes x_{n+1,p}^T \\ X_{n,p} & Z_{n,p} \end{pmatrix}$$

This is the time update of $U_{n,p}$. The order update of $U_{n,p}$ is given by

$$U_{n,p+1} = [X_{n,p}|x_{n-p-1}|Z_{n,p}|\eta_{n,p}]$$

where

$$\eta_{n,p} = row[x[m-p-1]x^T_{m,p}|x[m-p-1]x^T_{m,p}|x[m-p-1]^2 : m = n, n-1, ..., 0]$$

We define

$$Q_{n,p} = \begin{pmatrix} X^T_{n,p}X_{n,p} & X^T_{n,p}Z_{n,p} \\ Z^T_{n,p}X_{n,p} & Z^T_{n,p}Z_{n,p} \end{pmatrix} \begin{pmatrix} X^T_{n,p} \\ Z^T_{n,p} \end{pmatrix}$$
$$= U^T_{n,p}U_{n,p}$$

Then, we can write with the obvious notations,

$$Q_{n,p} = \begin{pmatrix} Q^x_{n,p} & Q^{x,z}_{n,p} \\ Q^{z,x}_{n,p} & Q^z_{n,p} \end{pmatrix}$$

$$Q^x_{n,p+1} = X^T_{n,p+1}X_{n,p+1} = \begin{pmatrix} Q^x_{n,p} & X^T_{n,p}x_{n-p-1} \\ x^T_{n-p-1}X_{n,p} & x^T_{n-p-1}x_{n-p-1} \end{pmatrix}$$

$$Q^z_{n,p+1} = \begin{pmatrix} Q^z_{n,p} & Z^T_{n,p}\eta_{n,p} \\ \eta^T_{n,p}Z_{n,p} & \eta^T_{n,p}\eta_{n,p} \end{pmatrix}$$

where we have used

$$Z_{n,p+1} = [Z_{n,p}|\eta_{n,p}]$$

Further,

$$Q^{x,z}_{n,p+1} = X^T_{n,p+1}Z_{n,p+1} =$$
$$\begin{pmatrix} Q^{x,z}_{n,p} & X^T_{n,p}\eta_{n,p} \\ x^T_{n-p-1}Z_{n,p} & x^T_{n-p-1}\eta_{n,p} \end{pmatrix}$$
$$Q^{z,x}_{n,p+1} = Q^{x,zT}_{n,p+1}$$

Combining all these gives

$$Q_{n,p+1}$$

$$Q_{n,p+1} = \begin{pmatrix} Q^x_{n,p} & X^T_{n,p}x_{n-p-1} & Q^{x,z}_{n,p} & X^T_{n,p}\eta_{n,p} \\ x^T_{n-p-1}X_{n,p} & x^T_{n-p-1}x_{n-p-1} & x^T_{n-p-1}Z_{n,p} & x^T_{n-p-1}\eta_{n,p} \\ Q^{z,x}_{n,p} & Z^T_{n,p}x_{n-p-1} & Q^z_{n,p} & Z^T_{n,p}\eta_{n,p} \\ \eta^T_{n,p}X_{n,p} & \eta^T_{n,p}x_{n-p-1} & \eta^T_{n,p}Z_{n,p} & \eta^T_{n,p}\eta_{n,p} \end{pmatrix}$$

By applying a permutation P that interchanges the second and third block columns and the second and third block rows, ie,

$$P = \begin{pmatrix} I & 0 & 0 & 0 \\ 0 & 0 & I & 0 \\ 0 & 1 & 0 & 0 \\ 0 & 0 & 0 & I \end{pmatrix}$$

we get

$$PQ_{n,p+1}P^T = \begin{pmatrix} Q^x_{n,p} & Q^{x,z}_{n,p} & X^T_{n,p}x_{n-p-1} & X^T_{n,p}\eta_{n,p} \\ Q^{z,x}_{n,p} & Q^z_{n,p} & Z^T_{n,p}x_{n-p-1} & Z^T_{n,p}\eta_{n,p} \\ x^T_{n-p-1}X_{n,p} & x^T_{n-p-1}Z_{n,p} & x^T_{n-p-1}x_{n-p-1} & x^T_{n-p-1}\eta_{n,p} \\ \eta^T_{n,p}X_{n,p} & \eta^T_{n,p}Z_{n,p} & \eta^T_{n,p}x_{n-p-1} & \eta^T_{n,p}\eta_{n,p} \end{pmatrix}$$

From this equation, the order update of $Q^{-1}_{n.p}$ can be easily computed. This is based on the fact that if

$$\begin{pmatrix} A & B \\ C & D \end{pmatrix} \begin{pmatrix} P & Q \\ R & S \end{pmatrix} = I,$$

then

$$AP + BR = I, AQ + BS = 0, CP + DR = 0, CQ + DS = I$$

so that

$$Q = -A^{-1}BS, P = -C^{-1}DR,$$
$$(B - AC^{-1}D)R = I, (D - CA^{-1}B)S = I$$

giving

$$R = (B - AC^{-1}D)^{-1}, S = (D - CA^{-1}B)^{-1},$$

$$Q = -A^{-1}B(D - A^{-1}B)^{-1},$$
$$P = -C^{-1}D(B - AC^{-1}D)^{-1}$$

Here, we are assuming that A, B, C, D are square matrices possibly of different sizes. If we assume that only A, D are square and nonsingular (but possibly of different sizes), then we would derive different formulas:

$$Q = -A^{-1}BS, (D - CA^{-1}B)S = I, R = -D^{-1}CP, (A - BD^{-1}C)P = I$$

and hence

$$P = (A - BD^{-1}C)^{-1}, S = (D - CA^{-1}B)^{-1}, R = -D^{-1}C(A - BD^{-1}C)^{-1},$$
$$Q = -A^{-1}B(D - CA^{-1}B)^{-1}$$

The order update of the $Q_{n,p}^{-1}$ can be obtained by applying the latter formulas. Specifically, we have expressed

$$PQ_{n,p+1}P = \begin{pmatrix} Q_{n,p} & A_{n,p+1} \\ A_{n,p+1}^T & B_{n,p+1} \end{pmatrix}$$

where $Q_{n,p}$ is of size $((p+1) + (p+1)^2) \times ((p+1) + (p+1)^2) = (p+1)(p+2) \times (p+1)(p+2)$ and hence $A_{n,p+1}$ is of size $(p+1)(p+2) \times (p+2)(p+3) - (p+1)(p+2) = (p+1)(p+2) \times 2(p+2)$. $B_{n,p+1}$ is of size $2(p+2) \times 2(p+2)$. Let

$$z_{n,p}^T = x_{n,p}^T \otimes x_{n,p}^T$$

Then,

$$\hat{y}[n] = x_{n,p}^T h + z_{n,p}^T g$$

where

$$h = h_{N,p}, g = g_{N,p}, n \leq N$$

$\hat{y}_{n,p}$ is the orthogonal projection of y_n onto the range of $U_{n,p} = [X_{n,p}|Z_{n,p}]$ where $X_{n,p}$ has rows $x_{k,p}^T, k \leq n$ and $Z_{n,p}$ has rows $z_{k,p}^T, k \leq n$. Now, the range of $U_{n,p+1}$ is the same as that of the sum of the ranges of $U_{n,p}$ and the range of $P_{n,p}^{\perp}[x_{n-p-1}|\eta_{n,p}]$ where $P_{n,p}$ is the orthogonal projection onto the range of $U_{n,p}$. Note that x_{n-p-1} is an $n+1 \times 1$ vector while $\eta_{n,p}$ is an $n+1 \times 2p+1$ matrix whose rows are $[x[k-p-1]x_{k,p}^T|x[k-p-1]x_{k,p}^T|x[k-p-1]^2]$ for $k = n, n-1, ..., 0$. We define the backward prediction error matrix

$$e_b[n|p] = P_{n,p}^{\perp}[x_{n-p-1}|\eta_{n,p}] \in \mathbb{R}^{n+1 \times 2p+4}$$

We also define the forward prediction error matrix $e_f[n|p]$ as follows:

$$e_f[n+1|p] = P_{n,p}^{\perp} x_{n+1}$$

Then, the filtering error involved in estimating y_n based on $x_n, x_{n-1}, ..., x_{n-p}$ and the columns of $\eta_{n,p}$ is given by

$$\hat{y}_{n,p+1} = y_n - P_{n,p+1} y_n$$

$$P_{n,p+1} y_n = P_{n,p} y_n + P_{e_b[n|p]} y_n$$

where P_A denotes orthogonal projection onto the range of A. So the filtering error at time n based on data at times $0, 1, ..., p+1$ is given by

$$f[n|p+1] = y_n - \hat{y}_{n,p+1} =$$
$$y_n - P_{n,p} y_n - P_{e_b[n|p]} y_n = f[n|p] - e_b[n|p](e_b[n|p]^T e_b[n|p])^{-1} e_b[n|p]^T y_n$$

Now, we compute along the same lines, the order updates of the forward and backward prediction errors:

$$e_f[n+1|p+1] = P_{n,p+1}^{\perp} x_{n+1}$$

$$= e_f[n+1|p] - e_b[n|p](e_b[n|p]^T e_b[n|p])^{-1} e_b[n|p]^T x_{n+1}$$

Now,

$$x_{n+1} = e_f[n+1|p] + P_{n,p} x_{n+1}$$

Now since $P_{n,p} e_b[n|p] = 0$ it follows that

$$e_f[n+1|p+1] = e_f[n+1|p] - e_b[n|p](e_b[n|p]^T e_b[n|p])^{-1} e_b[n|p]^T e_f[n+1|p]$$

[60] RLS lattice algorithms for quantum observable estimation: The evolution is governed by the Evans-Hudson flow j_t:

$$dj_t(X) = j_t(\theta^a_b(X))d\Lambda^b_a(t)$$

In discrete terms

$$j_{n+1}(X) - j_n(X) = j_n(\theta^a_b(X))(\Lambda^b_a[n+1] - \Lambda^b_a[n])$$
$$\Lambda^0_0(t) = t, \Lambda^k_0(t) = A_k(t), \Lambda^0_k(t) = A^*_k(t),$$
$$\Lambda^k_j(t) = \lambda(|e_j >< e_k|\chi_{[0,t]})$$

So

$$j_{n+1}(X) - j_n(X) = j_n(\theta^0_0(X)) + j_n(\theta^k_0(X))\Delta A_k[n+1] + j_n(\theta^0_k(X))\Delta A^*_k[n+1] +$$
$$j_n(\theta^k_j(X))\Delta\Lambda^j_k[n+1]$$

θ^0_0 is a linear map on the space of system obserbvables that has certain unknown parameters which are identified using the RLS lattice algorithm. The first term on the right is the signal part and the other three terms are the noise part. We write

$$\theta^0_0(X) = \sum_{k=1}^{p} a[k]\theta^0_{0k}(X)$$

with the $a[k]'s$ as the unknown parameters to be estimated. We have

$$< f\phi(u)|(j_{t+dt}(X) - j_t(X) - j_t(\theta^0_0(X))dt)^2|f\phi(u) >=$$
$$< f\phi(u)|(j_t(\theta^k_j(X))d\Lambda^j_k(t) + j_t(\theta^0_k(X))dA_k(t) + j_t(\theta^k_0(X))dA_k(t)^*)^2|f\phi(u) >$$
$$=< f\phi(u)|(j_t(\tilde{\theta}^a_b(X)d\Lambda^b_a(t))^2|f\phi(u) >$$

(where $\tilde{\theta}^0_0 = 0$ and $\tilde{\theta}^a_b = \theta^a_b$ if either $a > 0$ or $b > 0$).

$$=< f\phi(u)|j_t(\tilde{\theta}^a_b(X)\tilde{\theta}^c_d(X))|f\phi(u) > \epsilon^b_c u_d(t)\bar{u}_a(t)dt$$

Thus, the noise power in the process is

$$dt^{-1} < f\phi(u)|(j_{t+dt}(X) - j_t(X) - j_t(\theta^0_0(X))dt)^2|f\phi(u) >$$
$$=< f\phi(u)|j_t(\tilde{\theta}^a_b(X)\tilde{\theta}^c_d(X))|f\phi(u) > \epsilon^b_c u_d(t)\bar{u}_a(t)$$
$$=< f\phi(u)|j_t(\theta^k_j(X)\theta^j_p(X)|f(\phi(u) > u_p(t)\bar{u}_k(t)$$
$$+ < f\phi(u)|j_t(\theta^k_j(X)\theta^j_0(X))|f\phi(u) > \bar{u}_k(t)$$
$$+ < f\phi(u)|j_t(\theta^0_j(X)\theta^j_k(X))|f\phi(u) > u_k(t)$$

[66] Quantum scattering theory, the wave operators and the scattering matrix. A is the free particle Hamiltonian and $B = A + V$ is the Hamiltonian after the free particle starts interacting with the scatterer. Let $|\phi_i>$ be the "in state" ie the free particle state at time $t = 0$ after it arrives from an infinite distance from the scatterer at time $t = -\infty$. Then, we see that if $|\psi_i>$ is the scattered state at time $t = 0$ ie after the particle interacts with the scatterer, we must have

$$lim_{t \to -infty} exp(-itB)|\psi_i > -exp(-itA)|\phi_i >= 0$$

so

$$|\psi_i >= \Omega_-|\phi_i >$$

where

$$\Omega_- = lim_{t \to -\infty} exp(itB)exp(-itA)$$

Likewise let $|\phi_o>$ be the "out-state", ie the free particle state at time $t = 0$ which evolves under the free particle Hamiltonian A to the scattered state at time $t = \infty$. Let $|\psi_o>$ be the corresponding scattered state at time $t = 0$. Then we must have

$$lim_{t \to \infty} exp(-itB)|\psi_o > -exp(-itA)|\phi_o >= 0$$

and hence

$$|\psi_o >= \Omega_+|\phi_o >$$

where

$$\Omega_+ = lim_{t\to\infty} exp(itB)exp(-itA)$$

It should be noted that the wave operators $\Omega_\pm$ must be defined on appropriate domains. We have

$$< \psi_o|\psi_i >=< \phi_o|\Omega_+^*\Omega_-|\phi_i >$$

and the magnitude square of this quantity gives the probability that the free particle starting at $t = -\infty$ in the state $|phi_o >$ interacts with the scatterer and then gets scattered at time $t = \infty$ to the state $|\phi_o >$. This proability can be expressed as

$$P(\phi_i \to \phi_o) = | < \phi_o|S|\phi_i > |^2$$

where

$$S = \Omega_+^*\Omega_-$$

is the scattering matrix or scattering operator. We have the following interesting relations:

$$\Omega_+ - I = \int_0^\infty \frac{d}{dt}(exp(itB)exp(-itA))dt = i\int_0^\infty exp(itB)V.exp(-itA)dt$$

$$\Omega_- - I = -\int_{-\infty}^0 \frac{d}{dt}(exp(itB)exp(-itA))dt$$

$$= -i\int_{-\infty}^0 exp(itB)V.exp(-itA)dt$$

$$= -i\int_0^\infty exp(-itB)V.exp(itA)dt$$

We can derive some formal expressoins for the wave operators based on the above formulas: Let

$$A = \int_{\mathbb{R}} \lambda E_A(d\lambda), B = \int_{\mathbb{R}} \lambda E_B(d\lambda)$$

be the spectral representations of the self adjoint operators A and B respectively. Then, we get from the above

$$\Omega_+ - I = i\int_{[0,\infty)\times\mathbb{R}} exp(it(B - lambda + i\epsilon))VE_A(d\lambda)dt$$

$$= -\int_{\mathbb{R}} (B - \lambda + i\epsilon)^{-1}VE_A(d\lambda)$$

It follows that

$$\Omega_+ = \int_{\mathbb{R}} (I - (B - \lambda + i\epsilon)^{-1}VE_A(d\lambda)$$

$$= \int_{\mathbb{R}} (B - \lambda + i\epsilon)^{-1}((A - \lambda + i\epsilon)^{-1}E_A(d\lambda)$$

$$= (i\epsilon)^{-1}\int_{\mathbb{R}} (B - \lambda + i\epsilon)^{-1}E_A(d\lambda)$$

Example: Let A be self-adjoint and $B = A + V$ where V is a bounded operator. Consider

$$W(t) = exp(-itA).exp(itB) = exp(-itA).exp(it(A + V))$$

We know that $W(t)$ has the Dyson series

$$W(t) = I + \sum_{n\geq 1}\int_{0<t_n<...<t_1<t} U_0(-t_1)VU_0(t_2 - t_1)VU_0(t_3 - t_2)...U_0(t_n - t_{n-1})VU_0(-t_n)dt_1...dt_n$$

where

$$U_0(t) = exp(-itA)$$

We substitute the spectral representation of A:

$$A = \int_{\mathbb{R}} \lambda.E(d\lambda), U_0(t) = \int_{\mathbb{R}} exp(-it\lambda)E(d\lambda)$$

Then,

$$\int_{0<t_n<...<t_1<t} U_0(-t_1)VU_0(t_2-t_1)VU_0(t_3-t_2)...U_0(t_n-t_{n-1})VU_0(-t_n)dt_1...dt_n$$

$$= \int exp(i(\lambda_1 t_1 + \lambda_2(t_1-t_2) + ... + \lambda_n(t_{n-1}-t_n) + \lambda_0 t_n))E(d\lambda_1)VE(d\lambda_2)V...E(d\lambda_n)VE(d\lambda_0)$$

where the integral is over $0 < t_n < t_{n-1} < ... < t_1 < t$ and $\lambda_0, \lambda_1, ..., \lambda_n \in \mathbb{R}$. Now, consider

$$I = \int_{0<t_n<...<t_1<t} exp(i(\lambda_1 t_1 + \lambda_2(t_1-t_2) + ... + \lambda_n(t_{n-1}-t_n) + \lambda_0 t_n))$$

Substitute

$$t_n = s_0, t_{n-1} - t_n = s_1, ..., t_1 - t_2 = s_{n-1}$$

Then,

$$t_1 = s_0 + s_1 + ... + s_{n-1}$$

and

$$I = \int_{s_k>0\forall k, s_0+s_1+...+s_{n-1}<t} exp(i(\lambda_1(s_0+...+s_{n-1}) + \lambda_2 s_{n-1} + ... + \lambda_n s_1 + \lambda_0 s_0))ds_0...ds_{n-1}$$

$$= \int_{s_k>0\forall k, s_0+...+s_{n-1}<t} exp(i(\lambda_1-\lambda_0)s_0 + (\lambda_1-\lambda_n)s_1 + ... + (\lambda_1-\lambda_2)s_{n-1}))ds_0...ds_{n-1}$$

For large t ($t \to \infty$), this is

$$\Pi_{k\neq 1, k=0,...,n}[(exp(i(\lambda_1-\lambda_k)t) - 1)/(i(\lambda_1-\lambda_k)]$$

$$= \Pi_{k\neq 0} 2exp(i(\lambda_1-\lambda_k)t/2).sin((\lambda_1-\lambda_k)t/2)/(\lambda_1-\lambda_k)$$

This further approximates to (for large t)

$$exp(i\phi(\lambda,t))\pi^{n-1}\Pi_{k\neq 1}\delta(\lambda_1-\lambda_k)$$

where ϕ is a real phase factor. Then we get

$$I \approx < f| \int_{\mathbb{R}} exp(i\phi(\lambda,t))E(d\lambda)VE(d\lambda)VE(d\lambda)..E(d\lambda)VE(d\lambda)|g >$$

It follows that

$$|I| \leq \int_{\mathbb{R}} |< f|E(d\lambda)VE(d\lambda)VE(d\lambda)..E(d\lambda)VE(d\lambda)|g > |$$

$$\leq \int_{\mathbb{R}} \|\| V \|^n \| E(d\lambda)g \| . \| E(d\lambda)f \|$$

$$\leq \| V \|^n (\int \| E(d\lambda)g \|^2)^{1/2} (\int E(d\lambda)f \|^2)^{1/2}$$

$$= \| V \|^n \| f \| \| g \|$$

This approximation becomes exact in the limit $t \to \infty$ and it shows that the Dyson series for $W(t)$ given by

$$W(t) = I + \sum_{n\geq 1} \int_{0<t_n<...<t_1<t} U_0(-t_1)VU_0(t_2-t_1)VU_0(t_3-t_2)...U_0(t_n-t_{n-1})VU_0(-t_n)dt_1...dt_n$$

has the property that for all f, g,

$$lim_{t\to\infty} < f|W(t)|g >$$

converges if

$$\| V \| < 1$$

More precisely, if $\| V \| < 1$, the Dyson series for $< f|W(t)|g >$ is absolutely convergent for all sufficiently large t and the bound does not depend on t. It follows from the dominated convergence principle that for all f, g, the limit

$$lim_{t\to\infty} < f|W(t)|g >$$

exists.

[67] Quantum systems driven by Stroock-Varadhan martingales. Let $a(t,\omega), b(t,\omega)$ be progressively measurable functions with a taking values in the space of positive definite $d \times d$ real matrices and b taking values in $\mathbb{R}^d$. Consider the random partial differential operator

$$L_t = b(t,\omega)^T \nabla + \frac{1}{2} Tr(a(t,\omega)\nabla\nabla^T)$$

Then if is well known that if $X(t,\omega)$ satisfies the sde

$$dX(t,\omega) = b(t,\omega)dt + \sigma(t,\omega)dB_t(\omega)$$

where σ is any $d \times p$ progressively measurable matrix such that $a = \sigma\sigma^T$ and B is $\mathbb{R}^d$-valued Brownian motion, then the processes

$$M_t = f(X(t)) - \int_0^t (L_s f)(X(s))ds$$

and

$$N_t = f(X(t))exp(-\int_0^t (L_s f(X(s)))ds/f(X(s))$$

are Martingales. We are assuming that $f : \mathbb{R}^d \to \mathbb{R}$ is infinitely differentiable in the the first case and never assumes the value zero in the second case. Indeed, in the first case, we have

$$dM_t = L_t f(X(t))dt + f'(X(t))^T \sigma_t dB_t - L_t f(X(t))dt = f'(X(t))^T \sigma_t dB_t$$

which is indeed the differential of a Martingale. In the second case, we have

$$dN_t = exp(-\int_0^t L_s f(X(s))ds/f(X(s))L_t f(X(t))dt + (f'(X(t))^T \sigma_t dB_t)exp(\int_0^t Lf(X(s))ds/f(X(s))$$

$$-L_t f(X(t))dt.exp(-\int_0^t L_s f(X(s))ds/f(X(s)))$$

$$= exp(\int_0^t Lf(X(s))ds/f(X(s)))f'(X(t))^T \sigma_t dB_t$$

which is the differential of a Martingale. Stroock and Varadhan proved that even if we do not assume the Brownian motion driven sde, then N_t is a Martingale iff M_t is a Martingale. Suppose M_t is a Stroock-Varadhan Martingale. Consider the quantum sde

$$dU(t) = (-(iH_t + V_t^2 d < M >_t /2)dt - iV_t dM_t)U(t)$$

where H_t, V_t are self-adjoint operator valued processes. By the Ito formula, it is clear that $d(U(t)^*U(t)) = 0$ and hence if $U(0) = I$, then $U(t) = I$ for all t. The problem is, by taking measurements on an observable Y for example measuring its average value $h(t) = \mathbb{E}(Tr(\rho(0)U(t)^*YU(t)))$ and different times for a given initial state $\rho(0)$, can we estimate the function f in the definition of the Stroock-Varadhan martingales in both the cases. One approach would be to follow the method of moments, ie expand f as

$$f(x) = \sum_k \theta_k \phi_k(x)$$

where $\{\theta_k\}$ are unknown parameters and $\{\phi_k\}$ are linearly independent known test functions. In the first, case the Stroock-Varadhan martingale M_t is expressible as a linear combination of the parameters θ_k while in the second case, it is a more complicated function of the $\theta_k's$.

Appendix

A.1.Linear Algebra

Miscellaneous remarks and problems in linear algebra and functional analysis related to quantum mechanics

[1] Primary decomposition theorem

[2] Jordan decomposition theorem

First let N be a nilpotent operator in a finite dimensional complex vector space V. Let $M^m = 0, N^{m-1} \neq 0$. We can choose a vector v_{11} such that $N^{m-1}v_{1,1} \neq 0$. We choose another vector $v_{1,2}$ such that $N^{m-1}v_{1,2} \notin span\{N^{m-1}v_{1,1}\}$. In this way, we choose $v_{1,r}$ so that $N^{m-1}v_{1,r} \notin span\{N^{m-1}v_{1,k} : 1 \leq k \leq r-1\}$. Continuing this way, we get a set of vectors $\{v_{1,k} : 1 \leq k \leq r_1\}$ such that $B_1 = \{N^{m-1}v_{1,k} : 1 \leq k \leq r_1\}$ forms a basis for $Range(N^{m-1})$. Now $Range(N^{m-1}) \subset Range(N^{m-2})$ and $N(Range(N^{m-2})) = Range(N^{m-1}$. We also note that $\{N^{m-2}v_{1,k} : 1 \leq k \leq r_1\}$ is a linearly independent set in $Range(N^{m-2})$. So it can be extended to a basis $B_2 = \{N^{m-2}v_{1,k}, N^{m-2}v_{2,l} : 1 \leq k \leq r_1, 1 \leq l \leq r_2\}$ for $Range(N^{m-2})$. It follows that $\{N^{m-3}v_{1,k}, N^{m-3}v_{2,l} : 1 \leq k \leq r_1, 1 \leq l \leq r_2\}$ is a linearly independent set in $Range(N^{m-3})$ and hence can be extended to a basis $B_3 = \{N^{m-3}v_{1,k_1}, N^{m-3}v_{2,k_2}, N^{m-3}v_{3,k_3} : 1 \leq k_j \leq r_j, j = 1, 2, 3\}$ for $Range(N^{m-3})$. Continuing in this way, we finally arrive at a basis

$$B = B_m = \{N^{m-j}v_{j,k_j} : 1 \leq k_j \leq r_j, j = 1, 2, ..., m\}$$

for V. We leave it as an exercise to write down the matrix of N relative to B.

Now consider the basis $B_1 = \{w_{1,k} = N^{m-1}v_{1,k} : 1 \leq k \leq r_1\}$ for $Range(N^{m-1})$. We have $Nw_{1,k} = 0$. Let $w_{2,k} = N^{m-2}v_{1,k}$. Then $Nw_{2,k} = w_{1,k}, 1 \leq k \leq r_1$. Also consider $w'_{2,l} = N^{m-2}v_{2,l}$. We have

$$Nw'_{2,l} = \sum_k c(l,k)w_{1,k} = N\sum_k c(l,k)w_{2,k}$$

and so we can define

$$w''_{2,l} = w(2,l)' - \sum_k c(l,k)w_{2,k}$$

Then,

$$Nw''_{2,l} = 0$$

and clearly, $w''_{2,l}$ can be used in place of $w_{2,l}$ to get a basis $B'_2 = \{w_{2,k}, w''_{2,l} : 1 \leq k \leq r_1, 1 \leq l \leq r_2\}$ for $Range(N^{m-2})$. We note that

$$\sum c(k)w_{2,k} + \sum d(l)w''_{2,l} = 0$$

implies on operating with N that

$$\sum c(k)w_{1,k} = 0$$

and hence $c(k) = 0 \forall k$. Thus,

$$\sum d(l)w''_{2,l} = 0$$

and hence,

$$\sum d(l)(w'_{2,l} - \sum_r c(l,r)w_{2,k}) = 0$$

which implies that $\{w'_{2,l}, w_{2,k}\}_{l,k}$ form a linearly dependent set unless $d(l) = 0$ for all l. This proves the linear independence of the set B'_2 defined above. Continuing this process, we finally get a basis

$$B'_p = \{w_1(p,k_1), w_2(p,k_2), ..., w_p(p,k_p) : 1 \leq k_j \leq r_j, j = 1, 2, ..., p\}$$

for $Range(N^{m-p})$ for each p=1,2,...,m such that

$$Nw_j(p,k_j) = w_j(p-1,k_j), 1 \leq k_j \leq r_j, j = 1, 2, ..., p-1, Nw_p(p,k_p) = 0$$

Consider now the se ts

$$S(1,k_1) = \{w_1(1,k_1), w_1(2,k_1), ..., w_1(m,k_1)\}$$

$$S(2,k_2) = \{w_2(2,k_2), w_2(3,k_2), ..., w_2(m,k_2)\}$$

and in general,

$$S(j,k_j) = \{w_j(j,k_j), w_j(j+1,k_j), ..., w_j(m,k_j)\}$$

for $j = 1, 2, ..., m$. We observe that $span(S(j,k_j))$ is $N-$invariant and in fact

$$N(w_j(j,k_j)) = 0, N(w_j(j+1,k_j)) = w_j(j,k_j), ..., N(w_j(m,k_j)) = w_j(m-1,k_j)$$

So the restriction of N to $S(j,k_j)$ gives a Jordan block of size $m-j+1$. Suppose

$$\sum_{r \geq 0} c(r)w_j(j+r,k_j) = 0$$

for some j. Then operating on both sides with N^s gives using

$$N^{s+1}(w_j(j+r,k_j)) = w_j(j+r-s,k_j), s \leq r$$

and zero for $s > r$, that

$$\sum_{m\geq r\geq s} c(r)w_j(j+r-s,k_j) = 0$$

and hence with $s = m$ we get $c(m)w_j(j,k_j) = 0$ and hence $c(m) = 0$. Then with $s = m-1$ we get $c(m-1) = 0$ etc. Thus all the $c(r)'s$ are zero. This proves that $S(j,k_j)$ is a linearly independent set for each j and k_j. We now observe that

$$B'_m = \{w_1(m,k_1), w_2(m,k_2), ..., w_m(m,k_m) : 1 \leq k_j \leq r_j, j = 1,2,...,m\}$$

is a basis for $Range(N^0) = V$. Also, as noted above, each vector $w_j(m,k_j)$ generates the cyclic subspace

$$span\{w_j(j+r,k_j) : 0 \leq r \leq m-j\} = span(S(j,k_j))$$

Remark related to the Jordan decomposition: Let $\{N^{m-1}v_{1,k} : k = 1,2,...,r_1\}$ be a basis for $Range(N^{m-1})$, let $\{N^{m-2}v_{2,k} : 1 \leq k \leq r_2\} + Range(N^{m-2})$ be a basis for $Range(N^{m-1})/Range(N^{m-2})$ and in general, let $\{N^{m-j}v_{j,k} : 1 \leq k \leq r_j\} + Range(N^{m-j+1})$ be a basis for $Range(N^{m-j})/Range(N^{m-j+1})$ where. $j = 1,2,...,m$. Suppose we assume that $N^{m-j+1}v_{j,k} = 0, 1 \leq k \leq r_j, 1 \leq j \leq m$. Then, we claim that

$$B = \{N^p v_{j,k} : 1 \leq k \leq r_j, 1 \leq j \leq m, 0 \leq p \leq m-1\}$$

is a linearly independent set in V where V is the vector space on which N is defined. Indeed suppose

$$\sum_{0\leq p\leq m-1, 1\leq j\leq m, 1\leq k\leq r_j} c(p,j,k)N^p v_{j,k} = 0$$

Then applying N^{m-1} on both sides gives

$$\sum_k c(0,1,k)N^{m-1}v_{1,k} = 0$$

and hence $c(0,1,k) = 0 \forall k$. Hence,

$$\sum_{(p,j)\neq(0,1),k} c(p,j,k)N^p v_{j,k} = 0$$

and applying N^{m-2} to both sides gives

$$\sum c(0,2,k)N^{m-2}v_{2,k} = 0$$

which implies that $c(0,2,k) = 0 \forall k$. Continuing in this way gives $c(0,j,k) = 0 \forall j,k$. Thus,

$$\sum_{p\geq 1,j,k} c(p,j,k)N^p v_{j,k} = 0$$

Applying N^{m-2} to both sides gives

$$\sum_k c(1,1,k)N^{m-1}v_{1,k} = 0$$

and hence $c(1,1,k) = 0 \forall k$. Thus,

$$\sum_{p\geq 1,(p,j)\neq(1,1),j,k} c(p,j,k)N^p v_{j,k} = 0$$

Applying N^{m-3} to both sides gives

$$\sum_k c(1,2,k)N^{m-2}v_{2,k} = 0$$

and hence $c(1,2,k) = 0 \forall k$. Continuing in this way, we finally get $c(p,j,k) = 0 \forall p,j,k$ proving linear independence of the set B.

A neat proof of the canonical Jordan representation of a nilpotent matrix: Let $N^m = 0, N^{m-1} \neq 0$. Let $B_1 = \{N^{m-1}x(1,\alpha) : 1 \leq \alpha \leq d_1\}$ be a basis for $R(N^{m-1})$. Let

$$\{N^{m-2}x(1,\alpha_1), N^{m-2}(x(2,\alpha_2) : 1 \leq \alpha_1 \leq d_1, 1 \leq \alpha_2 \leq d_2\}$$

be a basis for $R(N^{m-2})$ etc. In general,

$$\{N^{m-s}x(1,\alpha_1), N^{m-s}x(2,\alpha_2), ..., N^{m-s}x(s,\alpha_s) : 1 \leq \alpha_k \leq d_k, k = 1,2,...,s\}$$

is a basis for $R(N^{m-s})$. We may assume that $N^{m-1}(x(2,\alpha_2)) = 0\forall\alpha_2, ...N^{m-s+1}x(s,\alpha_s) = 0\forall\alpha_s$, since $N^{m-1}(x(2,\alpha_2))$ is expressible as a linear combination of $N^{m-1}(x(1,\alpha_1), \alpha_1 = 1,2,...,d_1$, ie,

$$N^{m-1}x(2,\alpha_2) = \sum_k c(k)N^{m-1}x(1,k)$$

and hence $x(2,\alpha_2)$ can be replaced by $x(2,\alpha_2) - \sum_k c(k)x(1,k)$, ie $N^{m-2}x(2,\alpha_2)$ can be replaced by $N^{m-2}x(2,\alpha_2) - \sum_k c(k)N^{m-2}x(1,k)$ and these vectors for different α_2 along with the vectors $N^{m-2}x(1,k)$ for different k again form a basis for $R(N^{m-2})$. This argument can be continued. The final result is that the vectors $B = \{N^{m-s}x(k,\alpha_k) : 1 \leq \alpha_k \leq d_k, k = 1,2,...,s, s = 1,2,...,m-1\}$ forms a basis for V and the matrix of N relative to B has the Jordan canonical form of a Nilpotent matrix $\{Reference : T.Kato, "Perturbationtheoryforlinearoperators"\}$

[3] Evaluation of a function of a matrix using the Jordan canonical form. Let A be a matrix over $\mathbb{C}$. We know that

$$A = D + N$$

where D is diagonable and N is nilpotent. Thus, A can be written as a direct sum of Jordan blocks of the form

$$J_m(\lambda) = \begin{pmatrix} \lambda & 1 & 0 & 0 & ... & 0 \\ 0 & \lambda & 1 & 0 & ... & 0 \\ 0 & 0 & \lambda & .. & 0.. & 0 \\ 0 & ..0.. & 0.. & 0.. & \lambda & 1 \\ 0 & 0 & ..0 & ..0 & .. & \lambda \end{pmatrix}$$

This matrix belongs to $\mathbb{C}^{m \times m}$. We write

$$Z_m = ((\delta[j-i-1]))_{1 \leq i,j \leq m}$$

Then,

$$J_m(\lambda) = \lambda I_m + Z_m$$

We have for any function f that is infinitely differentiable at λ,

$$f(J_m(\lambda)) = f(\lambda)I_m + f'(\lambda)Z_m + f''(\lambda)Z_m^2/2! + ... + f^{(m-1)}(\lambda)Z_m^{m-1}/(m-1)!$$

since

$$Z_m^m = 0$$

Then, choosing a basis B such that the matrix A has the Jordan canonical form

$$A = \bigoplus_{k=1}^{r} \bigoplus_{j=1}^{p_k} J_{m_{k,j}}(\lambda_k)$$

where $\lambda_k, k = 1,2,...,r$ are the distinct eigenvalues of A. We can write

$$f(A) = \bigoplus_{k=1}^{r} \bigoplus_{j=1}^{p_k} f(J_{m(k,j)}(\lambda_k)$$

Another way to compute $f(A)$ is by using the Cauchy residue theorem. If A is diagonalble, then $T = TDT^{-1}$ where $D = diag[\lambda_1, ..., \lambda_n]$ and

$$(zI - A)^{-1} = T(zI - D)^{-1}T^{-1}$$

so if Γ_k is a contour enclosing only the eigenvalue λ_k, then

$$(2\pi i)^{-1}\int_{\Gamma_k}(zI-A)^{-1}dz = TE_kT^{-1} = P_k$$

say, where E_k is a diagonal matrix having a one at those points where D has the entry λ_k and zeroes at the other points. Thus, P_k is the projection onto $\mathcal{N}(T-\lambda_k)$ along $\bigoplus_{j\neq k}\mathcal{N}(T-\lambda_j)$. We clearly have $\sum_k E_k = I$ and hence $\sum_k P_k = I$. Clearly $E_kE_j = 0$ for $k\neq j$ and hence $P_kP_j = 0$ for $k\neq j$ and since $E_k^2 = E_k$, it follows that $P_k^2 = P_k$. Thus,

$$\sum_k P_k = I$$

the summation being over indices k corresponding to the distinct eigenvalues of T, defines a spectral resolution of identity. Note that we clearly have that if Γ is a contour enclosing all the eigenvalues of T, then $\int_\Gamma f(z)(zI-A)^{-1}dz$ can be replaced by

$$(2\pi i)^{-1}\sum_k\int_{\Gamma_k} f(z)(zI-A)^{-1}dz = (2\pi i)^{-1}\sum_k\int_{\Gamma_k} f(z)(zI-A)^{-1}dz = \sum f(\lambda_k)P_k = f(A)$$

provided that f has no singularity within Γ. By a standard continuity argument, this result is also valid for a non-diagonable matrix A. If A is non-diagonable, then since the set of diagonable matrices is dense in the space of matrices, it follows that there exists a sequence $\epsilon_k\to 0$ such that $A+\epsilon_kI$ is non-singular for all k and if we assume that f is continuous, then we get

$$f(A) = lim_k f(A+\epsilon_kI) = (2\pi i)^{-1}\int_\Gamma f(z)(zI-A-\epsilon_kI)^{-1}dz = (2\pi i)^{-1}\int_\Gamma f(z)(zI-A)^{-1}dz$$

where Γ encloses all the eigenvalues of A.

A.2.Functional Analysis

[1] Let S be a symmetric operator in a Hilbert space $\mathcal{H}$, ie $S\subset S^*$. (We are assuming that $D(S)$ is dense in $\mathcal{H}$). This means that $D(S)\subset D(S^*)$ and $S^*|_{D(S)} = S$. We wish to show that if $R(S+i)$ and $R(S-i)$ are both dense in $\mathcal{H}$, then S is essentially self-adjoint, ie $\bar{S}$ is self-adjoint. Further if $R(S+i)$ and $R(S-i)$ are both exactly $\mathcal{H}$, then S is self-adjoint. First we prove that S is closable. Indeed, suppose $x_n\in D(S)$ and $x_n\to 0, Sx_n\to y$. Then, to prove that S is closable, we must show that $y=0$. For any $z\in D(S)$, we have $z\in D(S^*)$ and hence,

$$< Sx_n, z > = < x_n, S^*z > = < x_n, Sz > \to 0, n\to\infty$$

Thus,

$$< y, z > = 0 \forall z\in D(S)$$

Since $D(S)$ is dense in $\mathcal{H}$, it follows that $y\perp\mathcal{H}$ and hence $y=0$, proving that S is closable.

Remark: S is closable iff $x_n, z_n\in D(S), x_n\to x, z_n\to x, Sx_n\to y, Sz_n\to w$ imply $y=w$. This is the same as requiring that $x_n-z_n\in D(S), x_n-z_n\to 0, S(x_n-z_n)\to v$ all imply that $v=0$ and this is the same as requiring that $x_n\in D(S), x_n\to 0, Sx_n\to y$ all imply $y=0$.

Now, let $x\in D(S^*)$. Then $(S^*-i)x = lim_n(S-i)z_n$ for some sequence $z_n\in D(S)$ because by hypothesis, $R(S-i)$ is dense in $\mathcal{H}$. Thus,

$$< (S^*-i)x, y > = < x, (S+i)y > = lim < (S-i)z_n, y > = lim < z_n, (S^*+i)y > = lim < z_n, (S+i)y > \forall y\in D(S)$$

where we have used $D(S)\subset D(S^*)$. Now, we show that z_n is a convergent sequence. We have that $lim(S-i)z_n = (S^*-i)x$ exists and hence $(S-i)z_n$ is a Cauchy sequence, ie

$$(S-i)(z_n-z_m)\to 0, n, m\to\infty$$

and hence

$$\| (S-i)(z_n-z_m) \|^2 \to 0$$

or equivalently,

$$\| S(z_n-z_m) \|^2 + \| (z_n-z_m) \|^2 + 2Im(< S(z_n-z_m), z_n-z_m >)\to 0$$

Now

$$< Sz, z > = < z, Sz > \forall z\in D(S)$$

since $S \subset S^*$. Thus, $Im(< Sz, z >) = 0 \forall z \in D(S)$. Thus, $Im(< S(z_n - z_m), z_n - z_m >) = 0$. This proves that

$$\| S(z_n - z_m) \|^2 + \| (z_n - z_m) \|^2 \to 0$$

and hence

$$\| z_n - z_m \| \to 0$$

Thus, z_n is Cauchy and hence convergent. Let $z_n \to z$. Then since $(S+i)z_n$ converges, it follows that Sz_n also converges. Thus, $z \in D(\bar{S})$ and we get

$$< x, (S+i)y >=< z, (S+i)y > \forall y \in D(S)$$

Thus,

$$x - z \perp R(S+i)$$

and since $R(S+i)$ is dense in $\mathcal{H}$, it follows that $x = z \in D(\bar{S})$. Thus, we have proved that

$$D(S^*) \subset D(\bar{S})$$

Now $S \subset S^*$ implies $\bar{S} \subset S^*$ since S^* is closed. Thus, $D(\bar{S}) \subset D(S^*) \subset D(\bar{S})$ from which we deduce that $D(S^*) = D(\bar{S})$ and therefore $\bar{S} = S^*$. Further, $S \subset S^*$ $impliesS^{**} \subset S^* = \bar{S}$ and since S^{**} is a closed extension of S, it follows that $S^{**} = \bar{S}$. Thus, $S^* = \bar{S} = S^{**}$ which proves that $\bar{S}$ is self-adjoint, ie, $(\bar{S})^* = \bar{S}$.

Remarks: (a) If A, B are operators in $\mathcal{H}$ such that $A \subset B$, then $B^* \subset A^*$. Indeed, let $x \in D(B^*)$. Then $< B^*x, y >=< x, By >=< x, Ay >, \forall y \in D(A) \subset D(B)$. Hence, $x \in D(A^*)$ and $< B^*x - A^*x, y >= 0 \forall y \in D(A)$. This proves that $B^*x - A^*x \perp D(A)$ and since A is densely defined, it follows that $B^*x - A^*x = 0$, ie, $B^*x = A^*x$, proving the claim. (Note that we are assuming without any loss in generality that all operators are densely defined).

(b) If A is any operator, then A^* is closed. Indeed, let $x_n \in D(A^*), x_n \to x, A^*x_n \to y$. Then for all $z \in D(A)$, we have

$$< A^*x_n, z > \to < y, z >, < A^*x_n, z >=< x_n, Az > \to < x, Az >$$

and hence,

$$< y, z >=< x, Az >$$

This proves that $y \in D(A^*)$ and $A^*x = y$, proving that A^* is closed.

(c) If A is any closable operator and B is a closed operator such that $A \subset B \subset \bar{A}$, then $B = \bar{A}$. Indeed, it suffices to show that $\bar{A} \subset B$. So let $x \in D(\bar{A})$. Then there exists a sequence $x_n \in D(A)$ such that $x_n \to x, Ax_n \to \bar{A}x$ and since $A \subset B$, we have $x_n \in D(B), x_n \to x, Bx_n = Ax_n \to \bar{A}x$ and since B is closed, it follows that $x \in D(B), Bx = \bar{A}x$. This proves the claim. A related statement is that if B is any closed operator such that $A \subset \bar{B}$, then $\bar{A} \subset \bar{B}$. This follows from the implications $A \subset \bar{B}$ implies $Gr(A) \subset \bar{G}r(B)$ and hence $\bar{G}r(A) \subset \bar{G}r(*B)$. Another related remark is that if A is any closable operator then A^{**} is a closed extension of A and hence by the above $\bar{A} \subset A^{**}$. For suppose $x_n \in D(A), x_n \to x, Ax_n \to y$. Then, $x \in D(\bar{A}), y = \bar{A}x$ and we have for $z \in D(A^*)$,

$$< y, z >= lim < Ax_n, z >= lim < x_n, A^*z >=< x, A^*z >$$

Hence, $x \in D(A^{**})$ and $y = A^{**}x$. This proves that $\bar{A} \subset A^{**}$.

Now we come to the last statement. Let $S \subset S^*$ and $R(S \pm i) = \mathcal{H}$. Then, we have to show that $S^* = S$. Indeed, let $x \in D(S^*)$. Then there exists a $y \in D(S)$ such that $(S^* - i)x = (S - i)y$ since $R(S - i) = \mathcal{H}$. Thus for any $z \in D(S) \subset D(S^*)$, it follows that

$$< (S^* - i)x, z >=< x, (S+i)z >=< (S-i)y, z >=< y, (S^*+i)z >=< y, (S+i)z >$$

and hence,

$$< x - y, (S+i)z >= 0, z \in D(S)$$

ie

$$x - y \perp \mathcal{H}$$

and therefore,

$$x = y \in D(S)$$

This proves that

$$D(S^*) \subset D(S) \subset D(S^*)$$

and hence

$$D(S) = D(S^*)$$

so that

$$S = S^*$$

The proof is complete.

A.3.Stochastic processes

Here we discuss various kinds of noise processes that arise in perturbed quantum systems and explain how to compute transition probabilities of quantum systems in the presence of such noise processes.

[1] Brownian motion

[2] Poisson process

[3] Compound Poisson processes

[4] Levy processes

[5] Reflected and absorbed Brownian motion

Let $B(t), t \geq 0$ be a standard Brownian motion starting at any given point and let

$$T_0 = inf(t \geq 0 : B(t) = 0)$$

Obviously, the statistics of T_0 will depend on $B(0)$. Absorbed Brownian motion is the Brownian motion process upto time T_0 and after time T_0 it is set equal to zero. We denote this process by $X(t), t \geq 0$. We can show that X is a Markov process by the following simple intuitive reasoning: If at time t_0 the process $X(t_0) = 0$, then obviously $X(t) = 0$ for all $t > t_0$. On the other hand, if at time t_0, $X(t_0) = x > 0$, then obviously $B(t_0) = x$ and by the Markovian property of $B(.)$, the statistics of $X(t)$ for all $t > t_0$ will depend only on (t_0, x). It remains to compute the transition probability of $X(.)$. For $x, y > 0$, we have

$$P(X(t) > x|X(0) = y) = P_y(B(t) > x, T_0 > t) = P_y(B(t) > x, min_{s \leq t}B(s) > 0)$$

$$= P_0(B(t) > x - y, min_{s \leq t}B(s) > -y)$$

$$= P_0(-B(t) > x - y, min_{s \leq t}(-B(s)) > -y)$$

$$= P_0(B(t) < y - x, -max_{s \leq t}(B(s)) > -y) = P_0(B(t) < y - x, S_t < y)$$

$$= P_0(B(t) < y - x) - P_0(B(t) < y - x, S_t > y)$$

where

$$S_t = max_{s \leq t}B(s)$$

By the reflection principle,

$$P_0(B(t) < y - x, S_t > y) = P_0(B(t) > y + x)$$

Hence,

$$P(X(t) > x|X(0) = y) = P_0(B(t) < y - x) - P_0(B(t) > y + x)$$

and hence the transition density of $X(t)$ (taking values in $[0, \infty)$) is given by

$$q_t(x|y) = -\frac{d}{dx}P(X(t) > x|X(0) = y) = (2\pi t)^{-1/2}(exp(-(x - y)^2/2t) - exp(-(x + y)^2/2t)), x, y > 0$$

Also

$$P(X(t) = 0|X(0) = y) = P_y(T_0 < t) = P_y(min_{s \leq t}B(s) \leq 0)$$

$$= P_0(min_{s \leq t}B(s) \leq -y)$$

$$= P_0(S_t > y) = 2P_0(B(t) > y)$$

the last step following from the reflection principle. We thus verify that

$$P(X(t) \geq 0|X(0) = y) = \int_0^\infty q_t(x|y)dx + 2P(B(t) > y)$$

$$= P(B(t) > -y) - P(B(t) > y) + 2P(B(t) > y) = P(B(t) > -y) + P(B(t) > y)$$

$$= P(B(t) < y) + P(B(t) > y) = 1$$

[6] Bessel processes

[7] The Brownian local time process

[1] Let $\theta(x)$ be the unit step function. Let $B(t)$ be Brownian motion and $L(t)$ its local time at zero, ie,

$$dL(t) = \delta(B(t))dt$$

Let

$$\tau(s) = inf\{t > 0 : L(t) \geq s\}$$

then, clearly $\tau(s)$ is a stopping time for each s. By Ito's formula,

$$d(B(t)\theta(B(t)) = \theta(B(t))dB(t) + \delta(B(t))dt + \frac{1}{2}B(t)\delta'(B(t))dt$$

Now,

$$x\delta'(x) = (x\delta(x))' - \delta(x) = -\delta(x)$$

Hence,

$$\theta(B(t))dB(t) = d(B(t)\theta(B(t)) - \frac{1}{2}dL(t)$$

and hence

$$\int_0^t \theta(B(s))dB(s) = B(t)\theta(B(t)) - \frac{1}{2}L(t)$$

Therefore applying Doob's optional stopping theorem to the exponential martingale

$$M(t) = exp(-a\int_0^t \theta(B(s))dB(s) - (a^2/2)\int_0^t \theta(B(s))ds)$$

and the stopping time $\tau(t)$ gives

$$1 = \mathbb{E}[M(\tau(t))] = exp(at - (a^2/2)\int_0^{\tau(t)} \theta(B(s))ds)$$

(Note that $B(\tau(t)) = 0, L(\tau(t)) = t$). Thus,

$$\mathbb{E}[exp(-s\int_0^{\tau(t)} \theta(B(s))ds)] = exp(-\sqrt{2s}t)$$

(Reference: Marc Yor, Some aspects of Brownian motion, Birkhauser).

[8] Stochastic integration with respect to continuous semi-Martingales.
Let X_t be continuous semimartingale. By the Doob-Meyer theorem, it can be decomposed as

$$X_t = A_t + M_t$$

where A_t is a process of bounded variation and M_t is a martingale. We can write

$$A_t = A_t^+ - A_t^-$$

where A_t^+ and A_t^- are increasing processes. The integral of a bounded adapted process Y_t w.r.t A_t is defined as an ordinary Riemann-Stieltjes integral or equivalently as a Lebesgue integral. Since M_t is not a process of bounded variation but has a well defined finite quadratic variation, its integral must be defined in the Ito sense. Specifically, defining $\int_0^T Y_t dM_t$ as the mean square limit of partial sums of the form $S_n = \sum_{k=1}^n Y_{t_{n,k}}(M_{t_{n,k+1}} - M_{t_{n,k}})$ as $max_k|t_{n,k+1} - t_{n,k}| \to 0$. We have for $n > m$,

$$\mathbb{E}(S_n - S_m)^2 = \sum_k \mathbb{E}(Y_{t_{n,k}}^2)\mathbb{E}(M_{t_{n,k+1}} - M_{t_{n,k}})^2$$

$$+\sum_k \mathbb{E}(Y_{t_{m,k}}^2)(\mathbb{E}(M_{t_{m,k+1}} - M_{t_{m,k}})^2)$$

$$-2\mathbb{E}(S_n S_m)$$

Suppose we assume that the partition $\{t_{n,k}\}$ is a refinement of the partition $\{t_{m,k}\}$. Consider for example a term like $Y_{s_1}((M_{s_2} - M_{s_1})$ in S_m and a term like $\sum_{k=1}^3 Y_{t_k}(M_{t_{k+1}} - M_{t_k})$ in S_n where $s_1 = t_1 < t_2 < t_3 < t_4 = s_2$. The expected value of their product is an example of a term in $\mathbb{E}(S_n S_m)$. This term can be expressed as

$$\mathbb{E}(Y_{t_1}^2(M_{t_2} - M_{t_1})^2) + \mathbb{E}(Y_{t_1}Y_{t_2}(M_{t_3} - M_{t_2})^2)$$

$$+\mathbb{E}(Y_{t_1}Y_{t_3}(M_{t_4}-M_{t_3})^2)$$

From this observation, it is clear that if the quantity

$$\mathbb{E}\int_0^T Y_t^2 d<M>_t<\infty$$

then,

$$\mathbb{E}((S_n-S_m)^2)\to 0, n,m\to\infty$$

and hence by the fact that $L^2(\Omega,\mathcal{F},P)$ is a Hilbert space, it follows that there exits a random variable S_∞ which we denote as $\int_0^T Y_t dM_t$ and call it the stochastic integral of Y w.r.t M over the interval $[0,T]$. We are assuming that the increasing function $t\to<M>_t=\int_0^t(dM_s)^2$ defines a finite measure on a bounded interval like $[0,T]$ and that the adapted process Y_t is Riemann integrable w.r.t this measure. More precisely, the stochastic integral w.r.t a martinagle can be defined for almost surely progressively measurable processes Y_t that are integable w.r.t the random measure $d<M>_t$ over finite intervals. For a detailed discussion of this construction see the book by Karatzas and Shreve on "Brownian motion and stochastic calculus" or the book by Revuz and Yor on "Continuous martingales and Brownian motion.

Remark: Consider the quantum system

$$dU(t)=[-(iH(t)dt+\frac{1}{2}\sum_{k,m=1}^{p}V_k(t)V_m(t)d<M_k,M_m>(t))-i\sum_{k=1}^{p}V_k(t)dM_k(t)]U(t)$$

where $M_k(t), k=1,2,...,p$ are Martingales. It is easily verified using Ito's formula that $U(t)$ is unitary for all t if $U(0)=I$. We introduce a perturbation parameter ϵ into the martingale terms and obtain

$$dU(t)=[-(iH(t)dt+\epsilon^2\frac{1}{2}\sum_{k,m=1}^{p}V_k(t)V_m(t)d<M_k,M_m>(t))-i\epsilon\sum_{k=1}^{p}V_k(t)dM_k(t)]U(t)$$

We solve for $U(t)$ using perturbation theory:

$$U(t)=\sum_{m\geq 0}\epsilon^m U_m(t)$$

Equating same powers of ϵ gives successively

$$dU_0(t)=-iH(t)U_0(t)dt,$$

$$dU_1(t)=-iH(t)U_1(t)-i\sum_k V_k(t)U_0(t)dM_k(t),$$

$$dU_2(t)=-iH(t)U_2(t)-\frac{1}{2}\sum_{k,m}V_k(t)V_m(t)U_0(t)d<M_k,M_m>(t)-i\sum_k V_k(t)U_1(t)dM_k(t)$$

and in general,

$$dU_r(t)=-iH(t)U_r(t)-\frac{1}{2}\sum_{k,m}V_k(t)V_m(t)U_{r-2}(t)d<M_k,M_m>(t)-i\sum_k V_k(t)U_{r-1}(t)dM_k(t), r\geq 2$$

After calculating

$$U_0(t)+\sum_{r=1}^{N}\epsilon^r U_r(t)$$

we calculate the transition probablity from an initial state $|i>$ to a final state $|f>$ in time T:

$$\mathbb{E}[|<f|U_0(T)+\sum_{r=1}^{N}\epsilon^r U_r(T)|i>|^2]$$

where the average is taken over the probability distribution of the martingale processes over $[0,T]$. Examples of Martingales and the Ito formula for them (a)

$$M(t)=\int_0^t f(s,\omega)dB(s,\omega)+\int_{s\leq t,x\in E}g(s,x,\omega)(N(ds,dx,\omega)-\lambda(s)dF(x)ds)$$

where f, g are progressively measurable functions with B being Brownian motion and $N(t, ., \omega)$ a Poisson field with $\mathbb{E}N(ds, E) = \lambda(s)dsF(E)$. Ito's formula for this martingale is

$$(dM(t))^2 = d < M > (t) = f(t)^2 dt + \int_{x \in E} g(t, x)^2 N(dt, dx)$$

A.4.Syllabus for a short course on Linear algebra and its application to classical and quantum signal processing.

[1] Vector space over a field, linear transformations on a vector space.

[2] Finite dimensional vector spaces, basis for a vector space, matrix of a linear transformation relative to a basis, Similarity transformation of the matrix of a linear transformation under a basis change.

[3] Examples of finite and infinite dimensional vector spaces.

[4] range, nullspace, rank and nullity of a linear transformation.

[5] Subspace, direct sum decompositions of vector spaces, projection operators.

[6] Linear estimation theory in the language of orthogonal projection operators.

[7] Statistics of the estimation error of a vector under a small random perturbation of the data matrix, statistics of the perturbation of the orthogonal projection operator under a small random perturbation of the data matrix.

[8] Primary decomposition theorem, Jordan decomposition theorem, functions of matrices.

[9] Cauchy's residue theorem in complex analysis and its approach to the computation of functions of a matrix.

[10] Norms on a vector space, norms on the space of matrices, Frobenius norm, spectral norm.

[11] Notions of convergence in a vector space, calculating the exponential function and inverse of a perturbed matrix using a power series.

[12] Recursive least squares lattice algorithms for time and order updates of prediction error filters based on appending rows and columns to matrices and computing functions of the appended matrices, RLS lattice for second order Volterra systems, Statistical properties of the prediction filter coefficients under the addition of a small noise process to the signal process: A statistical perturbation theory based approach.

[13] The MUSIC and ESPRIT algorithms for direction of arrival estimation based on properties of signal and noise eigensubspaces.

[14] Computing the solution to time varying linear state variable systems using the Dyson series. Convergence of the Dyson series.

[15] Computing the approximate solution to nonlinear state varable systems using Dyson series applied to the linearized system.

[16] Dyson series in quantum mechanics.

[17] Computing transition probabilities for quantum systems with random time varying potentials using the Dyson series.

[18] Approximate solution to stochastic differential equations driven by Brownian motion and Poisson fields using linearization combined with Dyson series. Mean and variance propagation equations based on linearization.

[19] The spectral theorem for finite dimensional normal operators and infinite dimensional unbounded self-adjoint operators in a Hilbert space.

[20] Properties of spectral families in finite and infinite dimensional Hilbert spaces.

[21] The general theory of estimating parameters in linear models for Gaussian and non-Gaussian noise.

[22] The quantum stochastic calculus of Hudson and Parthasarathy and its application to the modeling of a quantum system coupled to a photon bath.

[23] Kushner nonlinear filter and its linearized EKF version.

[24] The Belavkin quantum filter based on non-demolition measurements and its application to quantum control.

Application to quantum control: The Belavkin equation can be expressed as

$$d\rho_t = L_t(\rho_t)dt + M_t(\rho_t)dW_t$$

where W_t is a classical Wiener process arising from the innovations process of the measurement

$$dW_t = dY_t - \pi_i(S_t + S_t^*)dt = dY_t - Tr(\rho_t(S_t + S_t^*))dt$$

Note that ρ_t can be viewed as a classical random process with values in the space of signal space density matrices. S_t is a system space linear operator. The Belavkin equation is a commutative equation since all the terms appearing in it like $\rho_t, L_t(\rho_t), W_t$ etc. are signal space operator valued functionals of the commutative noise processs $\{Y_y\}$. Now let $U_c(t)$ be the control unitary satisfying the sde

$$dU_c(t) = (-(iH_1(t) + Q_1(t))dt - iK(t)dY_t)U_c(t)$$

We have

$$Y(t) = U(t)^* Y_i(t) U(t) = U(T)^* Y_i(t) U(T), T \geq t$$

Here $Y_i(t)$ is an operator on the Boson Fock space and is thus independent of the system Hilbert space operators. We have taking $Y_i(t) = A(t) + A(t)^*$,

$$dY(t) = dY_i(t) + j_t(Z_t)dt$$

where Z_t is a system space operator and $j_t(Z) = U(t)^* Z U(t)$. Thus,

$$dU_c(t) = (-(iH_1(t) + Q_1(t))dt - iK(t)(j_t(Z_t)dt + dY_i(t)))U_c(t)$$

We have

$$d(U_c^* U_c) = dU_c^* U_c + U_c^* dU_c + dU_c^* dU_c$$
$$= U_c(t)^*(-(Q_1^* + Q_1)dt + idY_t K(t) - iK(t)dY_t + dY(t)K(t)^2 dY(t))U_c(t)$$

If $K(t)$ commutes with $dY(t)$ and $Q_1 + Q_1^* = K(t)^2$, then we would get $d(U_c^* U_c) = 0$ and U_c will be a control unitary operator. Taking $K(t) = j_t(P_t)$ where P_t is a system operator, we have

$$K(t)dY(t) = j_t(P_t)(j_t(Z_t)dt + dY_i(t)) = j_t(P_t Z_t)dt + j_t(P_t)dY_i(t)$$

and

$$dY(t)K(t) = j_t(Z_t P_t)dt + j_t(P_t)dY_i(t)$$

So for $U_c(t)$ to be unitary, we require that $[Z_t, P_t] = 0$ for all t. Note that Z_t, P_t are system Hilbert space operators. We can now define

$$\rho_c(t) = U_c(t)\rho(t)U_c(t)^*$$

Now,

$$d\rho_c(t) = dU_c(t)\rho(t)U_c(t)^* + U_c(t)d\rho(t).U_c(t)^* + U_c(t)d\rho(t)dU_c(t)^* + dU_c(t)d\rho(t)U_c(t)^*$$
$$+U_c(t)d\rho(t).dU_c(t)^* + dU_c(t)\rho(t)dU_c(t)^*$$

[25] Linear algebra applied to the study of the linearized Einstein field equations in the presence of matter and radiation for the study of galactic evolution as the propagation of small non-uniformities in matter and radiation propagating in an expanding universe.

$g_{\mu\nu}^{(0)}(x)$ is the background Robertson Walker (RW) metric. Its perturbation is

$$g_{\mu\nu} = g_{\mu\nu}^{(0)} + \delta g_{\mu\nu}$$

The coordinate system can be chosen so that

$$\delta g_{0\mu} = 0$$

Then, the linearized Ricci tensor is

$$\delta R_{\mu\nu} = \delta\Gamma^{\alpha}_{\mu\alpha,\nu} - \delta\Gamma^{\alpha}_{\mu\nu,\alpha}$$
$$-(\delta\Gamma^{\alpha}_{\mu\nu})\Gamma^{\beta(0)}_{\alpha\beta} - \Gamma^{\alpha(0)}_{\mu\nu}\delta\Gamma^{\beta}_{\alpha\beta}$$
$$+\Gamma^{\alpha(0)}_{\mu\beta}\delta\Gamma^{\beta}_{\nu\alpha} + (\delta\Gamma^{\alpha}_{\mu\beta})\Gamma^{\beta(0)}_{\alpha\beta}$$

This expression can be expressed as

$$\delta R_{\mu\nu} = (\delta\Gamma^{\alpha}_{\mu\alpha})_{:\nu} - (\delta\Gamma^{\alpha}_{\mu\nu})_{:\alpha}$$

where the covariant derivatives are computed using the unperturbed RW metric. The energy momentum tensor of the matter field is

$$T_{\mu\nu} = (\rho + p)v_\mu v_\nu - p g_{\mu\nu}$$

and that of the radiation field is

$$S_{\mu\nu} = \frac{1}{4}F_{\alpha\beta}F^{\alpha\beta}g_{\mu\nu} - F_{\mu\alpha}F^{\alpha}_{\nu}$$

For example, computing in a flat space-time

$$S_{00} = \frac{1}{4}F_{\alpha\beta}F^{\alpha\beta} - F_{0\alpha}F^{\alpha}_{0}$$

Now

$$F_{\alpha\beta}F^{\alpha\beta} = -2F_{0r}F_{0r} + F_{rs}F_{rs} = -2|E|^2 + 2|B|^2$$

$$F_{0\alpha}F_0^\alpha = -F_{0r}F_{0r} = -|E|^2$$

so

$$S_{00} = \frac{1}{2}(-|E|^2 + |B|^2) + |E|^2 = \frac{1}{2}(|E|^2 + |B|^2)$$

which is the correct expression for the energy density of the electromagnetic field. Likewise, $S^{0r} = S^{r0}$ defines the energy flux as well as the momentum density and $S^{rs} = S^{sr}$ defines the momentum flux. We can write the expression for $\delta R_{\mu\nu}$ in the general form

$$\delta R_{\mu\nu} = C_1(\mu\nu\alpha\beta\rho, x)\delta g_{\alpha\beta,\rho}(x) + C_2(\mu\nu\alpha\beta\rho\sigma, x)\delta g_{\beta\rho,\sigma}(x)$$

where C_1, C_2 are functions of x determined completely from the background gravitational field $g_{\mu\nu}^{(0)}(x)$ The perturbation to the energy momentum tensor of matter is given by

$$\delta T_{\mu\nu} = (\delta\rho + \delta p)V_\mu^{(0)}V_\nu^{(0)} + (\rho^{(0)} + p^{(0)})(V_\mu^{(0)}\delta v_\mu + V_\nu^{(0)}\delta v_\mu)$$

$$-\delta p g_{\mu\nu}^{(0)} + p^{(0)}\delta g_{\mu\nu}$$

We note that

$$V_\mu^{(0)} = (1, 0, 0, 0),$$

and $p^{(0)}, \rho^{(0)}$ are functions of time only. Further,

$$g_{00}^{(0)} = 1, g_{11}^{(0)} = -S^2(t)f(r), g_{22}^{(0)} = -S^2(t)r^2, g_{33}^{(0)} = -S^2(t)r^2 sin^2(\theta),$$

$$f(r) = \frac{1}{1 - kr^2}$$

with $k = 1$ for a spherical universe, $k = 0$ for a flat universe and $k = -1$ for a hyperbolic universe.

We now consider a flat unperturbed universe for which the metric has the form

$$d\tau^2 = dt^2 - S^2(t)(dx^2 + dy^2 + dz^2)$$

Thus,

$$g_{00} = 1, g_{rr} = -S^2(t), r = 1, 2, 3$$

The Ricci tensor components are:

$$R_{00} = \Gamma^\alpha_{0\alpha,o} - \Gamma^\alpha_{00,\alpha} - \Gamma^\alpha_{00}\Gamma^\beta_{\alpha\beta}$$

$$+\Gamma^\alpha_{0\beta}\Gamma^\beta_{0\alpha}$$

Now,

$$\Gamma^\alpha_{0\alpha} = \Gamma^r_{0r} = \frac{1}{2}g^{rr}g_{rr,0} = S'/S, r = 1, 2, 3$$

So

$$\Gamma^\alpha_{0\alpha,0} = (S'/S)'$$

$$\Gamma^\alpha_{00,\alpha} = 0$$

$$\Gamma^\alpha_{00}\Gamma^\beta_{\alpha\beta} = 0$$

$$\Gamma^\alpha_{0\beta}\Gamma^\beta_{0\alpha} =$$

$$\sum_r(\Gamma^r_{0r})^2 = \sum_r(g^{rr}g_{rr,0}/2)^2 = 3S'^2/S^2$$

So

$$R_{00} = (S'/S)' + 3S'^2/S^2 = S''/S + 2S'^2/S^2$$

$$R_{kk} = \Gamma^\alpha_{k\alpha,k} - \Gamma^\alpha_{kk,\alpha} - \Gamma^\alpha_{kk}\Gamma^\beta_{\alpha\beta} + \Gamma^\alpha_{k\beta}\Gamma^\beta_{k\alpha}$$

$$\Gamma^0_{kk,0} - \Gamma^0_{kk}\sum_r \Gamma^r_{0r} + 2\Gamma^0_{kk}\Gamma^k_{k0}$$

(No summation over k)

$$= -g_{kk,00}/2 + (g_{kk,0}/2)(S'/S) + 2(-g_{kk,0}/2)(g_{kk,0}/2g_{kk})$$

$$= S''/2 - S^{'2}/2S + S'(S'/2S) = S''/2$$

We have in fact

$$R_{km} = (S''/2)\delta_{km}$$

Now let us study the perturbed Einstein field equations w.r.t. the above flat space-time metric. First note that we can choose our coordinate system so that $\delta g_{0\mu} = 0$ and hence $\delta g^{0\mu} = 0$. Raising and lowering of indices are carried out w.r.t. the above flat space time metric which is diagonal. The unperturbed space-time is comoving, ie,

$$v^{\mu(0)} = (1, 0, 0, 0)$$

define geodesics in the unperturbed space-time. The unperturbed pressure and density $p^{(0)}(t), \rho^{(0)}(t)$ are functions of t only. The unperturbed energy momentum tensor of matter is

$$T^{\mu\nu(0)} = (\rho^{(0)} + p^{(0)})v^{\mu(0)}v^{\nu(0)} - p^{(0)}g^{\mu\nu(0)}$$

so that

$$T^{00(0)} = \rho^{(0)}, T^{kk(0)} = p^{(0)}/S^2$$

with the other components of the energy momentum tensor being zero. The unperturbed Einstein field equations

$$R^{(0)}_{\mu\nu} = K.(T_{\mu\nu(0)} - \frac{1}{2}T^{(0)}g_{\mu\nu(0)}), K = -8\pi G$$

thus give after noting that

$$T^{(0)} = g^{(0)}_{\mu\nu}T^{\mu\nu(0)} =$$

$$\rho^{(0)} - 3p^{(0)},$$

$$T^{(0)}_{kk} = -S^2T^{kk(0)} = -p^{(0)}$$

and hence the unperturbed field equations are

$$S''/S + 2S^{'2}/S^2 = K.(\rho^{(0)} - \frac{1}{2}(\rho^{(0)} - 3p^{(0)}),$$

$$S''/2 = K(-p^{(0)} + \frac{S^2}{2}(\rho^{(0)} - 3p^{(0)}))$$

These two equations along with an equation of state: $p^{(0)} = f(\rho^{(0)})$ determing the three functions of time $S(t), \rho^{(0)}(t), p^{(0)}(t)$. The perturbed equations are

$$\delta R_{\mu\nu} = K\delta T_{\mu\nu}$$

where

$$\delta T_{\mu\nu} = (\rho^{(0)} + p^{(0)})(v^{(0)}_\mu \delta v_\nu + v^{(0)}_\nu \delta v_\mu) + (\delta\rho + \delta p)v_{\mu(0)}v_{\nu(0)}$$

$$-\delta p g^{(0)}_{\mu\nu} - p^{(0)}\delta g_{\mu\nu}$$

Now,

$$\delta R_{km} = \delta\Gamma^\alpha_{k\alpha,m} - \delta\Gamma^\alpha_{km,\alpha} - \Gamma^{\alpha(0)}_{km}\delta\Gamma^\beta_{\alpha\beta}$$

$$-\Gamma^{\beta(0)}_{\alpha\beta}\delta\Gamma^\alpha_{km} + \Gamma^{\alpha(0)}_{k\beta}\delta\Gamma^\beta_{m\alpha}$$

$$+\Gamma^{\beta(0)}_{m\alpha}\delta\Gamma^\alpha_{k\beta}$$

The perturbed field equations taking into account contributions from the electromagneti field are expressible in the form

$$C_1(\mu\nu\alpha\beta\gamma, x)\delta g_{\alpha\beta,\gamma}(x) + C_2(\mu\nu\alpha\beta\gamma\sigma, x)\delta g_{\alpha\beta,\gamma\sigma}(x)$$

$$= C_3(\mu\nu\alpha, x)\delta v_\alpha(x) + C_4(\mu\nu, x)\delta\rho(x) + C_5(\mu\nu, x)\delta p(x)$$

$$+C_6(\mu\nu\alpha\beta, x)\delta A_{\alpha,\beta}(x) = 0$$

The equations implied by the Bianchi identity are

$$(T^{\mu\nu} + S^{\mu\nu})_{:\nu} = 0$$

This is the same as

$$(T^{\mu\nu} + S^{\mu\nu})_{,\nu} + \Gamma^{\mu}_{\alpha\nu}(T^{\alpha\nu} + S^{\alpha\nu}) + \Gamma^{\nu}_{\alpha\nu}(T^{\mu\alpha} + S^{\mu\alpha}) = 0$$

We calculate is first order perturbed version:

$$(\delta T^{\mu\nu} + \delta S^{\mu\nu})_{,\nu} + \Gamma^{\mu(0)}_{\alpha\nu}(\delta T^{\alpha\nu} + \delta S^{\alpha\nu}) +$$

$$\Gamma^{\nu(0)}_{\alpha\nu}(\delta T^{\mu\alpha} + \delta S^{\mu\alpha}) +$$

$$\delta\Gamma^{\mu}_{\alpha\nu}(T^{\alpha\nu(0)} + S^{\alpha\nu(0)}) +$$

$$\delta\Gamma^{\nu}_{\alpha\nu}(T^{\mu\alpha(0)} + \delta S^{\mu\alpha(0)}) = 0$$

This can be put in the form

$$C_7(\mu\alpha\beta, x)\delta v_{\alpha,\beta}(x) + C_8(\mu\alpha, x)\delta v_{\alpha}(x) + C_9(\mu\alpha\beta\gamma, x)\delta A_{\alpha,\beta\gamma}(x)$$

$$+ C_{10}(\mu\alpha\beta\gamma, x)\delta g_{\alpha\beta,\gamma}(x) + C_{11}(\mu, x)\delta\rho(x) + C_{12}(\mu, x)\delta p(x) = 0$$

We note that this last equation contains 3 equations for the velocity components $\delta v_r, r = 1, 2, 3$ and one equation for $\delta\rho(x)$. $\delta p(x)$ is determined from $\delta\rho(x)$ using the equation of state. Also δv_0 is determined from δv_r using

$$0 = \delta(g_{\mu\nu}v^{\mu}v^{\nu})$$

$$= (\delta g_{\mu\nu})v^{\mu(0)}v^{\nu(0)} + g^{(0)}_{\mu\nu}(v^{\mu(0)}\delta v^{\nu} + v^{\nu(0)}\delta v^{\mu}$$

using that the unperturbed dynamics is comoving,ie, $v^{\mu(0)} = (1, 0, 0, 0)$ we get from this equation

$$\delta g_{00} + 2\delta v^0 = 0$$

so

$$\delta v^0 = -\frac{1}{2}\delta g_{00}$$

and hence,

$$\delta v_0 = \delta(g_{0\mu}v^{\mu}) = \delta g_{00}$$

since $v^{r(0)} = 0$ and $g^{(0)}_{0r} = 0$.

[26] Perturbation theory applied to electromagnetic problems. Here, we discuss the rudiments of the theory of time independent perturbation theory in quantum mechanics and explain how the same techniques can be used to solve waveguide and cavity resonator problems having almost aribtrary cross sections by transforming the boundary into a simpler boundary using the theory of analytic functions of a complex variable.

Let D be a flat connected region parallel to the xy plane. D represents the cross section of a waveguide or a cavity resonator. The boundary ∂D is a closed curve that represents the boundary of the guide or resonator. The z direction is orthogonal to the D plane and for the guide, it represents the direction along which the em waves propagate. For the resonator case, we assume that $0 \leq z \leq d$ with the surfaces $z = 0$ and $z = d$ being perfectly conducting surfaces as is the case with the side walls. We choose a system (q_1, q_2, z) of coordinates so that (q_1, q_2) are functions of (x, y) alone. Being more specific, we assume that

$$w = q_1(x, y) + jq_2(x, y) = f(z) = f(x + jy)$$

with an inverse

$$z = x + jy = g(w) = g(q_1 + jq_2)$$

and assume that $f(z)$ is an analytic function of the complex variable z and its inverse $g(w)$ is also an analytic function of the complex variable w. We can regard $(z, \bar{z})$ as independent variables just as (x, y) are. The relation between the two pairs of variables is

$$z = x + jy, \bar{z} = x - jy, x = (z + \bar{z})/2, y = (z - \bar{z})/2j$$

We have

$$\frac{\partial}{\partial x} = z_{,x}\frac{\partial}{\partial z} + \bar{z}_{,x}\frac{\partial}{\partial \bar{z}}$$

$$= \frac{\partial}{\partial z} + \frac{\partial}{\partial \bar{z}}$$

and likewise,

$$\frac{\partial}{\partial y} = j(\frac{\partial}{\partial z} - \frac{\partial}{\partial \bar{z}})$$

Thus,

$$\nabla_{\perp}^2 = \frac{\partial^2}{\partial x^2} + \frac{\partial^2}{\partial y^2}$$

$$= (\frac{\partial}{\partial z} + \frac{\partial}{\partial \bar{z}})^2 - (\frac{\partial}{\partial z} - \frac{\partial}{\partial \bar{z}})^2$$

$$= 4\frac{\partial^2}{\partial z \partial \bar{z}}$$

Now,

$$\frac{\partial}{\partial z} = w'(z)\frac{\partial}{\partial w}$$

$$\frac{\partial}{\partial \bar{z}} = \bar{w}'(z)\frac{\partial}{\partial \bar{w}}$$

also, we clearly have since $w'(z)$ is an analytic function of z and $\bar{w}'(z)$ is an analytic function of $\bar{z}$ and hence that $1/w'(z) = dz/dw$ is an analytic function of w and $1/\bar{w}'(z)$ is an analytic function of $\bar{w}$ that

$$\nabla_{\perp}^2 = 4\frac{\partial^2}{\partial z \partial \bar{z}} = 4|z'(w)|^{-2}\frac{\partial^2}{\partial w \partial \bar{w}}$$

$$= |z'(w)|^{-2}(\frac{\partial^2}{\partial q_1^2} + \frac{\partial q_2^2}{)}$$

Thus, the eigenvalue problem $(\nabla_{\perp}^2 + h^2)\psi(x, y) = 0$ with the Dirichlet boundary condition $\psi = 0$ on ∂D becomes

$$(\nabla_{\perp,q}^2 + h^2 F(q_1, q_2))\psi(q_1, q_2) = 0$$

with

$$\psi(a, q_2) = 0$$

where the boundary ∂D is the same as $q_1 = a$. For example, if we take $w = log(z) = log(\rho) + j\phi$, then $q_1 = a$ is the circle $q_1 = \rho = e^a$ which is a circle and

$$(q_1, q_2) = |z'(w)|^2 = |g'(q_1 + jq_2)|^2$$

Also, we have defined

$$\nabla_{\perp,q}^2 = \frac{\partial^2}{\partial q_1^2} + \frac{\partial^2}{\partial q_2^2} = L$$

say. Solution by perturbation theory: Let $h^2 = -\lambda$. Then, we have to solve

$$(L - \lambda(1 + \epsilon G(q_1, q_2)))\psi(q_1, q_2) = 0$$

where we are assuming that

$$F(q) = 1 + \epsilon G(q)$$

so that the boundary is a small perturbation of the rectangular boundary.

[27] Perturbation theory applied to general nonlinear partial differential equations with noisy terms. The gravitational wave equations, fluid dynamical equations, Klein-Gordon and Dirac equations are special cases of this. The field $\phi : \mathbb{R}^n \to \mathbb{R}^p$ satisfies a nonlinear pde

$$\sum_{k=1}^{n} b_k(x)\phi_{l,k}(x) + \sum_{k,m=1}^{p} a_{km}(x)\phi_{l,km}(x) + \delta.F_l(\phi_l(x), \phi_{l,k}(x), \phi_{l,km}(x), x)$$

$$+\delta.\sum_m G_{lm}(\phi_l(x), \phi_{l,k}(x), \phi_{l,km}(x), x)w_m(x)$$

where $w_m(x)$ are Gaussian noise processes.

[28] The Knill-Laflamme theorem for quantum error correcting codes: Explicit construction of the recovery operators for a noisy quantum channel in terms of the code subspace and the noise subspace.

[29] Post-Newtonian equations of hydrodynamics. The perturbations are carried out in powers of the velocity. The mass parameter is of the order of the square of the velocity ($v^2 = GM/r$ for the orbital velocity) and the following expansions are valid

$$\rho = \rho_2 + \rho_4 + ..., p = p_4 + p_6 + ...,$$
$$v^r = v_1^r + v_3^r + ...,$$
$$v^0 = 1 + v_2^0 + v_4^0 + ...$$
$$g_{00} = 1 + g_{00(2)} + g_{00(4)} + ...,$$
$$g_{0r} = g_{0r(3)} + g_{(0r(5)} + ...$$
$$g_{rs} = -\delta_{rs} + g_{rs(2)} + g_{rs(4)} + ...$$
$$g^{00} = 1 + g_2^{00} + g_4^{00} + ...$$
$$g^{0r} = g_3^{0r} + g_5^{0r} + ...,$$
$$g^{rs} = -\delta^{rs} + g_2^{rs} + g_4^{rs} + ...$$
$$T^{\mu\nu} = (\rho + p)v^\mu v^\nu - pg^{\mu\nu}$$

The equation

$$g_{\mu\nu}v^\mu v^\nu = 1$$

can be expressed as

$$g_{00}v^{02} + 2g_{0r}v^0 v^r + g_{rs}v^r v^s = 1$$

so the $O(1)$ equation is

$$v_0^0 = 1$$

The $O(v^2)$ equation is

$$v_2^0 + g_{00(2)} - \sum_r v_1^{r2} = 0$$

or equivalently,

$$v_2^0 = -g_{00(2)} + \sum_r v_1^{r2}$$

We have

$$T^{00} = T_2^{00} + T_4^{00} + ...$$
$$T^{0r} = T_3^{0r} + T_5^{0r} + ...$$
$$T^{rs} = T_2^{rs} + T_4^{rs} + ...$$

where

$$T_2^{00} = \rho_2, T_4^{00} = 2\rho_2 v_2^0, T_6^{00} = \rho_2 v_2^{02} + 2\rho_4 v_2^0 + 2p_4 v_2^0 - p_4 g_2^{00}$$
$$T_3^{0r} = 2\rho_2 v_2^0 v_1^r, T_5^{0r} = 2\rho_2 v_2^0 v_1^r + 2\rho_2 v_3^r + 2\rho_4 v_1^r + 2p_4 v_1^r,$$
$$T_2^{rs} =$$

[30] Lab problems on linear algebra based signal processing:

[1] If X is an $m \times n$ matrix, then calculate δP_X upto $O(\| \delta X) \|)$ where P_X is the orthogonal projection onto $\mathcal{R}(X)$ and X gets perturbed to $X + \delta X$ where δX is a small random perturbation of X. Calculate using this formula, the second order statistics of δP_X, ie, $\mathbb{E}(\delta P_X \otimes \delta P_X)$ in terms of $\mathbb{E}(\delta X \otimes \delta X)$. Calculate δP_X upto $O(||\delta X||^2)$

[2] Take an $n \times n$ matrix A. Add a row and a column to this matrix at the end and express the inverse of this matrix B in terms of the inverse of A. Assume now that A gets perturbed to $A + \delta A$ and correspondingly, the appended row and column get perturbed by small amounts. Then calculate the inverse of the perturbed appended matrix in terms of A^{-1} and the appended rows and columns and their perturbations upto linear orders in the perturbations.

[3] Generate some functionals of the Brownian motion process $B(t)$ like $M(t) = max(B(s) : s \leq t), m(t) = min(B(s) : s \leq t), T_a = min(t > 0 : B(t) = a), |B(t)|$ (reflected Brownian motion), Absorbed Brownian motion $(B(min(t, T_0)) : t \geq 0$ where $B(0) = a > 0$ and $T_0 = min(t > 0 : B(t) = 0$, Local time process $L^a(t)$ of the Brownian motion process at the level a. This is defined as $L^a(t) = \int_0^t \delta(B(s) - a)ds$ and is approximated by $(2\epsilon)^{-1} \int_0^t \chi_{[a-\epsilon,a+\epsilon]}(B(s))ds$ where ϵ is a very small positive number. For a bivariate standard Brownian motion process $(B_1(t), B_2(t)), t \geq 0$, simulate the area process $A(t) = \int_0^t B_1(s)dB_2(s) - B_2(s)dB_1(s)$ and calculate its statistics. Simulate the Bessel process of order d, ie $X(t) = (\sum_{k=1}^d B_k(t)^2)^{1/2}, t \geq 0$ and verify by numerical simulations that it satisfies its standard stochastic differential equation.

[4] Verification of the Knill-Laflamme theorem for quantum error correcting codes. Generate a set of $r < n$ linearly independent column vectors $\{f_1, ..., f_r\}$ in $\mathcal{H} = \mathbb{C}^n$. Denote the subspace spanned by these r vectors by $\mathcal{C}$. Calculate the orthogonal projection P onto $\mathcal{C}$ using the standard formula $P = A(A^*A)^{-1}A^*$ where $A = [f_1, ..., f_r] = \mathbb{C}^{n\times r}$. Generate K $n \times n$ matrices having the block structure

$$N_k = \begin{pmatrix} C_{1k} & C_{2k} \\ C_{3k} & C_{4k} \end{pmatrix}, k = 1, 2, ..., K$$

where C_{1k} is an $r \times r$ matrix for each k such that for $1 \leq k, j \leq K$, we have

$$C_{1j}^* C_{1k} + C_{3j}^* C_{3k} = \lambda_{jk} I_r$$

for some complex numbers λ_{jk}. This can be achieved by choosing the $r \times n$ matrices $[C_{1j}^*|C_{3j}^*], 1 \leq j \leq K$ as non-overlapping rows of an $n \times n$ unitary matrix U and then multiply the resultant matrices by complex constants. Thus, we must have $K \leq [n/r]$. The matrices C_{2k}, C_{3k}, C_{4k} can be chosen arbitrarily. Now take the orthogonal projection P onto $\mathcal{C}$ as

$$P = \begin{pmatrix} I_r & 0 \\ 0 & 0 \end{pmatrix}$$

It is then easily seen that

$$PN_k^* N_j P = \lambda_{kj} P$$

and hence the quantum code P can correct the noise subspace $\{N_k : 1 \leq k \leq K\}$.

[5] (a) Study waves in a plasma influenced by a strong gravitational field and electromagnetic fields by the method of linearization: The Boltzmann particle distribution function $f(t, r, v)$ where $v = (v^r = dx^r/dt)r = 1^3$ satisfies (after approximating the collision term by a linear relaxation term

$$f_{,t}(t, r, v) + v^k f_{,x^k}(t, r, v) + v_{,0}^k f_{,v^k}(t, r, v) = (f_0(r, v) - f(t, r, v))/\tau(v)$$

Here the velocity v^k satisfies the geodesic equation in an electromagnetic field:

$$dx^k/d\tau = \gamma v^k, dv^k/d\tau = \gamma d/dt(dx^k/d\tau) = \gamma d/dt(\gamma v^k) = \gamma^2 v_{,0}^k + \gamma\gamma_{,0} v^k$$

where

$$\gamma = dt/d\tau = (g_{00} + 2g_{0r}v^r + g_{rs}v^r v^s)^{-1/2}$$

$$\gamma^2 v_{,0}^k + \gamma\gamma_{,0} v^k + \gamma^2 \Gamma_{00}^k + 2\gamma^2 \Gamma_{0m}^k v^m + \gamma^2 \Gamma_{mp}^k v^m v^p = e\gamma(F_0^m + F_s^m v^s) - - - (a)$$

or equivalently,

$$v_{,0}^k + (\gamma_{,0}/\gamma)v^k + \Gamma_{00}^k + 2\Gamma_{0m}^k v^m + \Gamma_{mp}^k v^m v^p = e\gamma^{-1}(F_0^m + F_s^m v^s)$$

This value of $v_{,0}^k$ is substituted into the above Boltzmann equation to get

$$f_{,t}(t, r, v) + v^k f_{,x^k}(t, r, v) - (\gamma_{,0}/\gamma)v^k + \Gamma_{00}^k + 2\Gamma_{0m}^k v^m + \Gamma_{mp}^k v^m v^p - e\gamma^{-1}(F_0^m + F_s^m v^s))f_{,v^k}(t, r, v) - (f_0(r, v) - f(t, r, v))/\tau(v) = 0$$

We note that $\gamma_{,0}$ is a function of $v_{,0}^k, v^k, x$. We need to get a Boltzmann equation that does not involve $v_{,0}^k$. For this purpose, we go back to the equation of motion of the charged particle (a). First observe that

$$\gamma_{,0} = (g_{00} + 2g_{0r}v^r + g_{rs}v^r v^s)_{,0}^{-1/2} = (-\gamma^3/2)(g_{00,0} + g_{00,k}v^k + 2g_{0r,0}v^r + 2g_{0r,s}v^r v^s + 2g_{0r}v_{,0}^r$$

$$+ g_{rs,0}v^r v^s + g_{rs,m}v^r v^s v^m + 2g_{rs}v_{,0}^s) - - - (b)$$

Substituting (b)into (a) gives us a linear algebraic equation for $(v_{,0}^k)_{k=1}^3$ which is inverted to get $v_{,0}^k$ as a function (v^k), x^μ, the electromagnetic field $F^{\mu\nu}(x)$ and of course the metric $g_{\mu\nu}(x)$ and its first order partial derivatives $g_{\mu\nu,\alpha}$. Thus, we get a well defined Boltzmann equation. Now given the particle distribution function $f(t,r,v)$, we need to calculate the energy momentum tensor of matter. We have

$$T^{\mu\nu} = (\rho + p)V^\mu V^\nu - pg^{\mu\nu}$$

where

$$\rho = m\int fd^3v,$$

$$U^r = \int v^r fd^3v / \int fd^3v, V^r = \gamma(U)U^r$$

V^0 is calculated using

$$g_{\mu\nu}V^\mu V^\nu = 1$$

Equivalently,

$$\gamma(U)^2 g_{00} + 2\gamma(U)^2 U^r g_{0r} + \gamma(U)^2 g_{rs}U^r U^s = 1$$

or

$$\gamma(U) = (g_{00} + 2g_{0r}U^r + g_{rs}U^r U^s)^{-1/2}, V^0 = \gamma(U)$$

The average internal kinetic energy per particle is

$$K(t,r) = (m/2)\int \sum(\sum_r (v^r - U^r)^2)f(t,r,v)d^3v$$

and the pressure field is given by

$$p(t,r) = nm < |v-U|^2 > /3 = (m/3)\int \sum_r (v^r - U^r)^2 f(t,r,v)d^3v$$

where m is the mass of a plasma particle and n is the number of plasma particles per unit volume. This energy momentum tensor of the plasma can be added to the energy momentum tensor of the electromagnetic field and substituted into the right side of the Einstein field equations. Thus, we get a couple system of pde's for $f, g_{\mu\nu}, A_\mu$.

An alternate way to define the energy momentum tensor is as

$$T^{\mu\nu} = m\int f(t,r,v)\gamma(t,r,v)v^\mu v^\nu d^3v - p(t,r)g^{\mu\nu}(t,r)$$

where

$$v^\mu = dx^\mu/dt$$

so that

$$v^0 = 1, v^r = dx^r/dt$$

and $\gamma = \gamma(t,r,v)$ is defined by the equation

$$\gamma = (g_{00} + 2g_{0k}v^k + g_{km}v^k v^m)^{-1/2}$$

The pressure p is as defined earlier.

[31] Quantum image processing: The image field is obtained by passing a quantum em field through a spatio temporal filter having impulse response $h(t,\tau,r,r')$. Assume the Coulomb gauge with zero charge density. Then the scalar potential $A^0 = 0$ and the vector potential is given by

$$A^r = \int [(2|K|)^{-1/2}a(K,\sigma)e^r(K,\sigma)exp(-i(|K|t - K.r)) + (2|K|)^{-1/2}a^*(K,\sigma)\bar{e}^r(K,\sigma)exp(i(|K|t - K.r))]d^3K$$

The Coulomb gauge condition $divA = A^r_{,r} = 0$ implies $K^r e^r(K,\sigma) = 0$ or equivalently, $(K, e(K,\sigma)) = 0$ which means that there are only two degrees of polarization which are indexed by $\sigma = 1,2$. The electric field is

$$E_r = -A^r_{,0} = i\int [(|K|/2)^{1/2}a(K,\sigma)e^r(K,\sigma)exp(-i(|K|t - K.r)) - (|K|/2)^{1/2}a^*(K,\sigma)\bar{e}^r(K,\sigma)exp(i(|K|t - K.r))]d^3K$$

or equivalently in three vector notation,

$$E = i\int [(|K|/2)^{1/2}a(K,\sigma)e(K,\sigma).exp(-i(|K|t-K.r)) - (|K|/2)^{1/2}a^*(K,\sigma)\bar{e}(K,\sigma)exp(i(|K|t-K.r))]d^3K$$

The magnetic field is given by

$$B = curlA = i\int [(2|K|)^{-1/2}a(K,\sigma)K\times e(K,\sigma)exp(-i(|K|t-K.r))-(2|K|)^{-1/2}a^*(K,\sigma)K\times e(K,\sigma)exp(i(|K|t-K.r))]d^3K$$

The em field of the image can be regarded as the output of a spatio-temporal filter with this free em field as input. Thus, the output field is

$$E^o(t,r) = \int h(t,\tau,r,r')E(\tau,r')d\tau d^3r', B^o(t,r) = \int h(t,\tau,r,r')B(\tau,r')d\tau d^3r'$$

and hence the output field can be expressed as

$$E^o(t,r) = \int (H_E(t,r,K,\sigma)a(K,\sigma) + \bar{H}_E(t,r,K,\sigma)a^*(K,\sigma))d^3K$$

$$B^o(t,r) = \int [H_B(t,r,K,\sigma)a(K,\sigma) + \bar{H}_B(t,r,K,\sigma)a^*(K,\sigma)]d^3K$$

where H_E, H_B are 3×1 functions constructed from the image impulse response $h(t,\tau,r,r')$. The energy of the em field coming from the image is

$$H_F = \int (|E^o|^2 + |B^o|^2)d^3r/2$$

and this can be expressed in the form after discretization of the spatial frequencies

$$H_F(\theta) = (1/2)\sum_{k,m=1}^{p}(Q_1(k,m,\theta)a_k^*a_m + Q_2(k,m,\theta)a_ka_m + \bar{Q}_2(k,m,\theta)a_k^*a_m^*)$$

where

$$\bar{Q}_1(k,m,\theta) = Q_1(m,k,\theta)$$

Here θ is a parameter vector upon which the image impulse response $h(t,\tau,r,r')$ depends. θ is the parameter which contains all information about the image field and is to be estimated by exciting an atom with the output image field and and taking measurements on the state of the atom at different times. We can simplify the form of the image field Hamiltonian as

$$H_F(\theta) = \sum_{k,m=1}^{p} Q(k,m,\theta)a_k^*a_m$$

Here,

$$[a_k, a_m^*] = \delta_{km}, [a_k, a_m] = 0, [a_k^*, a_m^*] = 0$$

The Hamiltonian of the atom assumed to be an N state system, is an $N\times N$ Hermitian matrix H_A and the interaction Hamiltonian between the image field and the atom has the form

$$H_{int}(t) = \sum_{k=1}^{p}(F_k(t,\theta)\otimes a_k + F_k(t,\theta)^*\otimes a_k^*)$$

[32] Performance analysis of the MUSIC algorithm.

$$X = AS + W, X\in\mathbb{C}^{N\times K}, A\in\mathbb{C}^{N\times p}, S\in\mathbb{C}^{p\times K}, W\in\mathbb{C}^{N\times K}$$

$$R_{xx} = K^{-1}\mathbb{E}(XX^*), R_{ss} = K^{-1}\mathbb{E}(SS^*), \sigma_w^2 I = K^{-1}\mathbb{E}(WW^*), \mathbb{E}(S\otimes W) = \mathbb{E}(S\otimes W^*) = 0$$

All signals are complex Gaussian. Thus,

$$\mathbb{E}(X\otimes X) = 0, \mathbb{E}(S\otimes S) = 0, \mathbb{E}(W\otimes W) = 0$$

The stochastic perturbation in the array signal correlation matrix is given by

$$\delta R_{xx} = K^{-1}XX^* - R_{xx} =$$

$$K^{-1}ASS^*A^* + K^{-1}WW^* + K^{-1}ASW^* + K^{-1}WS^*A^* - AR_{ss}A^* - \sigma_w^2 I$$

$$= A\delta R_{ss}A^* + \delta R_{ww} + K^{-1}(ASW^* + WS^*A^*)$$

where

$$\delta R_{ss} = K^{-1}SS^* - R_{ss}, \delta R_{ww} = K^{-1}WW^* - \sigma_w^2 I$$

To calculate the mean and covariance of the DOA estimates, we need the mean and covariance of the statistical perturbation δR_{xx} of R_{xx}. Now,

$$\mathbb{E}(\delta R_{ss} \otimes \delta R_{ss}) = K^{-2}\mathbb{E}(SS^* \otimes SS^*) - R_{ss} \otimes R_{ss}$$

Now,

$$\mathbb{E}(SS^* \otimes SS^*) = \mathbb{E}[(S \otimes S)(S^* \otimes S^*)]$$

This is equivalent to calculating

$$\mathbb{E}(s_i s_j \bar{s}_k \bar{s}_m) = \mathbb{E}(s_i \bar{s}_k)\mathbb{E}(s_j \bar{s}_m) + \mathbb{E}(s_i \bar{s}_m)\mathbb{E}(s_j \bar{s}_k)$$

$$= [R_{ss}]_{ik}[\mathbb{R}_{ss}]_{jm} + [R_{ss}]_{im}[\mathbb{R}_{ss}]_{jk}$$

Also since S and W are independent random matrices, it follows that δR_{ss} and δR_{ww} are independent zero mean random matrices. The second order moments of δR_{xx} are thus computed as

$$\mathbb{E}(\delta R_{xx} \otimes \delta R_{xx}) =$$

$$\mathbb{E}[A\delta R_{ss}A^* + \delta R_{ww} + K^{-1}(ASW^* + WS^*A^*))^{\otimes 2}]$$

$$= (A \otimes A)\mathbb{E}(\delta R_{ss} \otimes \delta R_{ss})(A^* \otimes A^*)$$

$$+\mathbb{E}(\delta R_{ww} \otimes \delta R_{ww})$$

$$+K^{-2}(I + F)\mathbb{E}(ASW^* \otimes WS^*A^*)$$

where F denotes the flip operator:

$$F(x \otimes y) = y \otimes x$$

Computing the last expectation is equivalent to computing

$$\mathbb{E}(a_{ij}s_{jk}\bar{w}_{lk}w_{i'j'}\bar{s}_{k'j'}\bar{a}_{k'l'})$$

$$= a_{ij}\bar{a}_{k'l'}\mathbb{E}(s_{jk}\bar{s}_{k'j'})\mathbb{E}(w_{i'j'}\bar{w}_{lk})$$

Thus the second order moments of δR_{xx} are easily computed and this can be combined with matrix perturbation theory to obtain the covariance of the signal and noise eigenvalues and eigenvectors of $\hat{R}_{xx} = R_{xx} + \delta R_{xx}$. These covariances can in turn be used to calculate the error covariances in the DOA estimates using the MUSIC pseudospectrum.

[33] Estimating the quantum image parameters from measurements on the state of an atom excited by the quantum em field coming from the image in the interaction representation. The image em field interacts with an atom described by an $N \times N$ Hamiltonian matrix H_A. This interaction Hamiltonian can be expressed as

$$H_{int}(t|\theta) = \sum_{k=1}^{p}(G_k(t|\theta) \otimes a_k + G_k(t|\theta)^* \otimes a_k^*)$$

where $G_k(t|\theta) \in \mathbb{C}^{N \times N}$, θ is the image parameter vector. The joint density of the atom and the image em field at time t can be expressed using the Glauber-Sudarshan representation:

$$\rho(t) = C\int exp(-|z|^2)A(t,z) \otimes |e(z)><e(z)|d^{2p}z$$

where

$$|e(z)>= \sum_{n \geq 0} z^n a^{*n}|0>/n! = \sum_{n \geq 0} z^n|n>/\sqrt{n!}$$

since

$$|n>=a^{*n}|0>/\sqrt{n!}$$

is the normalized state of the field in which $a_k^*a_k$ has the eigenvalue n_k and $n=(n_k)$. Further

$$C=\pi^{-p}$$

We have

$$a_k|e(z)>=z_k|e(z)>, a_k^*|e(z)>=\frac{\partial}{\partial z_k}|e(z)>$$

Thus,

$$a_k|e(z)><e(z)>=z_k|e(z)><e(z)|, a_k^*|e(z)><e(z)|=\frac{\partial}{\partial z_k}|e(z)><e(z)|,$$

$$|e(z)><e(z)|a_k=\frac{\partial}{\partial \bar{z}_k}|e(z)>,e(z)|,$$

$$|e(z)><e(z)|a_k^*=\bar{z}_k|e(z)><e(z)|$$

Note that $z\rightarrow|e(z)>$ is an analytic function of z and so $\bar{z}\rightarrow<e(z)|$ is an analytic function of $\bar{z}$. Now, the joint density $\rho(t)$ satifies Schrodinger's equation in the interaction picture:

$$i\rho'(t)=[H_{int}(t|\theta),\rho(t)]$$

and this translates to

$$i\int A_{,t}(t,z)\otimes|e(z)><e(z)|exp(-|z|^2)d^{2p}z=$$

$$\sum_k(\int(G_k(t|\theta)A(t,z)\otimes a_k|e(z)><e(z)|+G_k(t|\theta)^*A(t,z)\otimes a_k^*|e(z)><e(z)|$$

$$-A(t,z)G_k(t|\theta)\otimes|e(z)><e(z)|a_k-A(t,z)G_k(t|\theta)^*\otimes|e(z)><e(z)|a_k^*)exp(-|z|^2)d^{2p}z)$$

$$=\sum_k(\int(G_k(t|\theta)A(t,z)z_k\otimes|e(z)><e(z)|+G_k(t|\theta)^*A(t,z)\otimes(\frac{\partial}{\partial z_k}|e(z)><e(z)|)$$

$$-A(t,z)G_k(t|\theta)\otimes(\frac{\partial}{\partial \bar{z}_k}|e(z)><e(z)|)-A(t,z)G_k(t|\theta)^*\bar{z}_k\otimes|e(z)><e(z)|)exp(-|z|^2)d^{2p}z)$$

$$=\sum_k(\int(z_kG_k(t|\theta)A(t,z)-(G_k(t|\theta)^*(\frac{\partial}{\partial z_k}-\bar{z}_k)A(t,z)$$

$$+(\frac{\partial}{\partial \bar{z}_k}-z_k)A(t,z)G_k(t|\theta)-\bar{z}_kA(t,z)G_k(t|\theta))\otimes|e(z)><e(z)|d^{2p}z)$$

Thus, we get

$$iA_{,t}(t,z)=(T(t|\theta)A)(t,z)$$

where $T(t|\theta)$ is a differential operator acting on the space of matrix valued functions of the complex variable $z\in\mathbb{C}^p$ defined by

$$T(t|\theta)X(z)=\sum_k(z_kG_k(t|\theta)X(z)-\bar{z}_kX(z)G_k(t|\theta)-G_k(t|\theta)^*(\frac{\partial}{\partial z_k}-\bar{z}_k)X(z)$$

$$+(\frac{\partial}{\partial \bar{z}_k}-z_k)X(z)G_k(t|\theta))$$

The formal solution to this partial differential equation is

$$A(t,z)=\tau\{exp(-i\int_0^t T(s|\theta)ds)\}(A(0,z))$$

where τ is the time ordering operator. The atomic (system) density at time t is given by

$$\rho_A(t)=Tr_2(\rho(t))=\pi^{-p}\int A(t,z)d^{2p}z$$

$A(t,z)$ and $\rho_A(t)$ are $N\times N$ matrices.

[34] Existence and uniqueness of solutions to stochastic differential equations.
(a) Kolmogorov's inequality for discrete time sub-martingales. Let $M_k, k = 0, 1, 2, ...$ be a non-negative sub-martingale. Let

$$\tau_a = min(k \geq 0 : M_k \geq a), a > 0$$

Then,

$$\{\tau_a = m\} = \{M_0 < a, M_1 < a, ..., M_{m-1} < a, M_m \geq 0\}$$

Thus, if $\mathcal{F}_n$ is the underlying filtration for $\{M_n\}$, it follows that

$$\{\tau_a = m\} \in \mathcal{F}_m$$

In particular, τ_a is a stop-time. Thus, by the submartinagle property, for $n \geq m$,

$$\mathbb{E}(M_n \chi_{\tau_a = m}) \geq \mathbb{E}(M_m \chi_{\tau_a = m}) \geq a.P(\tau_a = m)$$

and summing over m gives

$$\mathbb{E}(M_n) \geq \mathbb{E}(M_n \chi_{\tau_a \leq n}) \geq a.P(\tau_a \leq n)$$

or

$$P(\tau_a \leq n) = P(max_{0 \leq k \leq n} M_k \geq a) \leq \mathbb{E}(M_n)/a$$

This can be generalized to the continuous time scenario. Specifically, if $M_t, t \geq 0$ is a continuous martingale, then

$$P(max_{0 \leq s \leq t} |M_s| \geq a) \leq \mathbb{E}|M_t|/a$$

since $|M_t|$ is a submartingale by Jensen's inequality. Using this, we can deduce a version Doob's inequality:

$$\mathbb{E}(max_{0 \leq t \leq T} |M_t|^2) \leq C_1(\mathbb{E}(M_T^2)$$

Now consider the sde

$$dX_t = f(t, X_t)dt + g(t, X_t)dM_t, X_0 = x$$

where f, g satisfy appropriate Lipshitz conditions which will be specified later. We wish to prove the existence and uniqueness of the solution to this sde. We define the processes $X_t^{(n)}, n = 1, 2, ...$ recursively as

$$X_t^{(n+1)} = x + \int_0^t f(s, X_s^{(n)})ds + \int_0^t g(s, X_s^{(n)})dM_s$$

These processes are all adapted to the underlying filtration on which the process M is defined and we have by Doob's in

$$\mathbb{E}(max_{0 \leq t \leq T} |X_t^{(n+1)} - X_t^{(n)}|^2) \leq KT\mathbb{E} \int_0^T |X_s^{(n)} - X_s^{(n-1)}|^2 ds + KC_1(\mathbb{E} \int_0^T |X_t^{(n)} - X_t^{(n-1)}|^2 d < M >_t)$$

If we assume that the measure $< M >$ is absolutely continuous w.r.t. the Lebesgue measure, with bounded Radon-Nikodym derivative, then we derive from the above that

$$\mathbb{E}(max_{0 \leq t \leq T} |X^{(n+1)} - X_t^{(n)}|^2) \leq (K_0 T + K_1) \int_0^T |X_t^{(n)} - X^{(n-1)})_t|^2 dt$$

from which the existence of a solution to the sde can be inferred using standard arguments based on Gronwall's inequality.

[35] Statistical analysis of the RLS lattice algorithm. Let $x[n]$ be a random process and we form the vector

$$\xi_n = [x[n], x[n-1], ..., x[0]]^T \in \mathbb{R}^{n+1 \times 1}$$

$$z^{-p}\xi_n = [x[n-p], x[n-p-1], ..., x[-p]]^T \in \mathbb{R}^{n+1 \times 1}$$

where $x[k] = 0 for k < 0$. Define the data matrix

$$X_{n,p} = [z^{-1}\xi_n, ..., z^{-p}\xi_n] \in \mathbb{R}^{n+1 \times p}$$

Let $P_{n,p} = P_{\mathcal{R}(X_{n,p})}$. The forward and backward prediction error sequences or order p at times n and $n-1$ are respectively defined by

$$e_f[n|p] = P_{n,p}^{\perp}\xi_n, e_b[n-1|p] = P_{n,p}^{\perp} z^{-p-1}\xi_n$$

We have the obvious formula based on orthogonal direct sum decompositions:

$$P^{\perp}_{n,p+1} = P^{\perp}_{n,p} - P_{sp\{e_b[n-1|p]\}}$$

and hence

$$e_f[n|p+1] = e_f[n|p] - K_f[n|p]e_b[n-1|p] - - - (1)$$

and likewise,

$$\tilde{e}_b[n|p+1] = e_b[n-1|p] - K_b[n|p]e_f[n|p - - - (2)]$$

where

$$\tilde{e}_b[n|p+1] = P^{\perp}_{sp\{\xi_n, z^{-1}\xi_n, \dots, z^{-p}\xi_n\}} z^{-p-1}\xi_n$$

It is easy to see that

$$e_b[n|p+1] = \begin{pmatrix} \tilde{e}_b[n|p+1] \\ 0 \end{pmatrix}$$

Here, the forward reflection coefficient is

$$K_f[n|p] =< e_b[n-1|p], e_f[n|p> /E_b[n-1|p]$$

and the backward reflection coefficient is

$$K_b[n|p] =< e_b[n-1|p], e_f[n|p] > /E_f[n|p]$$

where

$$E_f[n|p] =\| e_f[n|p] \|^2, E_b[n-1|p] =\| e_b[n-1|p] \|^2$$

We thus have

$$0 \leq K_f[n|p]K_b[n|p] = | < e_b[n-1|p], e_f[n|p] > |^2/E_f[n|p]E_b[n-1|p] \leq 1$$

From (1) and (2), we get

$$E_f[n|p+1] = E_f[n|p] + K_f^2[n|p]E_b[n-1|p] - 2K_f[n|p]^2E_b[n-1|p]$$

$$= E_f[n|p] - K_f^2[n|p]E_b[n-1|p]$$
$$= (1 - K_f[n|p]K_b[n|p])E_f[n|p]$$

and likewise,

$$E_b[n|p+1] = (1 - K_f[n|p]K_b[n|p])E_b[n-1|p]$$

The time update formulas for K_f and K_p and of E_f, E_p require time update formulas for $P_{n,p}$. We have

$$P_{n,p} = X^T_{n,p}(X^T_{n,p}X_{n,p})^{-1}X_{n,p}$$

Let

$$R_{n,p} = X^T_{n,p}X_{n,p}$$

Then since

$$X_{n+1,p} = \begin{pmatrix} \eta^T_{n,p} \\ X_{n,p} \end{pmatrix}$$

where

$$\eta_{n,p} = x[n], x[n-1], ..., x[n-p+1]$$

we get

$$R_{n+1,p} = \eta_{n,p}\eta^T_{n,p} + R_{n,p}$$

and hence

$$R^{-1}_{n+1,p} = R^{-1}_{n,p} - \frac{R^{-1}_{n,p}\eta_{n,p}\eta^T_{n,p}R^{-1}_{n,p}}{1 + \eta^T_{n,p}R^{-1}_{n,p}\eta_{n,p}}$$

Thus,

$$P_{n+1,p} = [\eta_{n,p}|X^T_{n,p}]^T(R^{-1}_{n,p} - \frac{R^{-1}_{n,p}\eta_{n,p}\eta^T_{n,p}R^{-1}_{n,p}}{1 + \eta^T_{n,p}R^{-1}_{n,p}\eta_{n,p}}).$$

$$.[\eta_{n,p}|X^T_{n,p}]$$

$$\begin{pmatrix} \frac{\eta_{n,p}^T R_{n,p}^{-1} \eta_{n,p}}{1+eta_{n,p}^T R_{n,p}^{-1} \eta_{n,p}} & \frac{\eta_{n,p}^T R_{n,p}^{-1} X_{n,p}^T}{1+\eta_{n,p}^T R_{n,p}^{-1} \eta_{n,p}} \\ \frac{X_{n,p} R_{n,p}^{-1}}{1+\eta_{n,p}^T R_{n,p}^{-1} \eta_{n,p}} \eta_{n,p} & P_{n,p} - \frac{X_{n,p} R_{n,p}^{-1} \eta_{n,p} \eta_{n,p}^T R_{n,p}^{-1} X_{n,p}^T}{1+\eta_{n,p}^T R_{n,p}^{-1} \eta_{n,p}} \end{pmatrix}$$

[36] Electric dipole moment and magnetic dipole moment of an atom with an electron in a constant electromagnetic field. The unperturbed Hamiltonian of the atom is given by

$$H_0 = p^2/2m - eV(r)$$

The perturbing Hamiltonian is $-eV_1 + e^2 V_2$ where

$$V_1 = -(r, E) + (L_z + g\sigma_z)B_0/2m$$

$$V_2 = (B_0 \hat{z} \times r)^2/2m = B_0^2(x^2 + y^2)/2m$$

To calculate the eigenfunctions of $H_0 - eV_1 + e^2 V_2$ upto $O(e^2)$, we need to develop second order time independent perturbation theory for degenerate unperturbed systems. Consider therefore a Hamiltonian

$$H = H_0 + \delta H_1 + \delta^2 H_2$$

with the eigenvalues of H_0 being $E_n^{(0)}, n = 1, 2, ...$ and an orthonormal basis for the eigenspace $\mathcal{N}(H_0 - E_n)$ being $|\psi_{nk}^{(0)} >, k = 1, 2, ..., d_n$. Let

$$|\psi_n^{(0)} >= \sum_{k=1}^{d_n} c(n, k)|\psi_{nk}^{(0)} >$$

be the unperturbed state of the system. We note that this state has an eigenvalue E_n for H_0. The constants $c(n, k)$ are yet to be determined. We write for the perturbed state

$$|\psi_n >= |\psi_n^{(0)} > +\delta|\psi_n^{(1)} > +\delta^2|\psi_n^{(2)} + O(\delta^3)$$

and correspondingly for the perturbed energy level

$$E_n = E_n^{(0)} + \delta E_n^{(1)} + \delta^2 E_n^{(2)} + O(\delta^3)$$

Substituting these expansions into the eigenequation and equating coefficients of $\delta^m, m = 0, 1, 2$ successively gives

$$(H_0 - E_n^{(0)})|\psi_n^{(0)} >= 0 - - - (1)$$

which is already known,

$$(H_0 - E_n^{(0)})|\psi_n^{(1)} > +H_1|\psi_n^{(0)} > -E_n^{(1)}|\psi_n^{(0)} >= 0 - - - (2)$$

$$(H_0 - E_n^{(0)})|psi_n^{(2)} > +H_1|\psi_n^{(1)} > +H_2|\psi_n^{(0)} > -E_n^{(1)}|\psi_n^{(1)} > -E_n^{(2)}|\psi_n^{(0)} >= 0 - - - (3)$$

From (2), we get on formijng the bracket with $< \psi_{mk}^{(0)}|$ from the left,

$$(E_m^{(0)} - E_n^{(0)}) < \psi_{mk}^{(0)}|\psi_n^{(1)} > + \sum_l < \psi_{mk}^{(0)}|H_1|\psi_{nl}^{(0)} c(n, l)$$

$$-E_n^{(1)} c(n, k)\delta_{mn} = 0 - - - (4)$$

Setting $m = n$ gives us the secular equation for the possible values of $E_n^{(1)}$ that lift the degeneracy of the unperturbed state:

$$\sum_l < \psi_{nk}^{(0)}|H_1|\psi_{nl}^{(0)} > c(n, l) = E_n^{(1)} c(n, k), 1 \leq k \leq d_n$$

Thus, $E_n^{(1)}$ assumes the values $E_{nj}^1, k = 1, 2, ..., d_n$ which are the eigenvalues of the $d_n \times d_n$ secular matrix $((< \psi_{nk}^{(0)}|H_1|\psi_{nl}^{(0)} >))_{1 \leq k,l \leq d_n}$ with the eigenvector corresponding to the eigenvalue $E_{nj}^{(1)}$ being denoted by $((c_j(n, k)))_{k=1}^{d_n}$. We may assume that these d_n eigenvectors form an orthonormal basis for $\mathbb{C}^{d_n}$. From (4) with $m \neq n$, we get

$$< \psi_{mk}^{(0)}|\psi_n^{(1)} >= \sum_l \frac{< \psi_{mk}^{(0)}|H_1|\psi_{nl}^{(0)} > c(n, l)}{E_n^{(0)} - E_m^{(0)}}$$

and hence the first orde perturbation to the eigenvector $|\psi_n^{(0)}>$, namely $\delta.|\psi_{nj}^{(1)}>$ corresponding to the perturbed eigenvalue $E_n^{(0)}+\delta.E_{nj}^{(1)}$ is given by

$$|\psi_{nj}^{(1)}>=\sum_{mkl,m\neq n}|\psi_{mk}^{(0)}><\psi_{mk}^{(0)}|H_1|\psi_{nl}^{(0)}>c_j(n,l)/(E_n^{(0)}-E_m^{(0)})$$

Turning now to (3), we assume $E_n^{(1)}=E_{nj}^{(1)}$ and $|\psi_n^{(1)}>=|\psi_{nj}^{(1)}>$. Taking the bracket of this equation with $<\psi_{mk}^{(0)}|$, we get

$$(E_m^{(0)}-E_n^{(0)})<\psi_{mk}^{(0)}|psi_n^{(2)}>+<\psi_{mk}^{(0)}|H_1|\psi_{nj}^{(1)}>+<\psi_{mk}^{(0)}|H_2|\psi_n^{(0)}>-E_{nj}^{(1)}<\psi_{mk}^{(0)}|\psi_{nj}^{(1)}>-E_n^{(2)}c_j(n,k)\delta_{mn}=0$$

For $m=n$, this gives

$$<\psi_{nk}^{(0)}|H_1|\psi_{nj}^{(1)}>+\sum_l<\psi_{nk}^{(0)}|H_2|\psi_{nl}>c_j(n,l)-E_{nj}^{(1)}<\psi_{nk}^{(0)}|\psi_{nj}^{(1)}>$$

$$-E_n^{(2)}c_j(n,k)=0---(5)$$

Actually, these equations are not all linearly independent. The only linearly independent equation for $E_n^{(2)}$ which emerges from this is obtained by forming the bracket of (3) with $+<\psi_n^{(0)}|=\sum_k\bar{c}_j(n,k)<\psi_{nk}^{(0)}|$. In fact, if we extend the conjugate of this vector to an onb for $\mathbb{C}^{d_n}$ and form the bracket with these vectors, then the term involving $E_n^{(2)}$ disappears. Thus we infer from from (5) that

$$E_n^{(2)}=E_{nj}^{(2)}=$$

$$(\sum_k|c_j(n,k)|^2)^{-1}[\sum_k\bar{c}_j(n,k)<\psi_{nk}^{(0)}|H_1|\psi_{nj}^{(1)}>+\sum_{k,l}<\psi_{nk}^{(0)}|H_2|\psi_{nl}^{(0)}>\bar{c}_j(n,k)c_j(n,l)$$

$$-E_{nj}^{(1)}\sum_k\bar{c}_j(n,k)<\psi_{nk}^{(0)}|\psi_{nj}^{(1)}>]$$

[37] Induced characters: Let G be a finite group and H a subgroup of G. Select one element x in each left coset of H. Let I denote the set of all such $x's$. Thus, we have $\bigcup_{x\in I}xH=G$ and $xH\bigcap yH=\phi,x\neq y,x,y\in I$. Let L be a representation of H acting in the vector space V. We shall formally write $x.V$ for the vector space V attached to the element x. The representation space for the representation $\pi=Ind_H^GL$ (ie π is the representation of G induced by the representation L of H) may be denoted by $U=\bigoplus_{x\in I}x.V$. $\pi(g)$ acts on this space by mapping $x.v$ to $[gx].v$ where $g\in G,x\in I,v\in V$ and $[gx]\in I$ is the element for which $[gx]H=gxH$. We may use the notation $gxV=[gx]V$. Let χ_U denote the character of π and χ_V that of L. It is clear that $g\in G$ will map $x.V$ onto itself iff $gxV=xV$ iff $x^{-1}gxV=V$ iff $x^{-1}gx\in H$. For a given $g\in G$, let X_g denote the set of all such $x's$. In other words,

$$X_g=\{x\in I:x^{-1}gx\in H\}$$

It follows that

$$\chi_U(g)=\sum_{x\in X_g}\chi_V(x^{-1}gx)$$

We note that for any $x,g\in G$, $x^{-1}gx\in H$ iff $y^{-1}gy\in H$ for all $y\in G(g).x$ and in this case, $x^{-1}gx=y^{-1}gy$, where $G(g)$ is the centralizer of g in G. Thus, we can write for $x\in X_g$,

$$\chi_V(x^{-1}gx)=\mu(G(g))^{-1}\sum_{y\in G(g)x}\chi_V(y^{-1}gy)$$

where $\mu(G(g))$ denotes the number of elements in $G(g)$. It follows that

$$\chi_U(g)=\mu(G(g))^{-1}\sum_{x\in X_g,y\in G(g)x}\chi_V(y^{-1}gy)$$

Further, it is clear that $h\in H$ and $y^{-1}gy\in H$ implies $(yh)^{-1}gyh=h^{-1}y^{-1}gyh\in H$ and hence $\chi_V((yh)^{-1}gyh)=\chi_V(y^{-1}gy)$. Thus, the above formula can also be expressed as

$$\chi_U(g)=\mu(G(g))^{-1}\mu(H)^{-1}\sum_{x\in X_gH,y\in G(g)x}\chi_V(y^{-1}gy)$$

$$= (\mu(G(g))\mu(H))^{-1} \sum_{y\in G(g)X_gH} \chi_V(y^{-1}gy)$$

Now, consider the set of conjugacy classes of $g \in G$:

$$C(g) = \{xgx^{-1} : x \in G\}$$

$C(g)$ contains

$$\mu(C(g)) = \mu(G)/\mu(G(g))$$

distinct elements.

Reference: Claudio Procesi, "Lie groups, an approach through invariants and representations", Springer.

A.5. Some more problems in group theory and quantum mechanics.

[1] The basic observables of non-relativistic quantum mechanics obtained from the unitary representations of the Galilean group based on Mackey's theory of semidirect products. Suppose N is a vector space regarded as an Abelian group under addition. Let H be a group that acts on N as $n \rightarrow \tau_h(n)$. Let $G = N \otimes_s H$. Thus, any element $g \in G$ can be uniquely expressed uniquely as $g = nh, n \in N, h \in H$ and the composition law in G is given by $n_1h_1n_2h_2 = n_1\tau_{h_1}(n_2)h_2$. We have

$$\tau_{h_1h_2} = \tau_{h_1}o\tau_{h_2}, \tau_h(n+n') = \tau_h(n) + \tau_h(n')$$

In other words, $h \rightarrow \tau_h$ is an homomorphism of H into $aut(N)$ with $aut(N)$ being the same as $End(N)$ ($aut(N)$ is a group theoretic notation while $End(N)$ is a vector space theoretic notation. The composition of n_1h_1 and n_2h_2: Then we may regard N as being normalized by the action τ of H. Equivalently, via the construction of a group isomorphism, we can regard τ_h as begin given by $\tau_h(n) = hnh^{-1}$, so that

$$n_1h_1n_2h_2 = (n_1 + \tau_{h_1}(n_2)).h_2$$

Now let $B(n_1, n_2)$ be a skew symmetric real bilinear form on N that is H-invariant, ie,

$$B(\tau_h(n_1), \tau_h(n_2)) = B(n_1, n_2)$$

Then consider

$$\sigma(n_1h_1, n_2h_2) = exp(iB(n_1, \tau_{h_1}(n_2)))$$

We claim that σ satisfies the conditions for a multiplier on G, ie, for a projective unitary (pu) representation U of G, ie,

$$\sigma(g_1, g_2)\sigma(g_1g_2, g_3) = \sigma(g_1, g_2g_3)\sigma(g_2, g_3) - - - (1)$$

We leave this verification to the reader. This follows from the fact that if U satsifies by the definition of a pu representation

$$U(g_1)U(g_2) = \sigma(g_1, g_2)U(g_1g_2)$$

and hence by associativity of linear operator multiplication,

$$(U(g_1)U(g_2))U(g_3) = U(g_1)(U(g_2)U(g_3))$$

we get

$$\sigma(g_1, g_2)U(g_1g_2)U(g_3) = U(g_1)U(g_2g_3)\sigma(g_2, g_3)$$

or

$$\sigma(g_1, g_2)\sigma(g_1g_2, g_3)U(g_1g_2g_3) = \sigma(g_1, g_2g_3)\sigma(g_2, g_3)U(g_1g_2g_3)$$

ie (1).

[2] Let $f(r_1, ..., r_N)$ and $g(r_1, ..., r_N)$ be two functions on $(\mathbb{R}^3)^N$. Let $R \in SO(3)$ and $\sigma, \tau, \rho \in S_N$. Let $\chi_\lambda(\sigma)$ be an irreducible character of S_n corresponding to the Young tableaux (ie a partition of n) λ and consider

$$I(f, g, r_1, ..., r_N) = \sum_{\sigma,\tau\in S_n} f(r_{\sigma 1}, ..., r_{\sigma n})\bar{g}(r_{\tau 1}, ..., r_{\tau n})\chi_\lambda(\sigma\tau^{-1})$$

We have

$$I(f, g, r_{\rho 1}, ..., r_{\rho N}) = \sum_{\sigma,\tau} f(r_{\rho\sigma 1}, ..., r_{\rho\sigma n})\bar{g}(r_{\rho\tau 1}, ..., r_{\rho\tau n})\chi_\lambda(\sigma\tau^{-1})$$

$$= \sum_{\sigma,\tau} f(r_{\sigma 1},...,r_{\sigma n})\bar{g}(r_{\tau_1},...,r_{\tau n})\chi_\lambda(\rho\sigma\tau^{-1}\rho^{-1})$$

$$= I(f,g,r_1`,...,r_N)$$

since

$$\chi(\rho\sigma\rho^{-1}) = \chi(\sigma)$$

for any character χ of S_n. More generally, we can define

$$I(f,g,r_1,...,r_N,r'_1,...,r'_N) = \sum_{\sigma,\tau\in S_n} f(r_{\sigma 1},...,r_{\sigma n})\bar{g}(r'_{\tau 1},....,r'_{\tau n})\chi_\lambda(\sigma\tau^{-1})$$

Then, we get

$$I(f,g,r_{\rho 1},...,r_{\rho n},r'_{\rho 1},...,r'_{\rho n}) = \sum_{\sigma,\tau} f(r_{\rho\sigma 1},...,r_{\rho\sigma n})\bar{g}(r'_{\rho\tau 1},...,r'_{\rho\tau n})$$

$$.\chi_\lambda(\sigma\tau^{-1}) =$$

$$= \sum_{\sigma,\tau} f(r_{\sigma 1},...,r_{\sigma n})\bar{g}(r'_{\tau 1},...,r'_{\tau n})$$

$$\chi_\lambda(\sigma\tau^{-1}) = I(f,g,r_1,...,r_N,r'_1,...,r'_N)$$

$\forall r_1,...,r_N,r'_1,...,r'_N \in \mathbb{R}^3$. Now,let $R \in SO(3)$. The rotated and permuted image fields obtained from f and g are

$$f_1(r_1,...,r_N) = f(R^{-1}r_{\rho 1},...,R^{-1}r_{\rho N}), g_1(r'_1,...,r'_N) = g(R^{-1}r'_{\rho 1},...,R^{-1}r'_{\rho N})$$

and we get

$$I(f_1,g_1,r_1,...,r_N,r'_1,...,r'_N) = I(f,g,R^{-1}r_1,...,R^{-1}r_N,R^{-1}r'_1,...,R^{-1}r'_N) =$$

and hence if χ_l denotes an irreducible character of $SO(3)$, we get

$$\int_{SO(3)\times SO(3)} I(f_1,g_1,S_1r_1,...S_1r_N,S_2r'_1,...,S_2r'_N)\chi_l(S_1S_2^{-1})dS_1dS_2$$

$$= \int_{SO(3)\times SO(3)} I(f,g,R^{-1}S_1r_1,...,R^{-1}S_1r_N,R^{-1}S_2r'_1,...,R^{-1}S_2r'_N)\chi_l(S_1S_2^{-1})dS_1dS_2$$

$$= \int_{SO(3)\times SO(3)} I(f,g,S_1r_1,...,S_1r_N,S_2r'_1,...,S_2r'_N)\chi_l(RS_1S_2^{-1}R^{-1})dS_1dS_2$$

$$= \int_{SO(3)\times SO(3)} I(f,g,S_1r_1,...,S_1r_N,S_2r'_1,...,S_2r'_N)\chi_l(S_1S_2^{-1})dS_1dS_2$$

It follows that

$$I_0(f,g,r_1,...,r_N,r'_1,...,r'_N) = \int_{SO(3)\times SO(3)} I(f,g,S_1r_1,...S_1r_N,S_2r'_1,...,S_2r'_N)\chi_l(S_1S_2^{-1})dS_1dS_2$$

is invariant under permutations and rotations.

References for A.4 and A.5.:
[1] Hoffman and Kunze, Linear Algebra, Prentice Hall.
[2] T.Kato, Perturbation theory for linear operators, Springer.
[3] K.R.Parthasarathy, "An introduction to quantum stochastic calculus", Birkhauser.
[4] S.J.Orfanidis, "Optimum signal processing", Prentice Hall.
[5] C.R.Rao, "Linear statistical inference and its applications, Wiley.
[6] J.Gough and Koestler, "Quantum filtering in coherent states".
[7] Lec Bouten, "Filtering and control in quantum optics", Ph.D thesis.
[8] Leonard Schiff, "Quantum mechanics".
[9] W.O.Amrein, "Hilbert space methods in quantum mechanics".
[10] K.R.Parthasarathy, "Coding theorems of classical and quantum information theory". Hindustan Book Agency.
[11] Naman Garg, H.Parthasarathy and D.K.Upadhyay, "Estimating parameters of an image field modeled as a quantum electromagnetic field using its interaction with a finite state atomic system", Technical Report, NSIT, 2016.

[12] Naman Garg, H.Parthasarathy and D.K.Upadhyay, "MATLAB implementation of Hudson-Parthasarathy noisy Schrodinger equation and Belavakin's quantum filtering equation with analysis of entropy evolution", Technical Report, NSIT, 2016.

A.6. New Syllabus for a short course on transmission lines and waveguides.

[1] Study of non-uniform transmission lines by expanding the distributed parameters and the line voltage and current as Fourier series in the spatial variable z. The modes of propagation (propagation constants) appear as the eigenvalues of an infinite matrix defined in terms of the spatial Fourier series coefficients of the non-uniform line impedance and admittance.

[2] Study of the statistics of the line voltage and current (spatial correlations) when the distributed parameters of the line have small random fluctuations. The study is based on perturbation theory applied to matrix eigenvalue problems and is very similar to time independent perturbation theory used in quantum mechanics.

[3] Analysis of transmission lines when the distributed parameters are randomly fluctuating functions of both space and time. We focus on estimating the distributed parameter statistical correlations from measurements of the line voltage and current and applying the ergodic hypothesis for estimating the line voltage and current correlations and then matching these correlations with the theoretically derived correlations.

[4] Analysis of transmission lines with random loading along the line using infinite dimensional stochastic differential equations. The voltage and current loading along the line are assumed to be expandable in terms of basis functions of the spatial variable with the coefficients being white noise processes in time, ie, derivatives of Brownian motion processes. We then calculate the probability law of the line voltage and current

[5] Equivalence of transmission lines and waveguides obtained by expanding the guide electric and magnetic fields in terms of basis functions of (x, y) and the coefficients being functions of z. From the Maxwell equations, we derive an infinite series of first order linear differential equations for the coefficient functions of z and compare these equations with an infinite sequence of coupled transmission lines. The basis functions of (x, y) used in the expansion of the electric and magnetic fields must satisfy the boundary conditions, namely, that E_z and the normal derivative of H_z vanish on the boundary.

[6] Study of nonlinear hysteresis and nonlinear capacitive effects on the dynamics of a transmission line. Hysteresis is related to a nonlinear $B - H$ curve having memory and is a consequence of the Landau-Lifshitz theory of magnetism in which the magnetic moment of an atom precesses in an external magnetic field due to the $M \times H$ torque on it. The solution to this equation is a Dyson series for the magnetization M in terms of H and truncated upto second degree terms, this leads to a quadratic expression for the hysteresis voltage term as a function of the line current. In other words, we have a second order Volterra relation between the hysteresis voltage and current. This is a consequence of magnetic properties of the material of which the line is made. Nonlinear capacitive effects can be explained from the nonlinear-memory relation between the dipole moment/polarization of an electron relative to its nucleus when an external electric is incident on it. The binding of the electron to the atom has harmonic as well as anharmonic terms which causes the differential equation satisfied by the position of the electron to be nonlinear and hence solving this equation using perturbation theory, we obtain the dipole moment as a Volterra series in the external electric field. When applied to transmission lines, this manifests itself as a Volterra relation between the line charge (which is proportional to the electric displacement vecto $D = \epsilon_0 E + P$ where P is the polarization/dipole moment per unit volume). The time derivative of this charge is the capacitor current and this component is incorporated into the line current equation and analysis is done using perturbation theory.

[7] Quantization of the line equations using the Gorini-Kossakowski-Sudarshan-Lindblad (GKSL) formalism. The line equations for a lossless line are distributed parameter analogs of LC circuits. The lossless line equations like an LC circuit can be derived from a Lagrangian and hence from a Hamiltonian that is a quadratic function of the phase variables. The effect of noise on such a system is obtained by adding a GKSL term to the dynamics of states and observables. These GKSL terms can natually be obtained using the Hudson-Parthasarathy quantum stochastic calculus by tracing out over the bath variables. By choosing our GKSL operators L as complex linear functions of the phase variables, we obtain resistive damping terms in the dynamical equations and hence we are able to obtain a quantum mechanical model for a lossy line.

A.7. Creativity in the mathematical, physical and the engineering sciences.

In this section, we give a brief history of the various intellectual achievements in the mathematical, physical and engineering sciences showing how creativity in these sciences very often comes from a need to understand nature and the working of the world around us. The examples we choose are Newton, Maxwell, Einstein, Planck, Bose, Rutherford, Bohr, Heisenberg, Schrodinger, Dirac, Dyson, Feynman, Schwinger, Tomonaga, Weinberg, Salam, Glashow and Hawking in the physical sciences, Faraday and Edison and the inventor Esaki of the tunnel diode in the engineeering sciences and in the mathematical sciences, Euler, Gauss, Fourier, Fermat, Galois, Abel, Cauchy, Hilbert, Poincare, Kolmogorov,

Ramanujan, Harish-Chandra and more recently, Edward Witten, a pioneer in mathematical string and superstring theory.

The creation of quantum electrodynamics: After the creation of the quantum theory of atoms and molecules, there remained several gaps in our understanding of the physical world. For example, it was not clear how to explain various experimental observations like the force of an electron on itself, the electron self energy ie, the movement of an electron produces an em field which acts back on the electron, the phenomenon of vacuum polarization according to which a photon propagates in vacuum to produce an electron-positron pair which propagate and again annihilate each other to produce once again a photon, the anomalous magnetic moment of the electron which gives radiative corrections to the magnetic moment caused once again by the the em field generated by the electron acting back on itself. Compton scattering of an electron/positron by a photon also remained unexplained. In other words, a satisfactory quantum theory describing the interactions of electrons, positrons and photons and how to calculate probabilities of scattering processes of these particles remained to be carried out. Thus because of the need to understand these unexplained physical processes, Feynman, Schwinger and Tomonaga created new mathematical tools which eventually were sharpened by Dyson and the succeeding generation of theoretical physicists like Wienberg, Salam, Glashow, Witten, and more recently by the Indian physicists Sen and Ashtekar. Feynman proposed a path integral formulation for calculating the scattering matrix for particles. This involved identifying the Lagrangian density $\mathcal{L}_0[\phi]$ of the electron-positron Dirac field without their interactions and that of the electromagnetic field and then evaluating the action $S[\phi] = \int \mathcal{L}d^4x$ for these fields. Feynman followed it with evaluation of the Gaussian path integrals $\int exp(iS[\phi])\mathcal{D}\phi$ taking into account the Berezin change for Fermionic path integrals. Then the interaction term $S_{int}[\phi] = \int J^\mu A_\mu d^4x = \int \psi^* \gamma^0 \gamma^\mu \psi A_\mu d^4x$ $(\phi = (A_\mu, \psi))$ is considered and its contribution was evaluated by expanding the exponential $exp(iS_{int}[\phi])$ in as a power series in $S[\phi]$ and Feynman associated a diagram with each term in this series explaining how each term gives rise to a different order term in the scattering matrix and how these terms can be calculated easily by a diagrammatic algorithm. On the other hand Schwinger and Tomonaga proposed an operator theoretic approach to calculating the scattering matrix. Their algorithm was based on first quantizing the electromagnetic field using creation and annihilation operators. This had already been observed by Paul Dirac when he wrote down the energy of the electromagnetic field as a quadaratic function of the four vector potential in the spatial frequency domain and deduced that this quadratic structure meant that the em field should be considered as an ensemble of harmonic oscillators with two oscillators associated with each spatial frequency (two degrees of polarization arise from the fact that in the coulomb gauge, in the absence of charges, the electric scalar potential is zero while $divA = 0$ for the Coulomb gauge implies that in the spatial frequency domain, the magnetic vector potential is orthogonal to the wave vector). The next idea of Schwinger was to substitute this quantum electromagnetic field consisting of a superposition of operators into the atomic Hamiltonian described by position and momentum operators and obtain an interaction term between the position-momentum pair for the atom with the quantum electromagnetic field operators. This interaction Hamiltonian was then used to calculate the scattering matrix elements and deduce corrections to the magnetic moment of the electron. Schwinger and Tomonaga also proposed a Lorentz invariant scheme for writing down the Schrodinger equation (which is not Lorentz covariant) for field theories. The idea basically involved replacing the time t variable by a three dimensional surface variable σ. In other words, just as the $t = constt$ surface is the three dimensional Euclidean space $\mathbb{R}^3$ as a subspace of $\mathbb{R}^4$, likewise, $\sigma = constt.$ could be an arbitrary three dimensional submanifold of $\mathbb{R}^4$. The Schrodinger equation $i\frac{\partial\psi(t)}{\partial t} = H(t)\psi(t)$ was replaced by the Lorentz covariant equation

$$i\frac{\delta\psi(\sigma)}{\delta\sigma} = H(\sigma)\psi(\sigma)$$

and this formalism was used with great power by Schwinger and Tomonaga to calculate the scattering matrix element between two three dimensional surfaces. Unification of the Feynman and Schwinger-Tomonaga theory was performed by Freeman Dyson who simply showed why one should obtain the same results using Feynman diagrams and the operator theoretic approach. For a very long time, Dyson's notes on this was the standard textbook for all courses in quantum field theory all over the world. Dyson's work led to renormalization theory developed by himself and subsequently othe researchers. This involved getting rid of the infinities in quantum field theory by renormalizing charge, mass and fields, ie, scaling these quantities with numbers depending on an ultraviolet and infrared cutoff while integrating in the four frequency domain. After the great riddle of construcing a cogent quantum theory of electrodynamics was solved almost completely by these four powerful mathematical physicists, the problem of understanding nuclear weak and strong forces remained and also how to unify these with quantum electrodynamics. The problem of unifying quantum electrodynamics with the weak forces, called the Electroweak theory was successfully solved by Weinberg, Salam and Glashow using group theoretic formalism, more precisely the $SU(2) \times U(1)$ formalism. Both the weak forces and electromagnetic forces appeared as gauge fields in this theory. Principles of symmetry breaking were used in this unification giving rise to mass of electrons and other nuclear particle. Goldstone had a say in this unification when he proposed the idea of how when a Lagrangian of fields that is invariant under a group G has a vacuum state that is not G-invariant because of degeneracies of the vacuum state, the symmetry of the Lagrangian as viewed from the vacuum state is broken to s smaller group $H \subset G$ and associated with each degree of broken symmetry is a massless particle called a massless Goldstone boson. The unbroken symmetries correspond to massive particles. Symmetry can also be broken by adding a term to the G-invariant Lagrangian density. This is what happens in the electroweak theory. The electroweak-strong unification was achieved by Gell-Mann and Nee-Mann who based their theory on the group $SU(3) \times SU(2) \times U(1)$. The idea is to derive all the coupling constants of electrdynamics, the weak and the strong theories from one unified theory. The entire idea of unifying gauge fields is based on the basic principle of Yang and Mills, namely that one can construct a covariant derivative $\nabla_\mu = \partial_\mu + iA_\mu(x)$ acting on a vector space $(\mathbb{C}^n)$ valued function of x in such a way that the gauge field $A_\mu(x)$ takes values in a Lie algebra $\mathfrak{g}$ of a subgroup G of the unitary group $U(n)$. The wave function on which ∇_μ acts is $\mathbb{C}^n$. Further, this covaraiant derivative satisfies the property that under a local G-transformation by $g(x) \in G$, the gauge field $A_\mu(x)$ which takes values in $\mathfrak{g}$ transforms in such a way to $A'_\mu(x)$ so that

$$g(x)(\partial_\mu + A'_\mu(x)) = (\partial_\mu + A_\mu(x))g(x)$$

or equivalently as

$$g(x)\nabla'_\mu = \nabla_\mu g(x)$$

where both sides are regarded as first order differential operators acting on wave functions $\psi(x) \in \mathbb{C}^n$. This gives

$$iA'_\mu(x) = g(x)^{-1}\partial_\mu g(x) + ig(x)^{-1}A_\mu(x)g(x)$$

This idea of gauge transformation in which the massive field wave function $\psi(x) \in \mathbb{C}^n$ transforms to $g(x)\psi(x)$ while the massless gauge field $A_\mu(x)$ transforms in the above way leads to the conclusion that given a Lagrangian density of the form $\mathcal{L}(\psi(x), \nabla_\mu\psi(x))$ that is $G-invariant$, ie, $\mathcal{L}(g\psi(x), g\nabla_\mu\psi(x)) = \mathcal{L}(\psi(x), \nabla_\mu\psi(x))$ for all $g \in G$, it follows that $\mathcal{L}$ is invariant also under local G-actions $g(x)$ provided that the $A_\mu(x)$ sitting inside the covariant derivative ∇_μ is also subject to the above gauge transformation. When the group G is the Abelian group $U(1)$, $A_\mu(x)$ is simply a real valued function for each μ and the above gauge transformation reduces to the Lorentz gauge transformation for the electromagnetic potentials. Thus, the non-commutative Yang-Mills theory provides a sweeping generalization of the commutative em field theory. It can also be applied to describe the interaction of the Dirac field $\psi(x)$ with a noncommutative gauge field with the electromagnetic potentials also coming as an extra component of the gauge potential. Although this idea of Yang and Mills is a purely group theoretic construction, it turned out to be one of the most fruitful constructs for unifying almost all the quantum particle fields. What remains now is the development of a quantum theory of gravity which would enable one to associate a particle which we may call a graviton and to describe its interaction with other quantum particles like the photon, electron, positron, and the propagators of the weak and strong forces. It should be borne in mind that the action of a classical gravitational field on any quantum particle described by a relativistic wave equation like the Dirac equation can be achieved using the Idea of Yang and Mills, namely by introducing a gravitational connection $\Gamma_\mu(x)$ which is a matrix. For example, if the gravitational field is described by a tetrad $V^\mu_a(x)$ so that the metric is $g^{\mu\nu}(x) = \eta^{ab}V^\mu_a(x)V^\nu_b(x)$ with η^{ab} being the Minkowski metric of flat space-time, then this tetrad can be understood as a local transformation of curved space-time to an inertial frame. We then construct a covariant derivative $\partial_\mu + \Gamma_\mu(x)$ and transform it locally to an inertial frame by using $V^\mu_a(x)(\partial_\mu + \Gamma_\mu(x))$ and setting up the Dirac equation in a gravitational field as

$$[i\gamma^a V^\mu_a(x)(\partial_\mu + \Gamma_{mu}(x)) - m]\psi(x) = 0$$

where $\gamma^a, a = 0, 1, 2, 3$ are the usual Dirac gamma matrices. To qualify as a valid general relativistic wave equation, this must be invariant under local Lorentz transformations $\Lambda(x)$. Under such a local Lorentz transformation, $\psi(x)$ transforms to $D(\Lambda(x))\psi(x)$ where D is the Dirac representation of the Lorentz group and its Lie algebra generators are $J^{ab} = \frac{1}{4}[\gamma^a, \gamma^b]$. Suppose that under such a local Lorentz transformation $\Gamma_\mu(x)$ which is a 4×4 matrix changes to $\Gamma'_\mu(x)$. Then we should have with $\psi'(x) = D(\Lambda(x))\psi(x)$, the equation

$$[i\gamma^a \Lambda^b_a(x)V^\mu_b(x)(\partial_\mu + \Gamma'_\mu(x) - m]\psi'(x) = 0$$

where the change of space-time coordinates has been accounted for by the factor matrix elements $\Lambda^b_a(x)$ of the local Lorentz transformation of the inertial frame index a in the tetrad frame $V^\mu_a(x)$. Using the identity

$$D(\Lambda)\gamma^b D(\Lambda)^{-1} = \Lambda^b_a\gamma^a$$

we get from the above

$$[D(\Lambda(x))(i\gamma^b V^\mu_b(x)(D(\lambda(x))^{-1}(\partial_\mu + \Gamma'_\mu(x)) - m]D(\Lambda(x))\psi(x) = 0$$

This is equivalent to

$$i\gamma^b V^\mu_b(x)((D(\Lambda(x))^{-1}\partial_\mu D(\Lambda(x))) + \partial_\mu + D(\Lambda(x))^{-1}\Gamma'_\mu(x)D(\Lambda(x))) - m]\psi(x) = 0$$

It follows that for this to coincide with the untransformed Dirac equation in curved space-time, the connection $\Gamma_\mu(x)$ of the gravitational field in the Dirac representation should transform to $\Gamma'_\mu(x)$ where

$$D(\Lambda(x))^{-1}\Gamma'_\mu(x)D(\Lambda(x)) + (D(\Lambda(x))^{-1}\partial_\mu D(\Lambda(x))) = \Gamma_\mu(x)$$

or equivalently,

$$\Gamma'_\mu(x) = D(\Lambda(x))\Gamma_\mu(x)D(\Lambda(x))^{-1} - (\partial_\mu D(\Lambda(x)))D(\Lambda(x))^{-1}$$

Such a connection has been constructed and is given by

$$\Gamma_\mu(x) = \frac{1}{2}J^{ab}V_a^\nu V_{b\nu:\mu}$$

(Reference: Steven Weinberg, "Gravitation and Cosmology, Principles and Applications of the General Theory of Relativity", Wiley.)

Gravity was unified with classical electromagnetism by Einstein in this beautiful theory the general theory of relativity. This theory said that gravity is not a force, it is simply a curvature of space-time and when matter moves in such a curved space time, follows geodesics which are shortest paths on the curved four dimensional manifold of space-time. These shortest paths are curved because any path on a curved surface is curved. By saying that geodesics are curved, we mean that the relation between the spatial and time coordinates of a moving particle is nonlinear and hence the motion appears to us as being accelerated motion.

A.8 Classification and representation theory of semisimple Lie algebras.

By Serre's theorem, a complex semisimple Lie algebra L is generated by $3n$ generators $e_i, h_i, f_i, i = 1, 2, ..., n$ satisyfying the commutation relations

$$[e_i, e_j] = [f_i, f_j] = [h_i, h_j] == [e_i, f_j] = 0,$$

$$[e_i, f_j] = \delta_{ij}h_i, [h_i, e_j] = a_{ij}e_j, [h_i, f_j] = -a_{ij}f_j$$

where a_{ij} are integers called the Cartan integers. This result of Serre follows from the Cartan's theory which says that L has a maximal Abelian subalgebra $\mathfrak{h}$ (Called a Cartan Algebra) and that any two maximal Abelian subalgebras are mutually conjugate. This result is not true for real semisimple Lie algebras where there can be more than one non-conjugate Cartan algebras. The reprsentation theory for real semisimple Lie algebras was developed almost single handedly by the great Indian mathematician Harish-Chandra who derived generalizations of the character formula of H.Weyl using the theory of distributions and also obtained the complete Plancherel formula for such algebras by introducing in addition to the principal and supplementary series of irreducible representations (introduced by Gelfand for obtaining the Plancherel formula for complex semisimple Lie groups), the discrete series of irreducible representations.

Coming back to the theme of complex semisimple Lie algebras, Cartan proved that the elements of $\mathfrak{h}$ in the adjoint representation, act in a semsimple way on L, ie, each operator $ad(H), H \in \mathfrak{h}$ acting on the vector space L is diagonable. It follows from basic Linear algebra, that the operators $ad(H), H \in \mathfrak{h}$ are simulatneously diagonable and hence we have the following direct sum decomposition of L:

$$L = \mathfrak{h} \otimes \bigotimes_{\alpha\in\Phi} L_\alpha$$

where Φ is the set of all roots of L, ie, for each $\alpha \in \Phi$ and $X \in L_\alpha$, we have

$$[H, X] = \alpha(H)X$$

We are here defining

$$L_\alpha = \{X \in L : [H, X] = \alpha(H)X\forall H \in \mathfrak{h}\}$$

α is a non-zero linear functional on $\mathfrak{h}$, ie, $\alpha \in \mathfrak{h}^*$ and all the $\alpha's$ are distinct linear functionals. Moreover, Cartan's theory states that there exists a subset $\Delta \subset \Phi$ called a set of simple roots such that any $\alpha \in \Phi$ is either a purely positive or purely negative integer linear combination of elements of Δ. This means that writing $\Delta = \{\alpha_1, ..., \alpha_p\}$, we have that for any $\alpha \in \Phi$, there exist integers $m_1, ..., m_p$ such that $m_j \geq 0\forall j$ or $m_j \leq 0\forall j$ and

$$\alpha = \sum_{j=1}^{p} m_j\alpha_j$$

and further, no $\alpha \in \Delta$ has such a decomposition. Actually, there exist many such sets Δ. Further, $dimL_\alpha = 1\forall\alpha \in \Phi$. This follows from the following elementary argument, From the Jacobi identity, $[L_\alpha, L_\beta] \subset L_{\alpha+\beta}, \alpha, \beta \in \Phi$. Thus

$[L_\alpha, L_{-\alpha}] \subset \mathfrak{h}$. Choose an $e_\alpha \in L_\alpha, f_\alpha \in L_{-\alpha}$ such that $B(e_\alpha, f_\alpha) = b(\alpha)$ where $B(X,Y) = Tr(ad(X)ad(Y))$ and $b(\alpha)$ is a constant to be chosen appropriately. Cartan proved that B defines an non-degenerate symmetric bilinear form on L (only if L is semisimple). Now define $t_\alpha = a(\alpha)[e_\alpha, f_\alpha]$ where $a(\alpha)$ is a constant to be chosen appropriately. Then, $t_\alpha \in \mathfrak{h}$ and

$$[t_\alpha, e_\alpha] = \alpha(t_\alpha)e_\alpha, [t_\alpha, f_\alpha] = -\alpha(t_\alpha)f_\alpha$$

and more generally, for any $\alpha, \beta \in \Phi$,

$$[t_\alpha, e_\beta] = \beta(t_\alpha)e_\beta,$$

$$[t_\alpha, f_\beta] = -\beta(t_\alpha)f_\beta$$

Consistency is checked as follows using $B([X,Y], Z) = B(X, [Y, Z])$:

$$= b(\beta)\beta(t_\alpha) = \beta(t_\alpha)B(e_\beta, f_\beta) = B([t_\alpha, e_\beta], f_\beta)$$

$$= -B(e_\beta, [t_\alpha, f_\beta]) = \beta(t_\alpha)B(e_\beta, f_\beta) = b(\beta)\beta(t_\alpha)$$

We further have

$$B(t_\alpha, t_\beta) = a(\alpha)B([e_\alpha, f_\alpha], t_\beta)$$

$$= a(\alpha)B(e_\alpha, [f_\alpha, t_\beta]) = a(\alpha)\alpha(t_\beta)B(e_\alpha, f_\alpha) = a(\alpha)b(\alpha)\alpha(t_\beta)$$

and we denote this by (α, β). The above formula implies that

$$(\alpha, \beta) = (\beta, \alpha) = a(\alpha)b(\alpha)\alpha(t_\beta) = a(\beta)b(\beta)\beta(t_\alpha)$$

Now define

$$h_\alpha = 2t_\alpha/(\alpha, \alpha)$$

Then,

$$[h_\alpha, e_\alpha] = 2\alpha(t_\alpha)e_\alpha/(\alpha, \alpha) = 2e_\alpha$$

provided that we choose the $a(\alpha)'s$ and the $b(\alpha)'s$ so that

$$\alpha(t_\alpha) = (\alpha, \alpha)$$

For such a choice of the constants, we also have

$$[h_\alpha, f_\alpha] = -2f_\alpha,$$

and

$$[e_\alpha, f_\alpha] = t_\alpha/a(\alpha) = (\alpha, \alpha)h_\alpha/(2a(\alpha)) = h_\alpha$$

provided that we choose the $a(\alpha)'s$ and the $b(\alpha)'s$ so that

$$(\alpha, \alpha) = 2a(\alpha) = \alpha(t_\alpha)$$

In other words, we have found $e_\alpha \in L_\alpha, f_\alpha \in L_{-\alpha}, h_\alpha \in \mathfrak{h}$ so that $\{e_\alpha, f_\alpha, h_\alpha\}$ satisfy the same commutation relations as the standard generators of $sl(2, \mathbb{C})$. We denote the Lie algebra generated by these three elements by $sl_\alpha(2, \mathbb{C})$. We note that in the adjoint representation, $sl_\alpha(2, \mathbb{C})$ is (a module for) an irreducible representation of the Lie algebra $sl_\alpha(2, \mathbb{C})$. We are assuming here that relative to the set of simple roots Δ, α is a positive root, ie it is expressible as a positive integer linear combination of the elements of Δ. We note that if $X \in L_\alpha, Y \in L_\beta$, then

$$B([H, X], Y) = -B(X, [H, Y])$$

implies

$$(\alpha(H) + \beta(H))B(X, Y) = 0, H \in \mathfrak{h}$$

and hence $B(X, Y) = 0$ unless $\beta = -\alpha$. It is therefore clear from the non-degeneracy of B that for each $X \in L_\alpha$, $B(X, Y) \neq 0$ for some $Y \in L_{-\alpha}$ It is also clear that

$$dim L_\alpha = 1$$

This can be seen as follows: Consider the sum

$$M_\alpha = \mathfrak{h} \oplus \bigoplus_c L_{c\alpha}$$

the sum being over all the irreducible representations of $sl_\alpha(2,\mathbb{C})$ in which the weights are multiples of α. (By weight β, here we mean that if X is a weight-vector with weight β where β is a linear functional on $\mathfrak{h}$, then $ad(H)(X) = \beta(H)X\forall H \in \mathfrak{h}$. We note that M_α is a module for $sl_\alpha(2,\mathbb{C})$. When $H = h_\alpha$, then $\beta(h_\alpha)$ becomes an eigenvalue of $ad(h_\alpha)$ and hence β can be regarded as a weight for the Lie algebra $sl_\alpha(2,\mathbb{C})$. Since $sl_\alpha(2,\mathbb{C})$ as a Lie algebra is isomorphic to $sl(2,\mathbb{C})$, in any irreducible representation of $sl_\alpha(2,\mathbb{C})$, either a weight zero or a weight one will occur with a unique weight vector. Clearly, $ad(h_\alpha)$ when acting on M_α has weight zero iff the weight vector is in $\mathfrak{h}$. M_α contains the module $\mathfrak{h} + sl_\alpha(2,\mathbb{C})$ of $sl_\alpha(2,\mathbb{C})$ and hence it cannot contain any irreducible even module that has zero intersection with $\mathfrak{h} + sl_\alpha(2,\mathbb{C})$ appearing in M_α as a direct summand for any even irreducible module for $sl_\alpha(2,\mathbb{C})$ must necessarily contain a zero weight vector which must be an element of $\mathfrak{h}$ and hence will intersect the module $\mathfrak{h} + sl_\alpha(2,\mathbb{C})$(Note that $\mathfrak{h} + sl_\alpha(2,\mathbb{C})$ is a module for $sl_\alpha(2,\mathbb{C})$, ie, it is left invariant by the adjoint action of the latter and hence this module can be decomposed into irreducible modlues for $sl_\alpha(2,\mathbb{C})$). Therefore, we must have

$$M_\alpha = \mathfrak{h} + sl_\alpha(2,\mathbb{C})$$

and in particular,

$$dimL_\alpha = 1, \alpha \in \Phi$$

By an even module $\mathfrak{g}$ of $sl_\alpha(2,\mathbb{C})$, we mean that $ad(h_\alpha)$ has an even eigenvalue when operating on the root vectors in $\mathfrak{g}$ in the adjoint representation. For example, if the eigenvalues $\{-2q, -2q+2, ..., 0, 2, 4, ..., 2p\}\}$ occured in an irreducible submodule of M_α for the Lie algebra $sl_\alpha(2,\mathbb{C})$ as a direct summand different from $sl_\alpha(2,\mathbb{C})$ in the adjoint representation, then the zero weight vector in this represenation would be $\mathfrak{h}_\alpha$ which is a contradiction. Further, an odd summand (ie in which a vector having weight one occurs) also cannot occur, for then $\alpha/2$ would be a root $((\alpha/2)(h_\alpha) = 1)$ and hence $\alpha = 2(\alpha/2)$ cannot be a root by the above argument(Note that in an irreducible representation of $sl_\alpha(2,\mathbb{C})$, only the weights from the set $2\mathbb{Z}$ or only weights from $2\mathbb{Z}+1$ can occur. We have thus proved that if α is a root and $c\alpha$ is also a root, then $c = \pm 1$.

Remark: If $\mathfrak{g}$ is a semisimple Lie algebra and, then we can decompose

$$\mathfrak{g} = \bigoplus_{i=1}^{N} \mathfrak{g}_i$$

as a direct sum of vector spaces $\mathfrak{g}_i$ where each $\mathfrak{g}_i$ is an ideal in $\mathfrak{g}$, ie $[\mathfrak{g}, \mathfrak{g}_i] \subset \mathfrak{g}_i \forall i$ and hence $[\mathfrak{g}_i, \mathfrak{g}_j] \subset \mathfrak{g}_i \bigcap \mathfrak{g}_j = \{0\}, i \neq j$.

A.8.Schrodinger wave equations for quantum general relativity.

A manifold specified by space-time coordinates $\tilde{x}^\mu$ is given. Another coordinate system for this manifold is X^μ. The metric tensor relative to the former is $\tilde{g}_{\mu\nu}$ and the metric tensor for the latter is $g_{\mu\nu}$. Thus, we have

$$g_{\mu\nu} X^\mu_{,\rho} X^\nu_{,\sigma} = \tilde{g}_{\rho\sigma}$$

By $X^\mu_{,\rho}$ we mean $\frac{\partial X^\mu}{\partial x^\rho}$. We denote the spatial coordinates of the former system by x^a, x^b,etc, where $a, b = 1, 2, 3$. Thus, the spatial components of the metric in the former system are

$$\tilde{g}_{ab} = g_{\mu\nu} X^\mu_{,a} X^\nu_{,b}$$

We define

$$q_{ab} = \tilde{g}_{ab}$$

We denote by $((q^{ab}))$ the 3×3 matrix that is the inverse of $((q_{ab}))$. Now, write

$$X^\mu_{,0} = T^\mu = N^\mu + Nn^\mu$$

where N^μ is purely spatial, ie, of the form

$$N^\mu = N^a X^\mu_{,a}$$

and the vectors N^μ and n^μ are orthogonal with n^μ normalized by the factor N. This means that N^a is selected so that

$$N^\mu = N^a X^\mu_{,a}, g_{\mu\nu} n^\mu X^\nu_{,a} = 0$$

The first is the condition for N^μ to be a spatial vector and the second is the condition for n^μ to be orthogonal to all spatial vectors. We can visualize this by saying that the three dimensional spatial manifold Σ_t defined by $x^0 = t = constt$ is embedded in the four dimensional manifold specified by the coordinates X^μ. The vector (n^μ) is the unit normal to the manifold Σ_t and the vectors $(X^\mu_{,a})_{\mu=0}^3, a = 1,2,3$ are tangential to the manifold Σ_t. We thus get the following equations for N^a:

$$g_{\mu\nu}(X^\mu_{,0} - N^a X^\mu_{,a})X^\nu_{,b} = 0$$

or

$$\tilde{g}_{0b} - N^a \tilde{g}_{ab} = 0$$

or equivalently,

$$q_{ab}N^b = \tilde{g}_{0a}$$

We now prove the following decomposition:

$$g^{\mu\nu} = q^{\mu\nu} + n^\mu n^\nu$$

where

$$q^{\mu\nu}n_\nu = 0$$

In other words, $g^{\mu\nu}$ can be decomposed as a sum of a purely spatial part and a purely normal part with regard to the surface Σ_t. To see this, we write

$$g^{\mu\nu} = \tilde{g}^{\alpha\beta}X^\mu_{,\alpha}X^\nu_{,\beta} =$$

$$\tilde{g}^{00}(N^\mu + Nn^\mu)(N^\nu + Nn^\nu) + 2\tilde{g}^{0a}(N^\mu + Nn^\mu)X^\nu_{,a} +$$

$$\tilde{g}^{ab}X^\mu_{,a} + X^\nu_{,b}$$

We have to show that the cross term in this expansion is zero, ie, terms involving products of spatial parts $X^\mu_{,a}$ and the normal part n^μ. The cross part here is

$$2N\tilde{g}^{00}N^\mu n^\nu + 2\tilde{g}^{0a}Nn^\mu X^\nu_{,a}$$

To prove that this is zero amounts to proving that

$$\tilde{g}^{00}N^\mu n^\nu + \tilde{g}^{0a}n^\mu X^\nu_{,a} = 0$$

To prove this it suffices to show that

$$\tilde{g}^{00}N^\mu + \tilde{g}^{0a}X^\mu_{,a} = 0$$

or equivalently,

$$\tilde{g}^{00}N^a + \tilde{g}^{0a} = 0$$

Proving this is equivalent to proving that

$$\tilde{g}_{ba}\tilde{g}^{0a} + \tilde{g}^{00}\tilde{g}_{ab}N^a = 0$$

which is the same as

$$-\tilde{g}_{b0}\tilde{g}^{00} + \tilde{g}^{00}\tilde{g}_{ab}N^a = 0$$

(since $\tilde{g}_{ba}\tilde{g}^{0a} + \tilde{g}_{b0}\tilde{g}^{00} = \delta^a_b = 0$). Thus, we have to show that

$$q_{ab}N^a = \tilde{g}_{b0}$$

But this has already been established using the orthogonality of the normal vector n^μ with the spatial vectors $X^\mu_{,a}$.

A.9. Time travel in the special and general theories of relativity and the revised notions of space-time in quantum general relativity.

Gravitational red-shift: Let $U(r)$ be the gravitational potential. Then the approximate (Newtonian) metric of space-time is given by

$$d\tau^2 = (1 + 2U(r)/c^2)dt^2 - c^{-2}(dx^2 + dy^2 + dz^2)$$

We are assuming that U depends only on the radial coordinate relative to a system. The radial null geodesic (radial propagation of photons) is given by

$$0 = d\tau^2 = (1 + 2U(r)/c^2)dt^2 - dr^2$$

or equivalently,

$$dr/dt = (1 + 2U(r)/c^2)^{1/2}$$

Thus assuming $r_1 < r_2$, a photon pulse starting from r_1 at time t_1 arrives at r_2 at time t_2 given by

$$t_2 - t_1 = \int_{r_1}^{r_2} (1 + 2U(r)/c^2)^{-1/2} dr$$

In this expression, t_1, t_2 are coordinate times, ie, times measured by a clock at a large distance from the gravitiational field, ie, at a point where the gravitatitional field is zero. Now if another pulse starts from r_1 at coordinate time $t_1 + \delta t_1$, then it will arrive at r_2 at time $t_2 + \delta t_2$ where

$$t_2 + \delta t_2 - t_1 - \delta t_1 = \int_{r_1}^{r_2} (1 + 2U(r)/c^2)^{-1/2} dr$$

Note that we are assuming a static gravitational field. It thus follows that

$$\delta t_2 = \delta t_1$$

The proper time intervals measured by clocks static at r_1 and r_2 for the pulses are respectively given by

$$d\tau_1 = (1 + 2U(r_1)/c^2)^{1/2} dt_1, d\tau_2 = (1 + 2U(r_2)/c^2)^{1/2} dt_2$$

It follows therefore that

$$d\tau_1/d\tau_2 = [\frac{1 + 2U_1/c^2}{1 + 2U_2/c^2}]^{1/2}$$

where

$$U_1 = U(r_1), U_2 = U(r_2)$$

Hence, if $U_2 < U_1$, we get

$$d\tau_1 > d\tau_2$$

or in terms of frequencies,

$$\nu_1 = 1/d\tau_1 < 1/d\tau_2 = \nu_2$$

or more precisely,

$$\frac{\nu_1}{\nu_2} = [\frac{1 + 2U_2/c^2}{1 + 2U_1/c^2}]^{1/2} < 1$$

This is the classic gravitational shift phenomenon: Light propagating from weak gravitational field to a strong gravitational field gets blue shifted and equivalently, light propagating from a strong gravitational field to a weak gravitational field gets red shifted. We have also in addition proved that since $d\tau_1 > d\tau_2$, it follows that clocks run slower in a strong gravitational field, ie when the gravitational potential is more negative.

A.10[a].Transmission lines with random fluctuations in the parameters and random line loading:

$$v_{,z}(t,z) + (R_0(z) + \delta R(t,z))i(t,z) + L_0(z)i_{,t}(t,z) + (\delta L(t,z)i(t,z))_{,t} = w_v(t,z)$$

$$i_{,z}(t,z) + (G_0(z) + \delta G(t,z))v(t,z)) + C_0(z)v_{,t}(t,z) + (\delta C(t,z)v(t,z))_{,t} = w_i(t,z)$$

In these equations, $\delta R(t,z), \delta L(t,z), \delta C(t,z), \delta G(t,z), w_v(t,z), w_i(t,z)$ are random Gaussian space-time Gaussian fields. We wish to solve this system of pde's approximately using first order perturbation theory and hence calculate the approximate statistical correlations of the fluctuations in the line voltage and line current in terms of the correlations in the parameter flucutations and the voltage and current loading terms w_v and w_i.

A.10[b]. Taking non-linear hysteresis and nonlinear capacitive effects into account, generalize the problem of A.9.

A.11. Estimating parameters in statistical image models described by linear and nonlinear partial differential equations:

First consider the linear case: The model for the image field $\phi(x,y), (x,y) \in D$ is given by

$$\sum_{a,b=1}^{p} A(a,b,\theta)\partial_x^a \partial_y^b \phi(x,y) = s(x,y) + w(x,y), (x,y) \in D$$

where $w(x,y)$ is zero mean coloured Gaussian noise. $\theta \in \mathbb{R}^m$ is the parameter vector to be estimated from measurements on ϕ. Here, $s(x,y)$ is a given input non-random signal field. We assume that an initial guess estimate θ_0 of θ is known and that the correction $\delta\theta$ to this estimate is to be made. We write

$$\phi(x,y) = \phi_0(x,y) + \delta\phi(x,y)$$

where ϕ_0 is the solution with the guess parameter θ_0 and zero noise and $\delta\phi$ is the first order correction to ϕ_0 arising from the parameter estimate correction term $\delta\theta$ and the noise w. We regard $\delta\phi, \delta\theta, w$ all as being of the first order of smallness. We define

$$H(\omega_1, \omega_2, \theta) = \sum_{a,b} A(a, b, \theta)(j\omega_1)^a (j\omega_2)^b$$

Then if two dimensional spatial Fourier transforms are denoted by placing a hat on top of a signal/noise field, we get

$$H(\omega_1, \omega_2, \theta)\hat{\phi}(\omega_1, \omega_2) = \hat{s}(\omega_1, \omega_2) + \hat{w}(\omega_1, \omega_2)$$

Thus to zeroth order, we get

$$H(\omega_1, \omega_2, \theta_0)\hat{\phi}_0(\omega_1, \omega_2) = \hat{s}(\omega_1, \omega_2),$$

$$H(\omega_1, \omega_2, \theta_0)\delta\hat{\phi}(\omega_1, \omega_2) + (H_r(\omega_1, \omega_2, \theta_0)\delta\theta_r)\hat{\phi}_0(\omega_1, \omega_2)$$
$$= \hat{w}(\omega_1, \omega_2)$$

where

$$H_r(\omega_1, \omega_2, \theta_0) = \frac{\partial H(\omega_1, \omega_2, \theta_0)}{\partial \theta_r}$$

Thus, if

$$g(x, y) = \mathcal{F}^{-1}(H(\omega_1, \omega_2, \theta_0)^{-1})$$

it then follows that

$$\phi_0(x, y) = g(x, y) * s(x, y) = \int g(x - x', y - y')s(x', y')dx'dy'$$

Likewise, if we write

$$h_r(x, y) = \mathcal{F}^{-1}(H_r(\omega_1, \omega_2, \theta_0))$$
$$= \sum_{a,b} A_r(a, b, \theta_0)\delta^{(a)}(x)\delta^{(b)}(y)$$

where

$$A_r(a, b, \theta_0) = \frac{\partial A(a, b, \theta_0)}{\partial \theta_r}$$

then we get

$$\delta\phi(x, y) = -g(x, y) * h_r(x, y)\delta\theta_r + g(x, y) * w(x, y) - - - (2)$$

where summation over the repeated index r is implied. We measure $\delta\phi(x, y)$ as follows: First since we know θ_0 and the input signal field $s(x, y)$, we can calculate $\phi_0(x, y)$ using (1) and then we measure the actual noise perturbed image field $\phi(x, y)$ and calculate $\delta\phi(x, y) = \phi(x, y) - \phi_0(x, y)$. Now using (2), we calculate the maximum likelihood estimator of $\delta\theta$ as

$$\hat{\delta\theta} =$$

$$argmin_{\delta\theta}(\int (\delta\phi(x, y) + g_r(x, y)\delta\theta_r)Q(x, y|x', y')(\delta\phi(x', y') + g_s(x', y')\delta\theta_s)dxdydx'dy')$$

where $Q(x, y|x', y')$ is the inverse Kernel of

$$R(x, y|x', y') = \mathbb{E}[(g(x, y) * w(x, y)).(g(x', y') * w(x', y'))]$$

$$\int g(x - x_1, y - y_1)g(x' - x_1', y' - y_1')\mathbb{E}(w(x_1, y_1)w(x_1', y_1'))dx_1dy_1dx_1'dy_1'$$

ie,

$$\int Q(x, y|x'y')R(x', y'|x'', y'')dx'dy' = \delta(x - x'')\delta(y - y'')$$

and

$$g_r(x, y) = g(x, y) * h_r(x, y) = \sum_{a,b} A_r(a, b, \theta_0)\partial_x^a\partial_y^b g(x, y)$$

We write

$$\mathbf{g}(x, y) = ((g_r(x, y)))$$

Then,

$$\hat{\delta\theta} = [\int Q(x,y|x',y')\mathbf{g}(x,y)\mathbf{g}(x',y')^T dxdydx'dy']^{-1}[\int Q(x,y|x',y')\mathbf{g}(x,y)\delta\phi(x',y')dxdydx'dy']$$

A simple computation gives us the covariance of this parameter vector estimator:

$$Cov(\delta\hat{\theta}) = [\int Q(x,y|x',y')\mathbf{g}(x,y)\mathbf{g}(x',y')^T dxdydx'dy']^{-1}$$

Wavelet based image parameter estimation: Let $\psi_n(x,y), n = 1,2,...$ be a wavelet orthonormal basis for $L^2(\mathbb{R}^2)$. Here, the index n consists of the scaling and translational index in both the dimensions, ie, n corresponds to four ordered integer indices. We define

$$c(n,\phi) =< \phi, \psi_n >= \int \phi(x,y)\psi_n(x,y)dxdy$$

Then,

$$\phi(x,y) = \sum_n c(n,\phi)\psi_n(x,y)$$

Substituting this into the image pde model gives

$$\sum_{n,a,b} c(n,\phi)A(a,b,\theta)\partial_x^a\partial_y^b\psi_n(x,y) = s(x,y) + w(x,y)$$

Taking the inner product on both sides with $\psi_m(x,y)$ gives

$$\sum_{n,a,b} c(n,\phi)A(a,b,\theta) < \psi_m, \partial_x^a\partial_y^b\psi_n >= c(m,s) + c(m,w)$$

Define

$$P(m,n|\theta) = \sum_{a,b} A(a,b,\theta) < \psi_m, \partial_x^a\partial_y^b\psi_n >$$

The above equation can then be expressed as

$$\sum_n P(m,n|\theta)c(n,\phi) = c(m,s) + c(m,w)$$

Writing

$$\theta = \theta_0 + \delta\theta, \phi(x,y) = \phi_0(x,y) + \delta\phi(x,y)$$

gives us on applying first order perturbation theory,

$$\sum_n P(m,n|\theta_0)c(n,\phi_0) = c(m,s),$$

$$\sum_n((\sum_r P_r(m,n|\theta_0)\delta\theta_r)(n,\phi_0) + P(m,n|\theta_0)\delta c(n)) = c(m,w)$$

where

$$P_r(m,n|\theta_0) = \frac{\partial P(m,n|\theta_0)}{\partial\theta_r}$$

and

$$\delta c(n) = c(n,\phi_0 + \delta\phi) - c(n,\phi_0)$$

We write

$$c_0(n) = c(n,\phi_0), P_r(m,n) = P_r(m,n|\theta_0), P_0(m,n) = P(m,n|\theta_0)$$

and thus get

$$\sum_n P_0(m,n)c_0(n) = c(m,s)$$

$$\sum_{n,r} P_r(m,n)\delta\theta_r c_0(n) + \sum_n P_0(m,n)\delta c(n) = c(m,w)$$

A.12 Mackey's theory on the construction of the basic quantum observables from unitary representations of the Galilean group. $(a, v) in V = \mathbb{R}^3 \times \mathbb{R}^3$. V is the Abelian group of translations and uniform velocity motions acting on the space-time manifold $\mathcal{M} = \{(t, x) : t \in \mathbb{R}, x \in \mathbb{R}^3\}$. This action is given by

$$(a, v)(t, x) = (t + x + vt + a)$$

$(\tau, g) \in \mathbb{R} \times SO(3)$. $\mathbb{R} \times SO(3)$ is the non-Abelian group of time translations and rotations acting on $\mathcal{M}$:

$$(\tau, g)(t, x) = (t + \tau, gx)$$

The Galiean group G is the semidirect product of V and H:

$$G = V \otimes_s H$$

where H acts on V as follows:

$$(\tau, g).(a, v).(\tau, g)^{-1} = (a', v')$$

or equivalently,

$$(\tau, g).(a, v) = (a', v').(\tau, g)$$

Acting both sides on $(t, x) \in \mathbb{R}^4$ gives

$$(\tau, g)(t, x + vt + a) = (a', v')(t + \tau, gx)$$

or equivalently,

$$(t + \tau, gx + tgv + ga) = (t + \tau, gx + v't + a' + v'\tau)$$

or equivalently,

$$v' = gv, a' = g(a - v\tau)$$

Thus, the semimdirect product structure is given by

$$(\tau, g).(a, v).(\tau, g)^{-1} = (g(a - v\tau), gv) \in V$$

Any element of the Galilean group G can be expressed in two ways, one as an element (a, v, τ, g) defined by its action on $\mathbb{R}^4$ by

$$(a, v, \tau, g)(t, x) = (t + \tau, gx + vt + a)$$

and in another way as the element $(a, v).(\tau, g)$. The action of this on $\mathbb{R}^4$ is given by

$$(a, v).(\tau, g)(t, x) = (a, v)(t + \tau, gx) = (t + \tau, gx + vt + v\tau + a)$$

It follows that the relationship between these two methods of expressing an element of the Galilean group is given by

$$(a + v\tau, v, \tau, g) = (a, v).(\tau, g)$$

or equivalently by

$$(a - v\tau, v).(\tau, g) = (a, v, \tau, g)$$

Let now $\mathfrak{h}$ be a Hilbert space (like $\mathbb{C}^{2j+1}$ for a spin j particle) and $\mathfrak{H} = L^2(\mathbb{R}^3, \mathfrak{h})$ the Hilbert space of all measurable functions $f : \mathbb{R}^3 \to \mathfrak{h}$ for which $\int_{\mathbb{R}^3} \| f(x) \|^2 d^3x < \infty$. The projective unitary representations of G are obtained by using the multiplier

$$m((a, v).(\tau, g), (a', v').(\tau', g')) = exp(iB((a, v), (\tau, g)[(a', v')])$$

where $B : V \times V \to \mathbb{R}$ (with $V = \mathbb{R}^3 \times \mathbb{R}^3$) being a skew symmetric bilinear form invariant under $H = \{(\tau, g) : \tau \in \mathbb{R}, g \in SO(3)\}$. that is invariant under H or equivalently under (τ, g). Note that the action of H on V is defined by

$$(\tau, g)[(a, v)] = (\tau, g).(a, v).(\tau, g)^{-1} = (g(a - v\tau), gv)$$

We thus find that

$$m((a, v).(\tau, g), (a', v').(\tau', g')) = exp(iB((a, v), (g(a - v\tau), gv)))$$
$$= exp(i\lambda((a, gv) - (v, g(a - v\tau))))$$

for some $\lambda \in \mathbb{R}$ where $(u, v) = u^T v$. Let U be a projective unitary representation of G with this multiplier. Then,

$$U((a, v).(\tau, g)) = U(a, v)U(\tau, g)$$

We can write

$$U(a,0)=V_1(a),U(0,v)=V_2(v),U(\tau,g)=W_1(\tau)W_2(g)$$

where V_1,V_2 are unitary representations of the Abelian group $V=\mathbb{R}^3\times\mathbb{R}^3$ and W_1 and W_2 are unitary representations of $\mathbb{R}$ and $SO(3)$. By Stone's theorem on unitary representations of Abelian groups, it follows that there exist Hermitian operators $Q=(Q_1,Q_2,Q_3)$ and $P=(P_1,P_2,P_3)$ in $L^2(\mathbb{R}^3,\mathfrak{h})$ such that

$$V_1(a)=exp(-ia.P),V_2(v)=exp(-iv.Q)$$

and also a Hermitian operator H in the same space such that

$$W_1(\tau)=exp(-i\tau H)$$

We have

$$U(a_1,v_1)U(a_2,v_2)=exp(i\lambda(a_1^Tv_2-a_2^Tv_1))U(a_1+a_2,v_1+v_2)$$

In particular,

$$V_1(a)V_2(v)=U(a,0)U(0,v)=exp(i\lambda a^Tv)U(a,v),$$
$$V_2(v)V_1(a)=U(0,v)U(a,0)=exp(-i\lambda a^Tv)U(a,v)$$

Thus, we get the Weyl commutation relations

$$V_1(a)V_2(v)=exp(i2\lambda a^Tv)V_2(v)V_1(a)$$

which can be expressed as

$$exp(-ia.P).exp(-iv.Q)=exp(i2\lambda a^Tv)exp(-iv.Q).exp(-ia.P)$$

and hence by considering infinitesimal a and v in $\mathbb{R}^3$. we get

$$-P_iQ_j+Q_jP_i=2i\lambda\delta_{ij}$$

or equivalently,

$$[Q_i,P_j]=2i\lambda\delta_{ij}$$

To get agreement with quantum mechanics that the momentum operators P generate translations and the position operators Q generate uniform velocities, we must take $\lambda=1/2$ and thus, we get

$$[Q_i,P_j]=i\delta_{ij}$$

The Stone-Von-Neumann theorem then implies that the Hilbert space (can be chosen so that the actions of the $Q_i's$ and the $P_j's$ in $L^2(\mathbb{R}^3,\mathfrak{h})$ are such that

$$(Q_jf)(x)=x_jf(x),(P_jf)(x)=-i\frac{\partial f(x)}{\partial x_j}$$

In other words, regarding $L^2(\mathbb{R}^3,\mathfrak{h})$ as $L^2(\mathbb{R}^3)\otimes\mathfrak{h}$ (This is a Hilbert space isomorphism), we have that $Q_j=x_j\otimes I_{\mathfrak{h}}$ and $P_j=-i\frac{\partial}{\partial x_j}\otimes I_{\mathfrak{h}}$. We note compute

$$U(\tau,g)U(a,v)=U((\tau,g).(a,v))$$

Now,

$$(\tau,g).(a,v)(t,x)=(\tau,g)(t,x+vt+a)=(t+\tau,gx+tgv+ga)$$
$$=(ga,gv,\tau,g)(t,x)$$

ie,

$$(\tau,g).(a,v)=(ga,gv,\tau,g)$$

On the other hand,

$$(a,v).(\tau,g)(t,x)=(a,v).(t+\tau,gx)=(t+\tau,gx+vt+v\tau+a)$$
$$=(a+v\tau,v,\tau,g)(t,x)$$

Thus,

$$(\tau, g).(a, v) = (g(a - v\tau), gv).(\tau, g)$$

So we get

$$U(\tau, g)U(a, v)U(\tau, g)^{-1} = U(g(a - v\tau), gv)$$

Taking $g = I$, this gives

$$W_1(\tau)U(a, v)W_1(-\tau) = U(a - v\tau, v)$$

Setting $a = 0$ in this formula gives

$$W_1(\tau)V_2(v)W_1(-\tau) = U(-v\tau, v)$$

while setting $v = 0$ gives

$$W_1(\tau)V_1(a)W_1(-\tau) = V_1(a)$$

The second equation implies

$$[H, P_j] = 0, j = 1, 2, 3$$

We now note that

$$V_1(-v\tau)V_2(v) = exp(-i\tau|v|^2/2)U(-v\tau, v)$$

Thus, we get from the first equation

$$W_1(\tau)V_2(v)W_1(-\tau) = exp(i\tau|v|^2/2)V_1(-v\tau)V_2(v) = exp(i\tau|v|^2/2)exp(i\tau v.P)V_2(v)$$

For infinitesimal τ, this gives

$$-i[H, exp(-iv.Q)] = (i|v|^2/2 + iv.P)exp(-iv.Q)$$

or equivalently,

$$H - exp(-iv.Q).H.exp(iv.Q) = -|v|^2/2 - v.P - - - (1)$$

The $O(v)$ term of this equation gives

$$i[v.Q, H] = -v.P$$

or equivalently,

$$i[H, Q_j] = P_j$$

Combining this with the equation

$$[H, P_j] = 0$$

we may conclude using the commutation relations $[Q_i, P_j] = i\delta_{ij}$ that

$$H = P^2/2 + E = \frac{1}{2}\sum_{j=1}^{3} P_j^2 + E - - - (2)$$

where E is an operator of the form $I \otimes E_1$ in $L^2(\mathbb{R}^3) \otimes \mathfrak{h}$, ie, for $f(x) \in \mathfrak{h}, x \in \mathbb{R}^3$, we have

$$(Ef)(x) = E_1 f(x)$$

By considering the $O(v^2)$ term in (1), we get

$$[v.Q, [v.Q, H]] = -v^2$$

or equivalently,

$$[[H, Q_i], Q_j] = -\delta_{ij}/2$$

This is verified by (2):

$$[P^2/2, Q_i] = -iP_i, [[P^2/2, Q_i], Q_j] = -i[P_i, Q_j] = -\delta_{ij}$$

We now consider the equation $U(\tau, g)U(a, v)U(\tau, g)^{-1} = U(g(a - v\tau), gv)$ with $\tau = 0$. We get

$$W_2(g)U(a, v)W_2(g)^{-1} = U(ga, gv)$$

In particular, we get

$$W_2(g)V_1(a)W_2(g)^{-1} = V_1(ga),$$
$$W_2(g)V_2(v)W_2(g)^{-1} = V_2(gv)$$

These equations are the same as

$$W(2(g)P_iW_2(g)^{-1} = \sum_{j=1}^{3} g_{ji}P_j,$$

$$W_2(g)Q_iW_2(g)^{-1} = \sum_{j=1}^{3} g_{ji}Q_j$$

Here, $g \in SO(3)$. Thus, $W_2(g)$ has the effect of rotating the position and momentum operators.

N-particle system: We assume that $\mathcal{H}_i$ is the Hilbert space for the ith particle and that the projective unitary representation U of the Galilean group G acting in $\mathcal{H} = \otimes_{i=1}^{N}\mathcal{H}_i$ has the form

$$U(a, v, 0, g) = \otimes_{i=1}^{N} U_i(a, v, 0, g)$$

In other words, as regards translation, motion with uniform velocities and rotations, the actions of these on each particle in the system is the same. The above discussion for a single particle implies that each $U_i(a, v, 0, g)$ acts in the same way on the corresponding particle. However, time evolution described by the operator $U(0, v, \tau, I)$ acts on the entire system and may not be factorizable into a tensor product of single particle operators. This is because, the3 generator of this group which is the energy/Hamiltonian H is the sum of the individual kinetic energies and an interaction potential energy and the latter depends on some complex combination of all the particle position operators. So for the present, we can let

$$U(0, 0, \tau, I) = exp(-i\tau H)$$

where H is a Hermitian operator acting in $\mathcal{H}$. We also assume the existence of velocity operators $V_i = (V_{i1}, V_{i2}, V_{i3}), i = 1, 2, ..., N$ acting in $\mathcal{H}$ satisfying the following properies:

$$U(0, v, 0, I)V_{ij}U(0, v, 0, I)^{-1} = V_{ij} + v_j, 1 \leq j \leq 3, i = 1, 2, ..., N,$$

Since

$$U_i(0, v, 0, I) = exp(-iv.Q)$$

it follows that

$$i[v.\sum_k Q_k, V_{ij}] = -v_j$$

where $Q_k = (Q_{k1}, Q_{k2}, Q_{k3})$ are the position operators acting in $\mathcal{H}_k$ and likewise $P_k = (P_{k1}, P_{k2}, P_{k3}$ are the momentum operators acting in $\mathcal{H}_k$. It should be noted that by the theory for one particle discussed above and the separability of $U(a, v, 0, g)$ we have that $U(a, 0, 0, I) = exp(-ia.P) = \otimes_k exp(-iaP_k), P = \sum_k P_k$ and likewise for Q. Thus, we get

$$i[\sum_k Q_{kl}, V_{ij}] = -\delta_{jl}$$

We also have from the one particle theory and separability of $U(a, v, 0, g)$ that

$$i[Q_{kl}, P_{ij}] = -\delta_{ki}\delta_{lj}$$

So, if we postulate that

$$[Q_{kl}, V_{ij}] = 0, k \neq i$$

(this is true if we assume that V_{ij} acts in $\mathcal{H}_i$), then we get

$$i[Q_{kl}, V_{ij}] = -\delta_{ki}\delta_{ij}$$

and hence we derive

$$[Q_{kl}, P_{ij} - V_{ij}] = 0$$

which implies that

$$P_{ij} - V_{ij} = A_{ij}(Q)$$

where $A_{ij}(Q)$ is a function of $Q = (Q_{ij} : 1 \leq i \leq N, 1 \leq j \leq 3)$ only. We define

$$H_0 = \frac{1}{2}\sum_{k=1}^{N} V_k^2 = \frac{1}{2}\sum_{k=1}^{N}\sum_{i=1}^{3} V_{ki}^2$$

From the composition theory of Galilean group representations developed above, we have that

$$U(0,0,\tau,I)U(a,0,0,I) = U(a,0,\tau,I) = U(a,0,0,I)U(0,0,\tau,I)$$

and hence

$$[H, \sum_k P_{ki}] = 0$$

Note that

$$U(a,0,0,I) = exp(-ia.\sum P_k) = exp(-i\sum_{k,i} a_i P_{ki})$$

This means that the total momentum of the system of N particles is conserved. Now, we consider

$$[H_0, Q_{ki}] = [\sum_{r=1}^{3} V_{kr}^2/2, Q_{ki}] = [\sum_{r=1}^{3}(P_{kr} - A_{kr}(Q))^2/2, Q_{ki}] =$$

$$-i(P_{ki} - A_{ki}(Q)) = -iV_{ki}$$

We also assume (by definition of the velocity as the time derivative of the position),

$$\frac{d}{dt}U(0,0,-\tau,I).Q_{ki}U(0,0,\tau,I)|_{\tau=0} = V_{ki}$$

This gives

$$[H, Q_{ki}] = -iV_{ki}$$

and hence

$$[H - H_0, Q_{ki}] = 0$$

and therefore,

$$H - H_0 = V_0(Q) + E$$

where E is of the form $I \otimes E_1$ with E_1 acting in $\mathfrak{h}$ and $V_0(Q)$ and arbitrary function of the positions $Q = (Q_{ki} : 1 \leq k \leq N, i = 1,2,3)$. Thus, we finally get the general form of the total system Hamiltonian:

$$H = \frac{1}{2}\sum_{k,i}(P_{ki} - A_{ki}(Q))^2 + V_0(Q) + E$$

(Ref: K.R.Parthasarathy, "Mathematical Foundations of Quantum Mechanics", Hindustan Book Agency)

A.13. A remark on quantum stochastic calculus.

Let $u_t \in \mathcal{H}, U_t \in \mathcal{U}(\mathcal{H}), t \geq 0$ Let P be a spectral measure on $[0,\infty)$ with values in $\mathcal{P}(\mathcal{H})$ and assume that P commutes with all the $U_t's$. We can write $U_t = exp(iH_t)$ where H_t is a Hermitian operator in $\mathcal{H}$. Suppose that the $H_t's$ commute with each other and that H_t acts in $P_t\mathcal{H}$ where $P_t = P[0,t]$ (For example we can choose a Hermitian operator H in $\mathcal{H}$ that commutes with P and then define $H_t = P_tH = HP_t$). More generally, we shall assume that for $s < t$, $H_t - H_s$ acts in $P[s,t]\mathcal{H}$. Then we have

$$dU_t = iU_t dH_t$$

and since dH_t acts in $dP_t\mathcal{H} = P[t, t+dt]\mathcal{H}$ while U_t acts in $P_t\mathcal{H}$, it follows that for $v, w \in \mathcal{H}$, we have with $\Gamma(U_t)$ denoting the second quantization of U_t (ie, $\Gamma(U_t) = W(0, U_t)$),

$$d < e(v)|\Gamma(U_t)|e(w) >=< e(v)|d\Gamma(U_t)|e(w) >=$$

$$i < e(v)|\Gamma(U_t)|e(w) > .d < v|H_t|w >$$

and hence $\Gamma(U_t)$ satisfies the qsde

$$d\Gamma(U_t) = \Gamma(U_t)d\Lambda(H_t)$$

where $\Lambda(X)$ is the second quantization of X defined by

$$< e(v)|exp(\Lambda(X))|e(w) >=< e(v)|e(exp(X)w) >= exp(< v|exp(X)|w >)$$

A.14. Scattering theory with time dependent interactions with the scattering centre.

The free particle Hamiltonian is H_0 and the Hamiltonian after the particle starts interacting with the scattering centre is $H(t) = H_0 + \delta.V(t)$. We assume that $V(t) = V_0$ is a constant operator for $|t| > T$. Let $|\phi_1 >$ be a free particle state evolving according to H_0 in the remote past (the "in state") Let $|\psi_1 >$ be the corresponding scattered state evolving according to $H(t)$. Likewise, let $|\phi_2 >$ be a free particle state evolving according to H_0 in the future, ie, as $t \to \infty$ (ie, the "out state") and $|\psi_2 >$ the corresponding scattered state evolving according to $H(t)$. Define for $t_2 > t_1$,

$$U_0(t_2 - t_1) = exp(-i(t_2 - t_1)H_0),$$

$$U(t_2, t_1) = T\{exp(-i \int_{t_1}^{t_2} H(t)dt)\}$$

where $T\{.\}$ denotes the time ordering operator. We must have

$$lim_{t \to \infty}(U(t,0)|\psi_2 > -U_0(t)|\phi_2 >) = 0$$

and hence

$$|\psi_2 >= lim_{t \to \infty} U(t,0)^{-1} U_0(t)|\phi_2 >$$

We write

$$\Omega_2(t) = U(t,0)^{-1}U_0(t), t \geq 0,$$

Then on an appropriate domain of out states, we have the operator

$$\Omega_2 = lim_{t \to \infty}\Omega_2(t)$$

Thus,

$$|\psi_2 >= \Omega_2|\phi_2 >$$

Likewise,

$$lim_{t \to -\infty}(U(0,t)^{-1}|\psi_1 > -U_0(-t)|\phi_1 >) = 0$$

or equivalently,

$$|\psi_1 >= lim_{t \to -\infty} U(0,t)U_0(-t)|\phi_1 >$$
$$= lim_{t \to \infty} U(0,-t)U_0(t)|\phi_1 >$$
$$= \Omega_1|\phi_1 >$$

where

$$\Omega_1 = lim_{t \to \infty}\Omega_1(t)|\phi_1 >$$

where

$$\Omega_1(t) = U(0,-t)U_0(t)$$

The scattering matrix is given by

$$S = \Omega_2^* \Omega_1 = lim_{t \to \infty}\Omega_2(t)^*\Omega_1(t)$$
$$= lim_{t \to \infty} U_0(-t)U(t,0)U(0,-t)U_0(t)$$
$$= lim_{t \to \infty} U_0(-t)U(t,-t)U_0(t)$$
$$= lim_{t \to \infty} exp(itH_0).T\{exp(-i \int_{-t}^{t} H(s)ds)\}.exp(-itH_0)$$

Note: In discussing scattering theory with noise, we assume that $\{V(t)\}$ is an operator valued random process and then compute the average value of the scattering matrix with respect to the probability distribution of $\{V(t)\}$. By the Dyson series expansion,

$$T\{exp(-i \int_{-t}^{t} H(s)ds)\} =$$

$$exp(-2itH_0) + exp(-itH_0)(\sum_{n=1}^{\infty}(-i)^n \int_{-t<s_n<...<s_1<t} \tilde{V}(s_1)...\tilde{V}(s_n)ds_1...ds_n)exp(-itH_0)$$

where

$$\tilde{V}(s) = exp(isH_0)V(s)exp(-isH_0)$$

Remarks: Assume V to be a constant potential so that $H_0, H = H_0 + V$ are respectively the Hamiltonians of the free particle and the particle with interactions with the scattering centre taken into account.

$$\Omega_+ = lim_{t\to\infty} exp(itH)exp(-itH_0) =$$

$$I + i\int_0^\infty exp(itH)V.exp(-itH_0)dt$$

This equation is to be interpreted in the strong sense, ie, the domain of Ω_+ contains all vectors f for which

$$\int_0^\infty \| V.exp(-itH_0)f \| dt < \infty$$

Note: The scattering matrix is not well defined in general for finite dimensional quantum systems. For example, suppose that H_0, H are $n \times n$ Hermitian matrices with spectral decompositions

$$H_0 = \sum_{k=1}^{p} \mu_k E_k, H = \sum_{k=1}^{q} \lambda_k F_k$$

where $\{\mu_1, ..., \mu_p\}$ is the set of distinct eigenvalues of H_0 and $\{\lambda_1, ..., \lambda_q\}$ is the set of distinct eigenvalues of H and $\{E_k : 1 \leq k \leq p\}$ form a spectral resolution of the identity and $\{F_k : 1 \leq k \leq q\}$ also forms a spectral resolution of the identity. Then,

$$exp(itH).exp(-itH_0) = \sum_{k,r} exp(i(\lambda_k - \mu_r)t)F_k E_r$$

As $t \to \infty$ this does not converge to any $n \times n$ matrix if there exist a k, r such that $F_k E_r \neq 0$ and $\lambda_k \neq \mu_r$. Suppose however that the set of all ordered pairs (k, r) is such that whenever $\lambda_k \neq \mu_r$, we have that $F_k E_r = 0$, then the above converges to $\sum_{(k,r)\in D} F_k E_r$ where D is the set of all ordered pairs (k, r) for which $\lambda_k = \mu_r$.

The Lippman-Schwinger equations:

A.15. Quantum image processing.

The state of pixel number $m \in \{1, 2, ..., N\}$ is $|k, m>$ if its grey scale amplitude and phase is specifiied by the index $k \in \{1, 2, ..., p\}$. Thus, if the image is in a pure state $|I>$, then this state can be expressed as

$$|I> = \sum_{k,m} I[k, m]|k, m>$$

where

$$\sum_{k,m} |I[k, m]|^2 = 1$$

This means that given that we measure pixel number m, the probability of getting the grey scale amplitude and phase level k is given by

$$P_I(k|m) = I[k, m]|^2$$

More generally, the image can be in a mixed state ρ_I defined by

$$\rho_I = \sum_{1\leq k,r\leq p, 1\leq m,s\leq N} I[k, m, r, s]|k, m><r, s|$$

where the condition

$$Tr(\rho_I) = 1$$

implies that

$$\sum_{k,m} I[k, m, k, m] = 1$$

Then if the image is in the mixed state ρ_I, the probability of getting the grey scale level amplitude specified by the index k given that the mth pixel is measured is given by

$$P_I(k|m) = \frac{<k, m|\rho_I|k, m>}{\sum_r <r, m|\rho_I|r, m>}$$

$$= \frac{I[k,m,k,m]}{\sum_r I[r,m,r,m]}$$

We shall now express ρ_I in the frequency domain of the grey scale amplitudes: The quantum Fourier transform of the grey scale state $|k>$ is given by

$$|\tilde{k}>= p^{-1/2}\sum_{r=1}^{p} exp(-i2\pi kr/p)|r>$$

The inverse quantum Fourier transform is thus given by

$$|k>= p^{-1/2}\sum_{r=1}^{p} exp(i2\pi kr/p)|\tilde{r}>$$

Then,

$$\rho_I = \sum I[k,m,r,s]|k,m><r,s| = p^{-1}\sum I[k,m,r,s]exp(i2\pi(kk'-rr')/p)|\tilde{k}'><\tilde{r}'|\otimes|m><s|$$

$$= \sum \hat{I}[k',m,s,r']|\tilde{k}'><\tilde{r}'|\otimes|m><s|$$

where

$$\hat{I}[k',m,s,r'] = p^{-1}\sum_{k,r} I[k,m,r,s]exp(i2\pi(kk'-rr')/p)$$

We may express this as

$$\rho_I = \sum \hat{I}[k,m,r,s]|\tilde{k}><\tilde{r}|\otimes|m><s|$$

$$= \sum \hat{I}[k,m,r,s] = |\tilde{k},m><\tilde{r},s|$$

The average image energy at pixel number m and frequency $\tilde{k}$ is thus given by

$$\frac{<\tilde{k},m|\rho_I|\tilde{k},m>}{\sum_r <\tilde{r},m|\rho_I|\tilde{r},m>} = \frac{\hat{I}[k,m,k,m]}{\sum_r \hat{I}[r,m,r,m]}$$

Now we consider the processing of the quantum image state ρ_I in the spatial domain and in the frequency domain using linear filters. Consider first a matrix of size $pN \times pN$ defined by

$$T = \sum_{m,s} T_{ms}\otimes|m><s|$$

where T_{ms} is a $p\times p$ matrix for each $m,s\in\{1,2,...,N\}$. We assume that T is a unitary matrix, ie,

$$T^*T = I_{pN}$$

This condition is equivalent to requiring that

$$\sum T^*_{ms}T_{m's'}\otimes|s><m|m'><s'| = I_{pN}$$

or equivalently,

$$\sum_{m,s,s'} T^*_{ms}T_{ms'}\otimes|s><s'| = I_{pN}$$

or equivalently,

$$\sum_m T^*_{ms}T_{ms'} = \delta_{ss'}I_p$$

Applying the unitary operator T to ρ_I gives a transformed image field specified by the density matrix

$$\rho'_I = T\rho_I T^* = \rho_I = \sum I[k,m,r,s]T(|k><r|\otimes|m><s|)T^*$$

$$= \sum \hat{I}[k,m,r,s]T(|\tilde{k}><\tilde{r}|\otimes|m><s|)T^*$$

Now,

$$T(|k><r|\otimes|m><s|) =$$

$$(\sum_{jl} T_{jl} \otimes |j><l|)(|k><r| \otimes |m><s|)(\sum_{j'l'} T^*_{j'l'} \otimes |l'><j'|)$$

$$= \sum T_{jl}|k><r|T^*_{j'l'} \otimes |j><l|m><s|l'>|m><j'|$$

$$= \sum_{jj'} T_{jm}|k><r|T^*_{j's} \otimes |j><j'|$$

A.16. Quantization of the em fields inside a rectangular waveguide.

$$(\nabla_\perp^2 + h^2)E_z = 0, (\nabla_\perp^2 + h^2)H_z = 0$$

$$E_\perp = -\frac{\gamma}{h^2}\nabla_\perp E_z - \frac{j\omega\mu}{h^2}\nabla_\perp H_z \times \hat{z},$$

$$H_\perp = -\frac{\gamma}{h^2}\nabla_\perp H_z + \frac{j\omega\epsilon}{h^2}\nabla_\perp E_z \times \hat{z}$$

We wish to select potentials Φ, A such that when the em fields satisfy the above, then

$$E = -nabla\Phi - j\omega A, \mu H = \nabla \times A$$

or equivalently,

$$E_z = \gamma\Phi - j\omega A_z, E_\perp = -\nabla_\perp\Phi - j\omega A_\perp,$$

$$\mu H_z\hat{z} = \nabla_\perp \times A_\perp,$$

$$\mu H_\perp = \nabla_\perp A_z \times \hat{z} - \gamma\hat{z} \times A_\perp$$

In addition, we wish that the potentials Φ, A satisfy the Lorentz gauge condition

$$divA = -j\omega\epsilon\mu\Phi,$$

or equivalently,

$$\nabla_\perp.A_\perp - \gamma A_z = -j\omega\epsilon\mu\Phi$$

It is easily seen that the most general potentials satisfying the above requirements are given by

$$A_\perp = -(\nabla_\perp\Phi + E_\perp)/j\omega,$$

$$A_z = (\gamma\Phi - E_z)/j\omega,$$

where Φ is any scalar field that satisfies the Helmholtz equation

$$(\nabla_\perp^2 + h^2)\Phi = 0$$

In particular, we can take $\Phi = 0$ and then

$$A = -E/j\omega$$

The general solution for the electromagnetic fields in the guide with the boundary conditions that E_z and the normal components of H vanish on the side boundaries is given by

$$E_z = \sum C(n,m)exp(-\gamma_{nm}z)u_{nm}(x,y), H_z = \sum D(n,m)exp(-\gamma_{nm}z)v_{nm}(x,y)$$

where

$$\gamma_{nm} = (h_{nm}^2 - \omega^2\mu\epsilon)^{1/2},$$

$$u_{nm}(x,y) = (2/\sqrt{ab})sin(n\pi x/a)sin(m\pi y/b),$$

$$v_{nm}(x,y) = (2/\sqrt{ab})cos(n\pi x/a)cos(m\pi y/b)$$

These functions are normalized:

$$\int_0^a\int_0^b u_{nm}^2 dxdy = \int_0^a\int_0^b v_{nm}^2 dxdy = 1$$

and further they are orthogonal

$$\int u_{nm}u_{n'm'}dxdy = 0, (n,m) \neq (n',m')$$

$$\int v_{nm}v_{n'm'}dxdy = 0, (n,m) \neq (n',m')$$

We thus find that

$$E =$$

$$\hat{z}\sum C(n,m)u_{nm}(x,y)exp(-\gamma_{nm}z) - \sum[(\gamma_{nm}/h_{nm}^2)C(n,m)\nabla_\perp u_{nm}(x,y) + (j\omega\mu/h_{nm^2})D(n,m)\nabla_\perp v_{nm}(x,y)\times\hat{z}]exp(-\gamma_{nm}z)$$

We note that

$$\int(\nabla_\perp u_{nm}, \nabla_\perp v_{n'm'} \times \hat{z})dxdy$$

$$= \int(\nabla_\perp u_{nm} \times \nabla_\perp v_{n'm'}, \hat{z})dxdy$$

$$= \int(u_{nm,x}v_{n'm',y} - u_{nm,y}v_{n'm',x})dxdy = 0$$

on integration by parts (we are left with only boundary terms which vanish). We note that

$$E_\perp = -\sum[C(n,m)\gamma_{nm}/h_{nm}^2)\nabla_\perp u_{nm}(x,y) + D(n,m)(j\omega\mu/h_{nm^2})\nabla_\perp v_{nm}(x,y) \times \hat{z}]exp(-\gamma_{nm}z)$$

and hence

$$\int_0^a\int_0^b |E|^2 dxdy$$

$$= \sum(|C(n,m)|^2(1+|\gamma_{nm}|^2/h_{nm}^2) + |D(n,m)|^2(\mu\omega)^2/h_{nm}^2)exp(-2\alpha_{nm}z)$$

where

$$\gamma_{nm}(\omega) = \alpha_{nm}(\omega) + j\beta_{nm}(\omega)$$

Thus,

$$\epsilon\int_0^a\int_0^b\int_0^d |E|^2 dxdydz =$$

$$\sum_{n,m}(\lambda(n,m)|C(n,m)|^2 + \mu(n,m)|D(n,m)|^2)$$

where

$$\lambda(n,m) = (1+|\gamma_{nm}|^2/h_{nm}^2)(1-exp(-2\alpha_{nm}d))/2\alpha_{nm}$$

$$\mu(n,m) = ((\mu\omega)^2/h_{nm}^2)(1-exp(-2\alpha_{nm}d))/2\alpha_{nm}$$

Note that the energy in the magnetic field is given by

$$\int |\nabla \times A|^2 dxdydx/2\mu =$$

and

$$(\nabla \times A, B) = \nabla.(A \times B) - (A, \nabla \times B)$$

The first term on the rhs is a perfect divergence and by Gauss' theorem, its volume integral over the guide volume is zero if assuming that the fields vanish on the surface. Further,

$$\nabla \times B = \nabla \times (\nabla \times A) = \nabla(divA) - nabla^2 A = -\nabla^2 A$$

since $divE = 0$ implies $divA = 0$. Further, $\nabla^2 A = -\nabla^2(E/j\omega) = \omega^2\epsilon\mu E/j\omega = -j\omega\epsilon\mu E$ and hence,

$$(2\mu)^{-1}\int |\nabla \times A|^2 dxdydz =$$

$$(\epsilon/2)\int |E|^2 dxdydz$$

In other words, the energy in the magnetic field is same as the energy in the electric field. Thus, the total field energy is given by

$$\epsilon \int |E|^2 dxdydz$$

$$= \sum_{n,m} (\lambda(n,m)|C(n,m,\omega)|^2 + \mu(n,m)|D(n,m,\omega)|^2)$$

and this energy can be quantized using creation and annihilation operators in place of $C(n,m,\omega), D(n,m\omega)$ and their conjugates.

A.17. Image processing for non-Gaussian noise models based on the Edgeworth expansion.

The Edgeworth expansion: Let $\phi(x) = exp(-x^2/2)/\sqrt{2\pi}$, the standard normal density. Define the Hermite polynomials by

$$H_n(x) = (-1)^n exp(x^2/2) D^n exp(-x^2/2), D = d/dx$$

The Edgeworth expansion of a density $f(x)$ for which all the moments $\int_{\mathbb{R}} |x|^k f(x)dx, k = 1, 2, ...,$ are finite is given by

$$f(x) = \phi(x)(1 + \sum_{n \geq 1} c[n] H_n(x)) = \phi(x) + (2\pi)^{-1/2} \sum_{n \geq 1} c[n] D^n exp(-x^2/2)$$

Generating function and orthogonality of the Hermite polynomials:

$$\sum_{n \geq 0} t^n H_n(x)/n! = exp(x^2/2)(\sum_{n \geq 0} (-t)^n D^n/n!) exp(-x^2/2) = exp(x^2/2) exp(-tD) exp(-x^2/2)$$

$$= exp(x^2/2) exp(-(x-t)^2/2) = exp(tx - t^2/2)$$

Thus,

$$\sum_{n,m \geq 0} t^n s^m \int_{\mathbb{R}} H_n(x) H_m(x) \phi(x) dx = \int \phi(x) exp((t+s)x - t^2/2 - s^2/2) dx$$

$$= exp(ts)$$

and hence

$$\int H_n(x) H_m(x) \phi(x) dx = n! \delta[n-m], n, m \geq 0$$

Thus $\{H_n(x)/\sqrt{n!} : n \geq 0\}$ forms an orthnormal basis for the Hilbert space $L^2(\mathbb{R}, \phi(x)dx)$. It therefore follows that the coefficients $c[n], n \geq 0$ for the Edgeworth expansion of $f(x)$ are given by

$$c[n] = \int_{\mathbb{R}} f(x) H_n(x) dx/n!, n \geq 0$$

Note that $H_0(x) = 1$. Now consider a multivariate Edgeworth pdf defined by

$$\psi(x) = |A| \Pi_{i=1}^M f((Ax)_i)$$

where A is an $M \times M$ matrix and $x \in \mathbb{R}^M$. We have

$$\psi(x) = (2\pi)^{-M/2} |A| exp(-x^T A^T Ax/2) \Pi_{i=1}^M (1 + \sum_{n \geq 1} c[n] H_n((Ax)_i)$$

We shall calculate is moment generating function: Let $X \in \mathbb{R}^M$ have ψ as its pdf. Then

$$\hat{\psi}(t) = \mathbb{E} exp(< t, X >) = \int_{\mathbb{R}^M} exp(< t, x >) \psi(x) dx$$

$$= (2\pi)^{-M/2} \int_{\mathbb{R}^M} exp(< t, A^{-1}y >) exp(-y^T y/2) \Pi_{i=1}^M (1 + \sum_{n \geq 1} c[n] H_n(y_i)) dy$$

$$= (2\pi)^{-M/2} \Pi_{i=1}^M \int exp((A^{-T}t)_i \xi) exp(-\xi^2/2)(1 + \sum_{n \geq 1} c[n] H_n(\xi)) d\xi$$

To calculate this integral, we first observe that

$$(2\pi)^{-1/2}\int_{\mathbb{R}} exp(t\xi)exp(-\xi^2/2)H_n(\xi)d\xi$$

$$= (2\pi)^{-1/2}(-1)^n\int exp(t\xi).D_\xi^n exp(-\xi^2/2)d\xi$$

$$= t^n\int \phi(\xi)exp(t\xi)d\xi = t^n exp(t^2/2)$$

where integration by parts has been used. Thus for the above multivariate case, we get

$$\hat{\psi}(t) = exp(t^T A^{-1}A^{-T}t/2)\Pi_{i=1}^M(1+\sum_{n\geq 1} c[n]((A^{-T}t)_i)^n)$$

$$= exp(t^T(A^TA)^{-1}t/2)\Pi_{i=1}^M(1+\sum_{n\geq 1} c[n](A^{-T}t)_i)^n)$$

The approximate maximum likelihood estimator for an Edgeworth distribution: Suppose

$$y = Ax + w$$

where w has a multivariate Edgeworth expansion and x also has a multivariate Edgeworth expansion. We wish to estimate x based on y by maximizing $p(x|y)$, ie, the MAP estimate. We have

$$p(x|y)p_y(y) = p(y|x)p_x(x) = p_w(y - Ax)p_x(x)$$

Discrete time non-linear filtering theory applied to real time image parameter estimation. The parameter vector $v[n]$ at time n satisfies the stochastic difference equation

$$v[n+1] = f(n, v[n]) + \epsilon_v[n+1]$$

where $\epsilon_v[n]$ is an iid sequence. Thus, $v[n]$ is a discrete time Markov process with transition density

$$p(v[n+1]|v[n]) = p_{\epsilon_v}(v[n+1] - f(n, v[n]))$$

The image vector $x[n]$ is partitioned into patches $P_ix[n]$ with each patch given by

$$P_ix[n] = v_i[n] + \epsilon_i[n], i = 1, 2, ..., L$$

or equivalently,

$$Px[n] = v[n] + \epsilon[n]$$

where P is a non-singular square matrix and $\epsilon[n]$ is an iid sequence independent of the sequence $\epsilon_v[n], n \geq 1$. Finally, the measurement model for the image vector is given by

$$y[n] = x[n] + w[n]$$

where $w[n]$ is again an iid sequence independent of both the sequences ϵ_v and ϵ. We assume that all the three random sequences ϵ_v, ϵ, w have multivariate Edgeworth probability densities with possibly different linear combination coefficients. The aim is to dynamically estimate $v[n]$ based on

$$Y_n = \{y[k] : k \leq n\}$$

We have

$$p(v[n+1]|Y_{n+1}) = \frac{\int p(y[n+1]|v[n+1])p(v[n+1]|v[n])p(v[n]|Y_n)dv[n]}{\int p(y[n+1]|v[n+1])p(v[n+1]|v[n])p(v[n]|Y_n)dv[n]dv[n+1]}$$

We now observe that

$$y[n] = P^{-1}(v[n] + \epsilon[n]) + w[n] = P^{-1}v[n] + d[n]$$

where

$$d[n] = P^{-1}\epsilon[n] + w[n]$$

is again an iid vector valued noise. Thus, the MAP estimate of $v[n+1]$ given $Y_{n+1}$is given by

$$\hat{v}[n+1] = argmax_{v'} \int p_d(y[n+1] - Av')p_{\epsilon_v}(v' - f(n,v))p(n,v|Y_n)dv$$

where $A = P^{-1}$. We assume that

$$\hat{v}[n+1] = f(n,\hat{v}[n]) + \delta v' = \hat{v}_0[n+1] + \delta v'$$

where $\hat{v}_0[n+1] = f(n,\hat{v}[n])$, expand the above integral upto $O((\delta v')^2)$ and then maximize this w.r.t $\delta v'$ to get the extra correction. We have

$$p_d(y[n+1] - A(\hat{v}_0[n+1] + \delta v')) =$$

$$p_d(y[n+1] - A\hat{v}_0[n+1]) - p_d'(y[n+1] - A\hat{v}_0[n+1])^T A\delta v' + \frac{1}{2}\delta v'^T A^T p_d''(y[n+1] - A\hat{v}_0[n+1])A\delta v'$$

with neglect of $O(|\delta v'|^3)$ terms. Likewise,

$$p_{\epsilon_v}(\hat{v}_0[n+1] + \delta v' - f(n,v)) = p_{\epsilon_v}(\hat{v}_0[n+1] - f(n,v)) + p_{\epsilon_v}'(\hat{v}_0[n+1] - f(n,v))^T\delta v'$$

$$+\frac{1}{2}\delta v'^T p_{\epsilon_v}''(\hat{v}_0[n+1] - f(n,v))\delta v'$$

with neglect of $O(|\delta v'|^3)$. We thus obtain upto $O(|\delta v'|^2)$,

$$\hat{v}[n+1] = \hat{v}_0[n+1] + argmax_{\delta v'} \int (p_d(y[n+1] - A\hat{v}_0[n+1]) - p_d'(y[n+1] - A\hat{v}_0[n+1])^T A\delta v'$$

$$+\frac{1}{2}\delta v'^T A^T p_d''(y[n+1]-A\hat{v}_0[n+1])A\delta v')(p_{\epsilon_v}(\hat{v}_0[n+1]-f(n,v))+p_{\epsilon_v}'(\hat{v}_0[n+1]$$

$$-f(n,v))^T\delta v'+\frac{1}{2}\delta v'^T p_{\epsilon_v}''(\hat{v}_0[n+1]-f(n,v))\delta v')p(n,v|Y$$

$$= \hat{v}_0[n+1]+argmax_{\delta v'}[(\int (p_d(y[n+1]-A\hat{v}_0[n+1])p_{\epsilon_v}'(\hat{v}_0[n+1]-f(n,v))-p_{\epsilon_v}(\hat{v}_0[n+1]$$

$$-f(n,v))A^T p_d'(y[n+1]-A\hat{v}_0[n+1]))p(n,v|Y_n$$

$$+\delta v'^T(\int (A^T p_d'(y[n+1]-A\hat{v}_0[n+1])p_{\epsilon_v}'(\hat{v}_0[n+1]-f(n,v))^T+\frac{1}{2}(p_{\epsilon_v}''(\hat{v}_0[n+1]$$

$$-f(n,v))+A^T p_d''(y[n+1]-A\hat{v}_0[n+1])A)p(n,v|Y_n)dv)\delta v$$

This equation is of the form

$$\hat{v}[n+1] = \hat{v}_0[n+1] + argmax_{\delta v'}[f(n,Y_{n+1},\hat{v}_0[n+1])^T\delta v' + \frac{1}{2}\delta v'^T F(n,Y_{n+1},\hat{v}_0[n+1])\delta v'$$

$$= \hat{v}_0[n+1] - F(n,Y_{n+1},\hat{v}_0[n+1])^{-1}(f(n,Y_{n+1},\hat{v}_0[n+1]))$$

and provides the desired recursion.

Remark: The following approximation provides an alternate technique for improving the speed of the recursion:

$$\int \psi(n,y[n+1],v)p(n,v|Y_n)dv \approx$$

$$\psi(n,y[n+1],\hat{v}[n]) + \frac{1}{2}Tr(\psi_v''(n,y[n+1]\hat{v}[n])Cov(v[n]|Y_n))$$

To apply this formula, we note that the vector and matrices $f(n,Y_{n+1},\hat{v}_0[n+1])$, $F(n,Y_{n+1},\hat{v}_0[n+1])$ can be expressed as integrals of the form

$$f(n,Y_{n+1},\hat{v}_0[n+1]) = int\psi_1(n,y[n+1],\hat{v}_0[n+1],v)p(n,v|Y_n)dv$$

$$F(n,Y_{n+1},\hat{v}_0[n+1]) = \int \psi_2(n,y[n+1],\hat{v}_0[n+1],v)p(n,v|Y_n)$$

A.18.Cartan's classification of the simple Lie algebras and the Weyl character formula for the irreducible representations of Compact semisimple Lie groups.

A scheme S is a finite set linearly independent vectors (elements) $\alpha_1, ..., \alpha_n$ in a real vector space with an inner product $(.,.)$ such that

$$a(\alpha,\beta) = 2(\alpha,\beta)/(\alpha,\alpha)$$

is a non-positive integer for all $\alpha \neq \beta, \alpha, \beta \in S$. The Cauchy Schwarz inequality then implies that

$$0 \leq a(\alpha, \beta)a(\beta, \alpha) \leq 3, \alpha, \beta \in S, \alpha \neq \beta$$

ie, the product $a(\alpha, \beta)a(\beta, \alpha)$ assumes only the values $0, 1, 2, 3$. It is known from the general theory of semisimple Lie algebras, that a set of simple positive roots of a semisimple Lie algebra forms a scheme. The numbers $a(\alpha, \beta)$ are called the Cartan integers. Obviously $a(\alpha, \alpha) = 2$. To pictorially display a scheme S having n elments, we arrange these elements as vertices with the weight of each vertex α marked by a number proportional to $(\alpha, \alpha) = |\alpha|^2$.

Theorem 1: A connected scheme with n elements cannot have more than $n - 1$ links. For suppose that the elments of the scheme are $\alpha_k, k = 1, 2, ..., n$. Then consider

$$0 < (\sum_{i=1}^{n} \alpha_i/|\alpha_i, \sum_{i=1}^{n} \alpha_i/|\alpha_i|) =$$

$$n + \sum_{i<j} 2(\alpha_i, \alpha_j)/|\alpha_i||\alpha_j| = n - \sum_{i<j} (a(\alpha_i, \alpha_j)a(\alpha_j, \alpha_i))^{1/2} --- (1)$$

Note that if no link connects two elements α and β, then it means that $(\alpha, \beta) = 0$ or equivalently, $a(\alpha, \beta) = a(\beta, \alpha) = 0$. We have that α and β are connected iff

$$a(\alpha, \beta)a(\beta, \alpha) \geq 1$$

or equivalentl, iff

$$(a(\alpha, \beta).a(\beta, \alpha))^{1/2} \geq 1$$

and hence, we get from (1) that if p is the total number of links in S, then

$$0 < n - p$$

or equivalently,

$$p \leq n - 1$$

Theorem 2: A scheme cannot have any cycle. For suppose S is a connected scheme with a cycle $C = \{\alpha_1, ..., \alpha_k, \alpha_1\}$. Then C is a subscheme of S having k elements and at least k links. This contradicts the previous theorem.

Theorem 3: The number of links connecting any element in a scheme cannot exceed 3. For suppose α is an element of the scheme S that is connected to p elements $\alpha_1, ..., \alpha_p$. Let $W = span\{\alpha_1, ..., \alpha_p\}$. It is clear that no two $\alpha_i's$ connected to each other since otherwise we would have a subscheme with a cycle $(\alpha, \alpha_i, \alpha_j, \alpha_i\}$. Thus, $(\alpha_i, \alpha_j) = 0, i \neq j, i, j = 1, 2, ..., p$. In other words, the $\alpha_i's$ are mutually orthogonal vectors. Let γ be the orthogonal projection of α onto $span\{\alpha_1, ..., \alpha_p\}$. Then, we have

$$\gamma = \sum_{i=1}^{p} (\alpha, \alpha_i)\alpha_i/(\alpha_i, \alpha_i)$$

and hence,

$$(\gamma, \gamma) = \sum_{i=1}^{p} (\alpha, \alpha_i)^2/(\alpha_i, \alpha_i) < (\alpha, \alpha)$$

This can be expressed as

$$\sum_{i=1}^{p} a(\alpha, \alpha_i)a(\alpha_i, \alpha) < 4$$

and this proves our claim.

Theorem 4: A connected scheme cannot have two distinct elements each of which is connected to more than two links. For suppose $\alpha_1, ..., \alpha_n$ in S are connected in that order to their successors by single links and that α_1 is connected to elements in $S - C$ by two or more links and α_n is connected to elements in $S - C$ by two or more links. Here, $C = \{\alpha_1, ..., \alpha_n\}$. We have by construction $2(\alpha_i, \alpha_{i+1})/(\alpha_i, \alpha_i) = -1 = (\alpha_i, \alpha_{i+1})/(\alpha_{i+1}, \alpha_{i+1}), i = 1, 2, ..., n - 1$ since both of these quantities are non-positive integers (Cartan integers) whose product is 1. Thus, we get

$$(\alpha_i, \alpha_i) = (\alpha_1, \alpha_1) = c, i = 1, 2, ..., p$$

say, and

$$(\alpha_i, \alpha_{i+1}) = -c/2, i = 1, 2, ..., n-1$$

Now define

$$\alpha = \sum_{i=1}^{n} \alpha_i$$

Then, we have

$$(\alpha, \alpha_i) = (\alpha_{i-1}, \alpha_i) + (\alpha_i, \alpha_i) + (\alpha_{i+1}, \alpha_i), 2 \le i \le n-2$$
$$= -c/2 + c + c/2 = 0$$

Also

$$(\alpha, \alpha_1) = (\alpha_1, \alpha_1) + (\alpha_2, \alpha_1) = c - c/2 = c/2$$

and likewise,

$$(\alpha, \alpha_n) = c/2$$

Thus,

$$(\alpha, \alpha) = c = (\alpha_i, \alpha_i), i = 1, 2, ..., n$$

Also since $\alpha_2, ..., \alpha_{n-1}$ do not connect to $S - C$, we get for any $\beta \in S - C$ which is not connected to α_n,

$$(\alpha, \beta) = (\alpha_1, \beta)$$

and likewise for any $\beta \in S - C$ which is not connected to α_1, we have

$$(\alpha, \beta) = (\alpha_n, \beta)$$

This shows that we can define another scheme

$$S' = (S - C) \bigcup \{\alpha\}$$

such that α connects to four or more elements of $S - C$ which contradicts Theorem 3. This proves the theorem.

Theorem 5: The only scheme having a triple link can consist of exactly two elements connected by a triple link. If such a scheme $S = \{\alpha_1, \alpha_2\}$ exists. Then,

$$4(\alpha_1, \alpha_2)^2/(\alpha_1, \alpha_1)(\alpha_2, \alpha_2) = 3$$

which is possible if

$$(\alpha_1, \alpha_1) = 1, (\alpha_2, \alpha_2) = 3, (\alpha_1, \alpha_2) = -3/2$$

Note that this is a consequence of the fact that $2(\alpha_1, \alpha_2)/(\alpha_1, \alpha_1)$ and $2(\alpha_1, \alpha_2)/(\alpha_1, \alpha_2)$ must be negative integers.

Now consider an element $\delta \in S$ having three links attached to it with each of the three links leading to three simple chains. Let the first chain be $\delta, \alpha_{p-1}, \alpha_{p-2}, ..., \alpha_1$, the second chain is $\delta, \beta_{q-1}, ...\beta_1$ and the third chain is $\delta, \gamma_{r-1}, ..., \gamma_1$. Then since there cannot be any cycle in the scheme, it follows that $(\alpha_i, \beta_j) = 0, (\alpha_i, \gamma_j) = 0, (\beta_i, \gamma_j) = 0$ for all i, j.

A.19. Definition of the quantum stochastic integral in Boson Fock space.

Let $A_t(m), A_t(m)^*, \Lambda_t(H)$ denote respectively the annihilation, creation and conservation processes in the Boson Fock space $\Gamma_s(\mathcal{H} \otimes L^2(\mathbb{R}_+))$. We can write for $0 = t_0 < t_1 < t_2 < ...$, the direct sum decomposition

$$\mathcal{H} \otimes L^2(\mathbb{R}_+) = \bigoplus_{n=0}^{\infty} \mathcal{H}_{(t_n, t_{n+1}]}$$

where

$$\mathcal{H}_{(s,t]} = \mathcal{H} \otimes L^2((s,t])$$

where $L^2((s,t])$ is identified with the Hilbert subspace of $L^2(\mathbb{R}_+)$ consisting of all square integrable functions on $\mathbb{R}_+$ which vanish outside the interval $(s,t]$. If $\mathcal{H}_n, n = 1, 2, ...$ is a sequence of mutually orthogonal subspaces of a Hilbert space $\mathcal{H}$, then there is a Hilbert space isomorphism between the Boson Fock space $\Gamma_s(\bigoplus_n \mathcal{H}_n)$ and $\otimes_n \Gamma_s(\mathcal{H}_n)$. This isomorphism T is defined by identifying $e(\sum_n u_n)$ with $\otimes_n e(u_n)$ where $u_n \in \mathcal{H}_n$ and $e(u)$ is the exponential vector. First we recall that for a Hilbert space $\mathcal{H}$, the exponential vectors $e(u_1), ..., e(u_p)$ in $\Gamma_s(\mathcal{H})$ are linearly independent whenever $u_1, ..., u_p$ are distinct vectors in $\mathcal{H}$. To see this, suppose

$$\sum_{k=1}^{p} c(k)e(u_k) = 0, c(k) \in \mathbb{C}$$

Then,

$$\sum_k c(k)exp(< u, u_k >) = 0, u \in \mathcal{H}$$

Replacing u by tu and taking derivatives w.r.t t at $t = 0$ gives

$$\sum_k c(k) < u, u_k >^n = 0, n = 0, 1, 2, ...$$

Now the set of $u's$ for which $< u, u_k > \neq < u, u_m >$ for all $k \neq m$ forms a dense subset of $\mathcal{H}$. Hence, there exists a vector u such that $< u, u_k >, k = 1, 2, ..., p$ are all distinct. Hence the Vand-der-Monde matrix $((< u, u_k >^n))_{1 \leq k,m \leq p}$ is non-singular implying that $c(k) = 0, k = 1, 2, ..., p$. Now, since

$$e(u) = 1 \oplus \bigoplus_{n \geq 1} t^n u^{\otimes n} / \sqrt{n!}$$

we get

$$\frac{d^n}{dt^n} e(tu)|_{t=0} = \sqrt{n!} u^{\otimes n}$$

and since any symmetric tensor can be expressed as a linear combination of tensors of the form $u^{\otimes n}$, it follows that the exponential vectors $e(u), u \in \mathcal{H}$ span a dense subspace of $\Gamma_s(\mathcal{H})$. An adapted process $X_t, t \geq 0$ is a family of operators in $\Gamma_s(\mathcal{H})$ such that $X_t|e(u) >= (\tilde{X}_t|e(u_{t]})) \otimes |e(u_{(t}) >$ for all $t \geq 0$ where we have used the isomorphism that identifies $|e(u \oplus v) >$ with $|e(u) > \otimes |e(v) >$. Here $u_{t]} = u \otimes \chi_{[0,t]}$ and $u_{(t} = u \otimes \chi_{(t,\infty)}$. Ideally speaking, if T denotes the isomorphism that identifies the vector $e(\bigoplus_i u_i)$ in $\Gamma_s(\bigoplus_i \mathcal{H}_i)$ with $\otimes_i e(u_i)$ in $\bigotimes_i \Gamma_s(\mathcal{H}_i)$, then we should write the definition of an adapted process as

$$T(X_t|e(u) >) = (\tilde{X}_t|e(u_{t]}) >) \otimes e(u_{(t})$$

Let $P : 0 = t_0 < t_1 < ... < t_n = T$ be a partition of $[0, T]$. Its size is defined as

$$|P| = max_{0 \leq k \leq n-1}(t_{k+1} - t_k)$$

and we define the partial sum

$$I(X, A, P) = \sum_{k=0}^{n-1} X_{t_k}(A_{t_{k+1}}(m) - A_{t_k}(m))$$

Note that $X(t)$ is adapted so it acts in the Fock space $\Gamma_s(\mathcal{H}_{t]})$ while $A_{t_{k+1}} - A_{t_k}$ acts in the Fock space $\Gamma_s(\mathcal{H}_{(t_k,t_{k+1}]}$ since

$$(A_{t_{k+1}}(m) - A_{t_k}(m))|e(u) >= (\int_{t_k}^{t_{k+1}} \bar{<} m(t), u(t) > dt)|e(u) >$$

Note that time unfolds in quantum stochastic calculus as a continuous tensor product of Hilbert spaces. This can be visualized also in the classical probabilisitc setting by noting that if $\mathcal{F}_t, t \geq 0$ is a filtration on a probability space $(\Omega, \mathcal{F}, P)$ generated by a stochastic process $X(t), t \geq 0$, then for any $t_1 < t_2 < t_3$, if $\mathcal{F}_{(t_1,t_2]}$ denotes the σ field $\sigma(X(t) : t_1 < t \leq t_2)$, we can write

$$L^2(\mathcal{F}_{(t_1,t_3]}) = L^2(\mathcal{F}_{(t_1,t_2]}) \otimes L^2(\mathcal{F}_{(t_2,t_3]})$$

in the sense that any measurable functional of $X(t), t_1 < t \leq t_3$ can be expressed as a sum (poissbily infinite) of products of functions of $\{X(t) : t_1 < t \leq t_2\}$ and of $\{X(t) : t_2 < t \leq t_3\}$. More generally, we can write

$$L^2(\mathcal{F}_{(0,\infty)} = \otimes_{i=1}^{\infty} L^2(\mathcal{F}_{(t_i,t_{i+1}]})$$

where

$$0 = t_0 < t_1 < ...., t_n \to \infty$$

We now have

$$I(X, A, P)|e(u) >= \sum_{k=0}^{n-1} X(t_k)|e(u) >< < m, u >> ((t_k, t_{k+1}])$$

where

$$<< m, u >> ((s,t]) = \int_s^t < m(t'), u(t') > dt', s \leq t$$

Note that $<< m, u >>$ can be extended to a complex measure on $(\mathbb{R}_+, \mathcal{B}(\mathbb{R}_+))$. If Q is a partition finer that P, then it is clear that for each $k = 0, 1, ..., n-1$, there exist integers $a(k) < b(k)$ such that

$$(t_k, t_{k+1}] = \bigcup_{l=a(k)}^{b(k)} (s_l, s_{l+1}]$$

and the points $s_l, l = 0, 1, ...m-1$ all form the partition Q. Thus, we have

$$X(t_k) << m, u >> ((t_k, t_{k+1}]) - \sum_{l=a(k)}^{b(k)} X(s_l) << m, u >> ((s_l, s_{l+1}])$$

$$= \sum_{l=a(k)}^{b(k)} (X(t_k) - X(s_l)) << m, u >> ((s_l, s_{l+1}])$$

So

$$\| (I(X, A, P) - I(X, A, Q))|e(u) > \| \leq$$

$$max_{|t-s| \leq |P|, s, t \in [0,T]} \| (X(t) - X(s))|e(u) > \| \, | << m, u >> ([0,T])|$$

Assume that $X(t)$ is strongly uniformly continuous on $[0, T]$. Then

$$lim_{\delta \to 0} max_{|t-s| \leq \delta, s, t \in [0,T]} \| (X(t) - X(s))|e(u) > \| = 0$$

and hence it follow that if $P_n, n = 1, 2, ...$ is an increasing sequence of partitions, ie, $P_{n+1} > P_n \forall n$ and $|P_n| \to 0$ as $n \to \infty$, then

$$|(I(X, A, P_{n+m}) - I(X, A, P_n))|e(u) > \| \to 0, n \to \infty, m = 1, 2, ...$$

which implies that $I(X, A, P_n)|e(u) >, n = 1, 2, ...$ is a Cauchy sequence in the Boson Fock space $\Gamma_s(\mathcal{H} \otimes L^2(\mathbb{R}_+))$. and hence converges to an element of this space. Further, the limit is independent of the sequence of partitions P_n for if $Q_n, n = 1, 2, ...$ is another increasing sequence of partitions such that $|Q_n| \to 0$, then

$$\| (I(X, A, P_n) - I(X, A, Q_n))|e(u) > \| \leq \| (I(X, A, P_n) - I(X, A, P_n \cup Q_n))|e(u) > \|$$

$$+ \| (I(X, A, P_n \cup Q_n), I(X, A, Q_n)|e(u) > \|$$

and by the above logic, both of the terms on the rhs converge to zero, proving that the lhs also converges to zero and hence the strong limits of $I(X, A, P_n)$ and of $I(X, A, Q_n)$ are the same.

A.20. Hartree-Fock equations: $\psi_a(x), x \in \mathbb{R}^3$ are Fermionic operator fields and they satisfy the standard anticommutation relations

$$\{\psi_a(x), \psi_b^*(x')\} = \delta_{ab}\delta^3(x - x'), \{\psi_a(x), \psi_b(x')\} = 0, \{\psi_a(x)^*, \psi_b(x)^*\} = 0$$

Let

$$T(x) = -\nabla_x^2/2m$$

and let $V(x, x') = V(x', x)$ be a scalar potential field. The Hartree Fock seconed quantized Hamiltonian is defined by

$$H = H_0 + H_1,$$

$$H_0 = \int \psi_a(x)^* T(x) \psi_a(x) d^3x$$

with summation over the repeated index a being implied,

$$H_1 = \int V(x, x') \psi_a(x)^* \psi_a(x) \psi_b(x')^* \psi_b(x') d^3x d^3x'$$

Note that

$$H_0^* = H_0$$

follows by integration by parts and

$$H_1^* = H_1$$

so that

$$H^* = H$$

and hence H is a valid Hamiltonian. The Fermionic fields at time t are given by the rules of Heisenberg's matrix mechanics:

$$\psi_a(t,x) = exp(itH)\psi_a(x)exp(-itH),$$

and its adjoint

$$\psi_a(t,x)^* = exp(itH)\psi_a(x)^* exp(-itH)$$

Now,

$$\frac{\partial\psi_a(t,x)}{\partial t} = iexp(itH)[H,\psi_a(x)]exp(-itH)$$

$$[H,\psi_a(x)] = [H_0,\psi_a(x)] + [H_1,\psi_a(x)]$$

We have

$$[H_0,\psi_a(y)] = \int [\psi_b(x)^* T(x)\psi_b(x),\psi_a(y)]d^3x$$

$$= -\int \{\psi_b(x)^*,\psi_a(y)\}T(x)\psi_b(x)d^3x$$

$$= \int \delta_{ab}\delta^3(x-y)T(x)\psi_b(x)d^3x = -T(x)\psi_a(x)$$

$$[H_1,\psi_a(y)] = \int V(x,x')[\psi_b(x)^*\psi_b(x)\psi_c(x')^*\psi_c(x'),\psi_a(y)]d^3xd^3x'$$

Now,

$$\psi_a(y)\psi_b(x)^*\psi_b(x)\psi_c(x')^*\psi_c(x') =$$

$$(\delta_{ab}\delta^3(y-x) - \psi_b(x)^*\psi_a(y))\psi_b(x)\psi_c(x')^*\psi_c(x') =$$

$$\delta^3(y-x)\psi_a(y)\psi_c(x')^*\psi_c(x') + \psi_b(x)^*\psi_b(x)\psi_a(y)\psi_c(x')^*\psi_c(x')$$

So

$$[\psi_b(x)^*\psi_b(x)\psi_c(x')^*\psi_c(x'),\psi_a(y)] =$$

$$-\psi_b(x)^*\psi_b(x)\{\psi_c(x')^*,\psi_a(y)\}\psi_c(x') - \delta^3(y-x)\psi_a(y)\psi_c(x')^*\psi_c(x')$$

$$= -\delta_{ac}\delta^3(y-x')\psi_b(x)^*\psi_b(x)\psi_c(x') - \delta^3(y-x)\psi_a(y)\psi_c(x')^*\psi_c(x')$$

$$= -\delta^3(y-x')\psi_b(x)^*\psi_b(x)\psi_a(y) - \delta^3(y-x)\psi_a(y)\psi_c(x')^*\psi_c(x')$$

$$= -\delta^3(y-x')\psi_b(x)^*\psi_b(x)\psi_a(y) - \delta^3(y-x)(\delta_{ac}\delta^3(y-x') - \psi_c(x')^*\psi_a(y))\psi_c(x')$$

$$= -\delta^3(y-x')\psi_b(x)^*\psi_b(x)\psi_a(y) - \delta^3(y-x)\psi_c(x')^*\psi_c(x')\psi_a(y) - \delta^3(y-x)\delta^3(y-x')\psi_a(y)$$

Thus, we get

$$\frac{\partial\psi_a(t,y)}{\partial t} =$$

$$= -iT(t,y)\psi_a(t,y) - 2i\int V(y,x)\psi_b(t,x)^*\psi_b(t,x)d^3x - iV(y,y)\psi_a(t,y)$$

We write this equation as

$$i\frac{\partial\psi_a(t,y)}{\partial t} = \tilde{H}(t,y)\psi_a(t,y)$$

where $\tilde{H}(t,y)$ is the effective Hamiltonian operator defined by

$$\tilde{H}(t,y) = T(t,y) + 2\int V(y,x)\psi_b(t,x)^*\psi_b(t,x)d^3x + V(y,y)$$

and

$$T(t,y) = exp(itH)T(y)exp(-itH)$$

This is the Hartree-Fock equation.

A.21. Quantum scattering theory. Explicit determination of the scattering operator. $H_0 = P^2/2m, H = H_0 + V$.

$$exp(-itH_0)\psi(Q) = \psi_t(Q)$$

say. Then,

$$id\psi_t(Q)/dt = H_0\psi_t(Q) = -\nabla_Q^2\psi_t(Q)/2m$$

$$\psi_t(Q) = exp(it\nabla_Q^2/2m)\psi(Q)$$

$$exp(it\nabla_Q^/2m) = (2\pi\sigma^2)^{-3/2}\int exp(-|x|^2/2\sigma^2)exp((x,\nabla Q))d^3x$$

Then,

$$exp(\sigma^2\nabla_Q^2/2) = exp(it\nabla_Q^2/2m)$$

Hence,

$$\sigma^2 = it/m, \sigma = \sqrt{it/m}$$

So

$$\psi_t(Q) = (2\pi it/m)^{-3/2}\int exp(-m|x|^2/2it)exp((x,\nabla_Q))\psi(Q)dx$$

$$= (2\pi it/m)^{-3/2}\int exp(-m|x|^2/2it)\psi(Q+x)d^3x$$

$$= (2\pi it/m)^{-3/2}\int exp(-m|Q-x|^2/2it)\psi(x)d^3x$$

Define the Kernel function

$$K_t(Q) = (2\pi it/m)^{-3/2}exp(-m|Q|^2/2it)$$

Let

$$W(t) = exp(itH)exp(-tH_0)$$

Then,

$$W'(t) = iexp(itH).V(Q).exp(-itH_0)$$

(V is assumed to be a function of Q only. Thus,

$$W'(t) = iW(t)exp(itH_0)V(Q)exp(-itH_0)$$

Define

$$Z(t) = exp(itH_0).V(Q).exp(-itH_0)$$

Then,

$$Z'(t) = iexp(itH_0)[H_0, V(Q)]exp(-itH_0)$$

and

$$[H_0, V(Q)] = [P^2, V(Q)]/2m = (2m)^{-1}([P_a, V(Q)]P_a + P_a[P_a, V(Q)])$$

$$= -i(2m)^{-1}((\nabla_Q V(Q), P) + (P, \nabla_Q V(Q)))$$

$$(-i/m)(V'(Q), P) - (1/2m)\nabla_Q^2 V(Q)$$

Another way to evaluate this is to note that

$$Z(t) = V(exp(itH_0)Q.exp(-itH_0))$$

Now,

$$exp(itH_0)Q.exp(-itH_0) = exp(itad(H_0))(Q) = Q + it[H_0, Q] + (it)^2[H_0, [H_0, Q]] + ... =$$

$$Q + tP/m$$

Thus,

$$Z(t) = V(Q + tP/m)$$

Thus,

$$W'(t) = iW(t)V(Q + Pt/m)$$

Now suppose $|f>, |u> \in \mathcal{H}_{ac}(H_0)$. Then, by definition of the absolutely continuous spectrum of an operator, the Radon-Nikodym derivatives $d<u|E_0(\lambda)|u>/d\lambda$ and $d<f|E_0(\lambda)|f>/d\lambda$ exist and are finite. We have for $V=|u><u|$ with $<u|u>=1$,

$$W(t)|f>=exp(itH)exp(-itH_0)|f>=(i\int_0^t exp(itH)V.exp(-itH_0)dt)|f>$$

and for $\Omega_+|f>=lim_{t\to\infty}W(t)|f>$ to exist, it is sufficient that

$$X=\int_0^\infty \| V.exp(-itH_0)|f>\| dt<\infty$$

Now,

$$V.exp(-itH_0)|f>=|u><u|exp(-itH_0)|f>$$

so

$$\| V.exp(-itH_0)|f>\|=|<u|exp(-itH_0)|f>|$$

$$=|\int_{\mathbb{R}}(d<u|E_0(\lambda)|f>/d\lambda)exp(-i\lambda t)d\lambda|$$

This is the magnitude of the Fourier transform of the function $\lambda\to d<u|E(\lambda)|f>/d\lambda$. We note that

$$|<u|dE(\lambda)|f>|\leq <u|dE_)(\lambda)|u>^{1/2}<f|dE_0(\lambda)|f>^{1/2}$$

Hence,

$$|d<u|E_0(\lambda)|f>/d\lambda|\leq (d\| E_0(\lambda)|u>\|^2/d\lambda)^{1/2}(d\| E_0(\lambda)|f>\|^2/d\lambda)^{1/2}$$

A necessary condition for the magnitude of the Fourier transform of a function to be integrable is that the Fourier transform be finite. Thus, a necessary condition for $X<\infty$ is satisfied since the Radon-Nikodym derivatives $d\| E(\lambda)|u>\|^2/d\lambda$ and $d\| E_0(\lambda)|f>\|^2/d\lambda$ are finite because both $|f>$ and $|u>$ belong to $\mathcal{H}_{ac}(H_0)$.

Suppose $H_0=P$ (in one dimension) and $V=V(Q)$. $H=H_0+V=P+V(Q)$. Then, $V.exp(itP)f(x)=V(x)f(x+t)$. So, $\Omega_+|f>$ will exist if the function $t\to(\int_{\mathbb{R}}V(x)^2|f(x+t)|^2dx)^{1/2}$ is integrable on $\mathbb{R}_+$.

Consider now two Hamiltonians $H_0, H=H_0+V$ and let $\phi:\mathbb{R}\to\mathbb{R}$. Consider now the Hamiltonians $\phi(H_0),\phi(H)$. Let

$$W_\phi(t)=exp(it\phi(H))exp(-it\phi(H_0))$$

Then,

$$W_\phi(t)=I+i\int_0^t exp(is\phi(H))(\phi(H)-\phi(H_0))exp(-is\phi(H_0))ds$$

So $W_\phi(\infty)|f>$ will exist if

$$\int_0^\infty \| (\phi(H)-\phi(H_0))exp(-is\phi(H_0))|f>\| ds<\infty$$

We note that

$$<u|(\phi(H)-\phi(H_0))exp(-is\phi(H_0))|f>=\int_{\mathbb{R}}exp(-is\phi(\lambda))d<u|\phi(H)E_0(\lambda)|f>$$

$$-\int_{\mathbb{R}}exp(-is\phi(\lambda))\phi(\lambda)d<u|E_0(\lambda)|f>$$

In particular, if ϕ is an invertible function we can write

$$<u|(\phi(H)-\phi(H_0))exp(-is\phi(H_0))|f>=$$

$$\int(exp(-is\lambda)d<u|\phi(H)E_0(\phi^{-1}(\lambda))|f>/d\lambda)d\lambda$$

$$-\int exp(-is\lambda)\lambda(d<u|E_0(\phi^{-1}(\lambda))|f>/d\lambda)d\lambda$$

provided we assume that the concerned Radon-Nikodym derivatives exist. These will exist provided that all $|u>$, $\phi(H)|u>$ and $|f>$ belong to the absolutely continuous parts of the spectral measure $E_0o\phi^{-1}$ ie they belong to $\mathcal{H}_{ac}(\phi(H_0))$.

A.22.Hartree Fock approximation to the two electron problem of the Helium atom. The Hamiltonian is

$$H = H_1 + H_2 + V_{12}$$

where

$$H_1 = -\nabla_1^2/2m - 2e^2/r_1, H_2 = -\nabla_2^2/2m - 2e^2/r_2, V_{12} = e^2/r_{12}$$

Let us try the wave function (antisymmetric because the two electrons form a Fermionic pair)

$$\psi = (\psi_1 \otimes \psi_2 - \psi_2 \otimes \psi_1)/\sqrt{2}$$

with the constraints

$$< \psi_1|\psi_1 >=< \psi_2|\psi_2 >= 1, < \psi_1|\psi_2 >= 0$$

As in all eigenvalue problems we extremize

$$S =< \psi|H|\psi > -2E_1(< \psi_1|\psi_1 > -1) - 2E_2(< \psi_2|\psi_2 >) - 2\lambda_1 Re(< \psi_1|\psi_2 >) - 2\lambda_2 Im(< \psi_2|\psi_1 >))$$

We first observe that taking into account the constraints,

$$S =< \psi_1 \otimes \psi_2 - \psi_2 \otimes \psi_1|(H_1 + H_2 + V_{12})|\psi_1 \otimes \psi_2 - \psi_2 \otimes \psi_1 >$$

$$-2E_1(< \psi_1|\psi_1 > -1) - 2E_2(< \psi_2|\psi_2 >) - 2\lambda_1 Re(< \psi_1|\psi_2 >) - 2\lambda_2 Im(< \psi_2|\psi_1 >))$$

$$=< \psi_1|H_1|\psi_1 > + < \psi_2|H_1|\psi_2 > + < \psi_1|H_2|\psi_1 > + < \psi_2|H_2|\psi_2 > + < \psi_1 \otimes \psi_2|V_{12}|\psi_1 \otimes \psi_2 >$$

$$< \psi_2 \otimes \psi_1|V_{12}|\psi_2 \otimes \psi_1 > - < \psi_1 \otimes \psi_2|V_{12}|\psi_2 \otimes \psi_1 >$$

$$- < \psi_2 \otimes \psi_1|V_{12}|\psi_1 \otimes \psi_2 > -2E_1(< \psi_1|\psi_1 > -1) - 2E_2(< \psi_2|\psi_2 >) - 2\lambda_1 Re(< \psi_1|\psi_2 >) - 2\lambda_2 Im(< \psi_2|\psi_1 >))$$

Now,

$$\delta S/\delta \bar{\psi}_1 = 0$$

gives

$$2H_1\psi_1(r_1) + 2 < I \otimes \psi_2|V_{12}|\psi_1 \otimes \psi_2 >$$

$$-2 < I \otimes \psi_2|V_{12}|\psi_2 \otimes \psi_1 >$$

$$-2E_1\psi_1 - \lambda_1|\psi_1 > -i\lambda_2|\psi_2 >= 0$$

and likewise another equation for $\delta S/\delta \bar{\psi}_2 = 0$. Expanding the above, we get

$$2(-\nabla_1^2/2m - 2e^2/r_1 - E_1 - \lambda_1)\psi_1(r_1) + 2\int \bar{\psi}_2(r_2)(e^2/r_{12})(\psi_1(r_1)\psi_2(r_2) - \psi_2(r_1)\psi_1(r_2))d^3r_2$$

$$-i\lambda_2\psi_2(r_1) = 0$$

We note that the second term can be expressed as

$$\int \bar{\psi}_2(r_2)(e^2/r_{12})(\psi_1(r_1)\psi_2(r_2) - \psi_2(r_1)\psi_1(r_2))d^3r_2$$

$$= [\int (|\psi_2(r_2)|^2(e^2/r_{12})d^3r_2]\psi_1(r_1) - (\int (e^2/r_{12})\bar{\psi}_2(r_2)\psi_1(r_2)d^3r_2)\psi_2(r_1)$$

The first term in this expression represents the potential energy produced by a smeared second electron charge on the first charge, the charge density of this smeared distribution being given by $e|\psi_2(r_2)|^2$. The second term in the above expression represents the effect of the spin interaction between the two electrons, the interaction caused by the fact that both the electrons cannot occupy the same state.

Remark: The constraint $< \psi_1|\psi_2 >= 0$ need not be introduced. Neither do the constraints $< \psi_1|\psi_1 >=< \psi_2|\psi_2 >= 1$ need to be imposed. We only need to introduce the constraint that the overall wave function be normalized, ie,

$$1 =< \psi|\psi >$$

which is equivalent to

$$2 =< \psi_1 \otimes \psi_2 - \psi_2 \otimes \psi_1|\psi_1 \otimes \psi_2 - \psi_2 \otimes \psi_1 >=$$

$$2 < \psi_1|\psi_1 >< \psi_2|\psi_2 > -2| < \psi_1|\psi_2 > |^2$$

or equivalently,

$$1 =< \psi_1|\psi_1 >< \psi_2|\psi_2 > -| < \psi_1|\psi_2 > |^2$$

This results in the following version of the Hartree Fock equation

$$2(-\nabla_1^2/2m - 2e^2/r_1)\psi_1(r_1) + \int \bar{\psi}_2(r_2)(e^2/r_{12})(\psi_1(r_1)\psi_2(r_2) - \psi_2(r_1)\psi_1(r_2))d^3r_2$$

$$+ \int \bar{\psi}_2(r_1)(e^2/r_{12})(\psi_2(r_1)\psi_1(r_2) - \psi_1(r_1)\psi_2(r_2))d^3r_2$$

$$-E(< \psi_2|\psi_2 > \psi_1(r_1) - < \psi_2|\psi_1 > \psi_2(r_1)) = 0$$

with another equation of the same type, ie, with the same value of E for ψ_2. We can generalize this to a system of N interacting Fermions. The Hamiltonian of such a system is given by

$$H = \sum_{i=1}^N H_i + \sum_{1\le i<j\le N} V_{ij}$$

where H_i acts in the Hilbert space $\mathcal{H}_i$ and V_{ij} acts in the Hilbert space $\mathcal{H}_i \otimes \mathcal{H}_j$. H acts in the Hilbert space $\bigotimes_{i=1}^N \mathcal{H}_i$. The test wave function chosen is the antisymmetrized form of a separable wave function:

$$\psi = \sum_{\sigma\in S_N} sgn(\sigma)\psi_{\sigma 1} \otimes ... \otimes \psi_{\sigma N}$$

The normalization condition reads

$$1 =< \psi|\psi >= N!det((< \psi_a|\psi_b >)) = N!\sum_\sigma sgn(\sigma)\Pi_{k=1}^N < \psi_k|\psi_{\sigma k} >= \Lambda[\psi]$$

say. We have

$$S_1[\psi] =< \psi|\sum_{k=1}^N H_k|\psi >= \sum_{\sigma,\tau\in S_N} sgn(\sigma\tau) < \psi_{\sigma 1} \otimes ...\psi_{\sigma N}|H_k|\psi_{\tau 1} \otimes ...\psi_{\tau N} >$$

$$= \sum_{\sigma,\tau,k} sgn(\sigma\tau)(\Pi_{j=1,j\ne k}^N < \psi_{\sigma j}, \psi_{\tau j} >) < \psi_{\sigma k}|H_k|\psi_{\tau k} >$$

$$= \sum_{\sigma,\tau,k} sgn(\sigma\tau)(\Pi_{j=1,j\ne k}^N < \psi_j, \psi_{\tau\sigma^{-1}j} >) < \psi_k, |H_{\sigma^{-1}k}|\psi_{\tau\sigma^{-1}k} >$$

$$= \sum_{\rho,\sigma,k} sgn(\rho)(\Pi_{j=1,j\ne k}^N < \psi_j, \psi_{\rho j} >) < \psi_k|H_{\sigma k}|\psi_{\rho k} >$$

and

$$S_2[\psi] =< \psi|\sum_{i<j} V_{ij}|\psi >=$$

$$\sum_{\sigma,\tau} sgn(sigma\tau)(\Pi_{k=1,k\ne i,j}^N < \psi_{\sigma k}|\psi_{\tau k} >) < \psi_{\sigma i} \otimes \psi_{\sigma j}|V_{ij}|\psi_{\rho i} \otimes \psi_{\rho j} >$$

$$= \sum_{\sigma,\rho,i,j:\sigma^{-1}i<\sigma^{-1}j} sgn(\rho)(\Pi_{k=1,k\ne i,j}^N < \psi_k|\psi_{\rho k} >) < \psi_i \otimes \psi_j|V_{\sigma^{-1}i,\sigma^{-1}j}|\psi_{\rho i} \otimes \psi_{\rho j} >$$

Let

$$S[\psi] = S_1[\psi] + S_2[\psi]$$

The variational principle

$$\delta_{\bar{\psi}_k}(S[\psi] - E(\Lambda[\psi] - 1)) = 0$$

thus gives

$$\sum_{\rho,\sigma} sgn(\rho)(\Pi_{j=1\ne k}^N < \psi_j|\psi_{\rho j} >)H_{\sigma k}|\psi_{\rho k} >$$

$$+ \sum_{\rho,\sigma,j} sgn(\rho)|\psi_{\rho k} > (\Pi_{l=1,l\ne k,j}^N < \psi_l|\psi_{\rho l} >) < \psi_j|H_{\sigma j}|\psi_{\rho j} >$$

$$+ \sum_{\sigma,\rho,j:\sigma^{-1}k<\sigma^{-1}j} sgn(\rho)(\Pi_{m=1,m\neq k,j}^{N} < \psi_m|\psi_{\rho m} >)|\psi_{\rho k} >< I \otimes \psi_j|V_{\sigma^{-1}k,\sigma^{-1}j}|I \otimes \psi_{\rho j} >$$

$$+$$

$$\sum_{\sigma,\rho,i:\sigma^{-1}i<\sigma^{-1}k} sgn(\rho)(\Pi_{m=1,m\neq i,k}^{N} < \psi_m|\psi_{\rho m} >)|\psi_{\rho k} >< \psi_i \otimes I|V_{\sigma^{-1}i,\sigma^{-1}k}|\psi_{\rho i} \otimes I >$$

$$+ \sum_{\sigma,\rho,\sigma^{-1}i<\sigma^{-1}j} sgn(\rho)(\Pi_{m=1,m\neq i,j,k}^{N} < \psi_m|\psi_{\rho m} >)|\psi_{\rho k} >< \psi_i \otimes \psi_j|V_{\sigma^{-1}i,\sigma^{-1}j}|\psi_{\rho i} \otimes \psi_{\rho j} >$$

$$= N!E \sum_{\sigma} sgn(\sigma)|\psi_{\sigma k} > (\Pi_{j=1,j\neq k}^{N} < \psi_j|\psi_{\sigma j} >), k = 1, 2, ..., N$$

A.23.Plasmonic waveguides.

A rectangular waveguide is filled with a plasma. The charge per particle of the plasma is q and the particle distribution function is $f(t, r, v)$. This distribution function is assumed to vanish on the boundaries of the guide and hence its Fourier transform w.r.t time can be expanded as

$$ff(\omega, r, v) = \sum_{n,m\geq 1} f_{nm}(\omega, v)u_{nm}(x, y)exp(-\gamma_{nm}(\omega)z)$$

where

$$u_{nm}(x, y) = (2/\sqrt{ab})sin(n\pi x/a)sin(m\pi y/b), 0 \leq x \leq a, 0 \leq y \leq b$$

For a fixed n, m, we write $f(\omega, v)$ for f_{nm} and γ for γ_{nm}. The Maxwell equations

$$curlE = -j\omega\mu H, curlH = J + j\omega\epsilon E$$

where

$$J(\omega, x, y) = \int qv\delta f(\omega, x, y, v)d^3v$$

where the $f(\omega, x, y, v)exp(-\gamma z)$ is a component of the particle density corresponding to propagation constant γ. In component form, we have

$$E_{z,y}(x, y) + \gamma E_y(x, y) = -j\omega\mu H_x(x, y), \gamma E_x + E_{z,x} = j\omega\mu H_y,$$

$$E_{y,x} - E_{x,y} = -j\omega\mu H_z,$$

$$H_{z,y} + \gamma H_y = J_x(x, y) + j\omega\epsilon E_x,$$

$$\gamma H_x + H_{z,x} = -J_y(x, y) - j\omega\epsilon E_y,$$

$$H_{y,x} - H_{x,y} = J_z(x, y) + j\omega\epsilon E_z(x, y)$$

where the frequency argument ω has been omitted. Here,

$$J_x(x, y) = \int qv_x\delta f(x, y, v)d^3v, J_y(x, y) = \int qv_y\delta f(x, y, v)d^3v, J_z(x, y) = \int qv_z\delta f(x, y, v)d^3v - - - (a)$$

These equations can be solved for E_x, E_y, H_x, H_y in terms of $J(x, y)$ and the partial derivatives of $E_z(x, y), H_z(x, y)$. Having done so, we substitute these into the Boltzmann kinetic transport equation with the unperturbed Maxwell distribution function $f_0(v)$ being taken in place of f where multiplication with the em fields is concerned. Thus, we get the approximate equation

$$j\omega\delta f(x, y, v) + v_x\delta f_{,x}(x, y, v) + v_y\delta f_{,y}(x, y, v) - \gamma v_z\delta f(x, y, v) + q(E + \mu v \times H, \nabla_v)f_0(v) = -\delta f(x, y, v))/\tau(v) - - - (b)$$

where the relaxation time approximation for the collision term has been used. Here, $\delta f(x, y, v) = f(x, y, v) - f_0(v)$. To proceed further, we first solve the above equations for E_x, E_y, H_x, H_y:

$$\begin{pmatrix} \gamma & -j\omega\mu \\ j\omega\epsilon & -\gamma \end{pmatrix} (E_x, H_y)^T = (-E_{z,x}, H_{z,y} - J_x(x, y))^T,$$

$$\begin{pmatrix} \gamma & j\omega\mu \\ j\omega\epsilon & \gamma \end{pmatrix} (E_y, H_x)^T = (-E_{z,y}, -H_{z,x} - J_y(x, y))^T,$$

Solving these gives

$$E_x = -(\gamma/h^2)E_{z,x} - (j\omega\mu/h^2)(H_{z,y} - J_x) - - - (1)$$
$$E_y = -(\gamma/h^2)E_{z,y} + (j\omega\mu/h^2)(H_{z,x} + J_y) - - - (2)$$
$$H_x = (j\omega\epsilon/h^2)E_{z,y} - (\gamma/h^2)(H_{z,x} + J_y) - - - (3)$$
$$H_y = -(j\omega\epsilon/h^2)E_{z,x} - (\gamma/h^2)(H_{z,y} - J_x) - - - (4)$$

where

$$h^2 = \gamma^2 + \omega^2\epsilon\mu$$

Substituting these into the third components of the Maxwell curl equations gives

$$(H_{z,xx} + J_{y,x}) + (H_{z,yy} - J_{x,y}) + h^2 H_z = 0$$

$$(E_{z,xx} + E_{z,yy}) - (\gamma/j\omega\epsilon)(J_{x,x} + J_{y,y}) + h^2 E_z = 0$$

or equivalently,

$$(\nabla_\perp^2 + h^2)E_z = (\gamma/j\omega\epsilon)(J_{x,x} + J_{y,y}),$$
$$(\nabla_\perp^2 + h^2)H_z = J_{x,y} - J_{y,x}$$

In accordance with the boundary conditions on the tangential components of E and the normal components of H, we have the expansions

$$E_z(x,y,z) = \sum_{n,m} E_{nm}u_{nm}(x,y)exp(-\gamma_{nm}z), H_z(x,y,z) = \sum_{n,m} H_{nm}w_{nm}(x,y)exp(-\gamma_{nm}z)$$

where

$$w_{nm}(x,y) = (2/\sqrt{ab})cos(n\pi x/a)cos(m\pi y/b)$$

Formally, we can write

$$E_z = (\nabla_\perp^2 + h^2)^{-1}(\gamma/j\omega\epsilon)(\nabla_\perp.J_\perp) - - - (5)$$
$$H_z = (\nabla_\perp^2 + h^2)^{-1}(J_{x,y} - J_{y,x}) - - - (6)$$

and express E_x, E_y, H_x, H_y in terms of J using eqns. (1)-(6). Finally, these expressions are substituted into the linearized Boltzmann eqn. (b) by making use of (a). The result is a linear integro-partial differential equation for the Boltzmann particle distribution function $f(\omega, x, y, v)$ with γ as a parameter. The solutions of this equation will generally lead to discrete values of the propagation constant γ.

A.24. Winding number of planar Brownian motion. Let $Z_t = X_t + iY_t$ be a complex Brownian motion, ie, X, Y are independent real standard Brownian motion processes. We write

$$\rho_t = \sqrt{X_t^2 + Y_t^2} = |Z_t|, \theta_t = Tan^{-1}(Y_t/X_t) = Arg(Z_t)$$

Then,

$$log(\rho_t) = Relog Z_t, \theta_t = Imlog(Z_t)$$

Away from the origin, $z \to log z$ is an analytic function of a complex variable and hence its Laplacian vanishes. Thus, from Ito's formula,

$$dlog(Z_t) = dZ_t/Z_t$$

and hence $log(Z_t)$ is a Martingale. Writing

$$log(z) = u(x,y) + iv(x,y), z = x + iy$$

it follows that $u(X_t, Y_t)$ and $v(X_t, Y_t)$ are both real Martingales. We have

$$u(X_t, Y_t) = log(\rho_t), v(X_t, Y_t) = \theta_t$$

Further, the quadratic variation of the Martingales $u(X_t, Y_t)$ and $v(X_t, Y_t)$ are the same processes. They are equal to

$$\int_0^t |\nabla u(X_s, Y_s)|^2 ds = \int_0^t |\nabla v(X_s, Y_s)|^2 ds = \int_0^t ds/|Z_s|^2$$

where we make use of the Cauchy-Riemann equations,

$$u_{,x} = v_{,y}, u_{,y} = -v_{,x}$$

and of course

$$u_{,x} + iu_{,y} = u_{,x} - iv_{,x}, v_{,x} + iv_{,y} = -u_{,y} + iv_{,y}$$

while on the other hand, writing $f(z) = logz = u + iv$, we have

$$f'(z) = u_{,x} + iv_{,x}, if'(z) = u_{,y} + iv_{,y}$$

Thus,

$$|f'(z)|^2 = |\nabla u|^2 = |\nabla v|^2 = 1/|z|^2 = 1/(x^2 + y^2) = 1/\rho^2$$

Thus, writing

$$C(t) = \int_0^t ds/|Z_s|^2 = \int_0^t ds/\rho_s^2$$

it follows that there exists a planar Brownian motion $\beta(t) + i\gamma(t)$ such that

$$log(\rho_t) + i\theta_t = logZ_t = \beta(C(t)) + i\gamma(C(t))$$

so that

$$\beta(C(t)) = log(\rho_t), \gamma(C(t)) = \theta_t$$

In fact, if $\tau()$ is the inverse function of C, then we have that $u(X_{\tau(t)}, Y_{\tau(t)})$ and $v(X_{\tau(t)}, Y_{\tau(t)})$ are independent Brownian motion processes which we denote by $\beta(t)$ and $\gamma(t)$ respectively. We have

$$f(Z_{\tau(t)}) = \beta(t) + i\gamma(t)$$

and

$$df(Z_{\tau(t)}) = f'(Z_{\tau(t)})dZ_{\tau(t)}$$

(since $(dZ)^2 = 0$). Thus,

$$(df(Z_{\tau(t)}))^2 = f'(Z_{\tau(t)})^2(dZ_{\tau(t)})^2 = 0$$

which proves independence of the Brownian motions β and γ. Note that we can write

$$\beta(t) = log(\rho_{\tau(t)}), \gamma(t) = \theta_{\tau(t)}$$

Remark: The real and imaginary parts $\beta(t)$ and $\gamma(t)$ of $f(Z_{\tau(t)})$ are continuous martingales with quadratic variation matrix equal to $\begin{pmatrix} dt & 0 \\ 0 & dt \end{pmatrix} = dt.I_2$ which proves by Levy's theorem for vector valued Martingales that these two processes are independent standard Brownian motion processes. Now, the angle turned by the planar Brownian motion process Z_t around the origin respectively inside and outside a circle of radius r are

$$\theta_{r-}(t) = Im\int_0^t \chi_{|Z_s|<r} dlog(Z_s)), \theta_{r+}(t) = Im\int_0^t \chi_{|Z_s|>r} dlog(Z_s)$$

and these can be expressed as

$$\theta_{r-}(t) = \int_0^t \chi_{\rho_s<r} d\theta_s$$

$$= Im\int_0^t \chi_{|Z_s|<r} dZ_s/Z_s,$$

$$\theta_{r+}(t) = \int_0^t \chi_{\rho_s>r} d\theta_s = Im\int_0^t \chi_{|Z_s|>r} dZ_s/Z_s$$

Using the above time change result,

$$\theta_{r-}(t) = \int_0^{C(t)} \chi_{\beta(s)<log(r)} d\gamma(s),$$

$$\theta_{r+}(t) = \int_0^{C(t)} \chi_{\beta(s)>log(r)} d\gamma(s)$$

Now define

$$T_a = min(t \geq 0 : \beta(t) = a)$$

and

$$\sigma_r = min(t \geq 0 : \rho_t = r)$$

Then,

$$\sigma_r = min(t : log(\rho_t) = log(r)) = min(t : \beta(C(t)) = log(r))$$

Thus,

$$C(\sigma_r) = min(C(t) : \beta(C(t)) = log(r)) = min(t : beta(t) = log(r)) = T_{log(r)}$$

Now for a set E in $\mathbb{C}$, consider

$$I(t) = \int_{\sigma\sqrt{t}}^{t} \chi_{Z(s)\in E} dZ_s/Z_s$$

We write

$$Z_t = a + \hat{Z}_t, a \neq 0$$

We assume that $Z_0 = a$ so that $\hat{Z}_0 = 0$. We have

$$I(t) = \int_{\sigma\sqrt{t}}^{t} \chi_{\hat{Z}_s \in E-1} d\hat{Z}_s/(a + \hat{Z}_s)$$

Now, the process $\hat{Z}_{ts}, s \geq 0$ has the same law as $\sqrt{t}\hat{Z}_s, s \geq 0$ and hence $I(t)$ has the same distribution as

$$\tilde{I}(t) = \int_{t^{-1}\sigma'_{\sqrt{t}}}^{1} \chi_{\hat{Z}_s \in (E-1)/\sqrt{t}} \sqrt{t} d\hat{Z}_s/(a + \sqrt{t}\hat{Z}_s)$$

where σ'_r is the same as σ_r but with the process $a + \sqrt{t}\hat{Z}_{s/t}, s \geq 0$ used in place of $\hat{Z}_s, s \geq 0$ (Note that these two processes have the same law). Now

$$t^{-1}\sigma'_{\sqrt{t}} = min(s/t : |a + \sqrt{t}\hat{Z}_{s/t}| = \sqrt{t})$$
$$= min(s : |a + \sqrt{t}\hat{Z}_s| = \sqrt{t})$$
$$= min(s : |a/\sqrt{t} + \hat{Z}_s| = 1)$$

This gives the result that $t^{-1}\sigma'_{\sqrt{t}}$ converges as $t \to \infty$ to the random variable $min(s : |\hat{Z}_s| = 1) = \sigma$ say. It follows on taking lim $t \to \infty$ that the limit in law of $I(t)$ as $t \to \infty$ is given by the random variable

$$\int_{\sigma}^{1} \chi_{\hat{Z}_s \in E_0} d\hat{Z}_s/\hat{Z}_s$$

where

$$E_0 = lim_{t\to\infty}(E - 1)/\sqrt{t}$$

which is a finite random variable since $d\hat{Z}_s/\hat{Z}_s = dlog(\hat{Z}_s)$ and σ equals the limit as $t \to \infty$ of the random variable $min(s : |1/\sqrt{t} + \hat{Z}_s| = 1)$. In particular, we get that

$$lim_{t\to\infty}|\theta_{r-}(t) - \theta_{r-}(\sigma_{\sqrt{t}})|/logt = 0$$

and likewise for θ_{r+}. Thus, it follows that if $lim_{t\to\infty}\theta_{r-}(\sigma_{\sqrt{t}})/log(t)$ exists in law, then this limit coincides in law with the limit in law of $\theta_{r-}(t)/log(t)$ and likewise for $\theta_{r+}(t)$. Now,

$$\theta_{r-}(\sigma_{\sqrt{t}}) = \int_0^{C(\sigma_{\sqrt{t}})} \chi_{\beta(s)<log(r)} d\gamma(s)$$

$$= \int_0^{T_{log\sqrt{t}}} \chi_{\beta(s)<log(r)} d\gamma(s)$$

and we have for any $u > 0$,

$$T_a = min(s : \beta(s) = a) = min(us : \beta(us) = a) = u.min(s) : \sqrt{u}\beta(s) = a)(inlaw)$$

$$= uT_{a/\sqrt{u}}$$

It follows that $\theta_{r-}(\sigma_{\sqrt{t}})$ has the same law as

$$\int_0^{uT_{log(\sqrt{t})/\sqrt{u}}} \chi_{\beta(s)<log(r)} d\gamma(s)$$

Thus, by changing the argument s to us in the above formula for $\theta_{r-}(\sigma_{\sqrt{t}})$, it follows that $\theta_{r-}(\sigma_{\sqrt{t}})$ has the same law as

$$\int_0^{T_{log(\sqrt{t})/\sqrt{u}}} \chi_{\beta(s)<log(r)/\sqrt{u}} \sqrt{u} d\gamma(s)$$

Taking $u = (log\sqrt{t})^2$ gives us the result that $\theta(\sigma_{\sqrt{t}})/logt$ has the same law as

$$(1/2)\int_0^{T_1} \chi_{\beta(s)<log(r)/log(\sqrt{t})} d\gamma(s)$$

We note that the process $\sqrt{a}\beta(s), s \geq 0$ has the same law as the process $\beta(as), s \geq 0$ and $\sqrt{a}\gamma(s), s \geq 0$ has the same law as $\gamma(as), s \geq 0$. Letting $t \to \infty$ gives the result that $\theta_{r-}(\sigma_{\sqrt{t}})/log(t)$ converges in law (distribution) as $t \to \infty$ to

$$(1/2)\int_0^{T_1} \chi_{\beta(s)<0} d\gamma(s)$$

We have thus deduced that $\theta_{r-}(t)/log(t)$ converges in law as $t \to \infty$ to the random variable

$$(1/2)\int_0^{T_1} \chi_{\beta(s)<0} d\gamma(s)$$

where

$$T_1 = min(s) : \beta(s) = 1)$$

Likewise, $\theta_{r+}(t)/log(t)$ converges in law as $t \to \infty$ to the random variable

$$(1/2)\int_0^{T_1} \chi_{\beta(s)>0} d\gamma(s)$$

A.25. Perturbation to the energy-momentum tensor of the electromagnetic field caused by the expansion of the universe in the Robertson-Walker metric. The metric is

$$d\tau^2 = dt^2 - S^2(t)(dx^2 + sy^2 + dz^2)$$

The Maxwell equations in this metric are

$$(F^{\mu\nu}\sqrt{-g})_{,\nu} = 0$$

Now,

$$-g = S^6(t), \sqrt{-g} = S^3(t)$$

$$F_{\alpha\beta} = A_{\beta,\alpha} - A_{\alpha,\beta}$$

Gauge condition chosen is

$$(A^\mu\sqrt{-g})_{,\mu} = 0$$

or equivalently,

$$A^\mu_{:\mu} = 0$$

This is the general relativistic version of the Lorentz gauge. Now,

$$F_{0r} = A_{r,0} - A_{0,r} = -(S^2A^r)_{,0} - A^0_{,r} = E_r$$

say.

$$F_{rs} = A_{s,r} - A_{r,s} = S^2(A^r_{,s} - A^s_{,r})$$

E_r is the general relativistic version of the electric field and

$$F_{12} = S^2(A^1_{,2} - A^2_{,1}) = -B_3,$$

$$F_{23} = S^2(A^2_{,3} - A^3_{,2}) = -B_1,$$

$$F_{31} = S^2(A^3_{,1} - A^1_{,3}) = -B_2$$

B_r is the general relativistic version of the magnetic field. The Maxwell equation

$$(F^{0r}\sqrt{-g})_{,r} = 0$$

gives

$$F_{0r,r} = 0$$

or equivalently,

$$E_{r,r} = divE = 0$$

This can be expressed as

$$(S^2A^r)_{,0r} + \nabla^2 A^0 = 0$$

or equivalently, with $A = (A^r)$ and $V = A^0$,

$$(S^2)'divA + S^2(divA)_{,0} + \nabla^2 V = 0$$

The equation

$$(F^{r\mu}\sqrt{-g})_{,\mu} = 0$$

is the same as

$$(F^{r0}S^3)_{,0} + S^3 F^{rs}_{,s} = 0$$

or equivalently,

$$(SF_{0r})_{,0} + S^{-1}F_{rs,s} = 0$$

or

$$(S(A_{r,0} - A_{0,r}))_{,0} + S^{-1}(A_{s,rs} - \nabla^2 A_r) = 0$$

or

$$-(S(S^2A^r)_{,0})_{,0} - (SA^0_{,r})_{,0} + S(\nabla^2 A^r - A^s_{,rs}) = 0$$

or equivalently,

$$(2SS'^2A^r + S^2A^r_{,0} + S^3A^r_{,00} + 4S^2S'A^r_{,0} + (S^2)''A^r) + S'A^0_{,r} + SA^0_{,r0}$$
$$+S(divA)_{,r} - S\nabla^2 A^r = 0$$

or

$$2SS'^2A + S^2A_{,0} + S^3A_{,00} + 4S^2S'A_{,0} + (S^2)''A + S'\nabla V + S\nabla V_{,0}$$
$$+S\nabla(divA) - S\nabla^2 A = 0$$

The gauge condition reads

$$S^3A^r_{,r} + (S^3A^0)_{,0} = 0$$

or equivalently,

$$S^3(divA + V_{,0}) + (S^3)'V = 0$$

or equivalently,

$$divA + V_{,0} + 3S'V/S = 0 - - - (1)$$

We record the other two equations below:

$$(S^2)'divA + S^2(divA)_{,0} + \nabla^2 V = 0 - - - (2),$$

$$2S'^2A + SA_{,0} + S^2A_{,00} + 4SS'A_{,0} + ((S^2)''/S)A + (S'/S)\nabla V + \nabla V_{,0}$$
$$+\nabla(divA) - \nabla^2 A = 0 - - - (3)$$

Substituting (1) into (2) gives

$$-(S^2)'(V_{,0}+3S'V/S)-S^2(V_{,00}+3(S'V/S)_{,0})+\nabla^2 V=0$$

which can be rearranged as

$$\nabla^2 - S^2 V_{,00} = 2SS'V_{,0}+6S'^2V+3S^2(S'V/S)_{,0} ---(4)$$

This is the modified general relativistic wave equation for V. Likewise, substituting (1) into (3) gives

$$\nabla^2 A - S^2 A_{,00} = 2S'^2A+4SS'A_{,0}+((S^2)''/S)A+(S'/S)\nabla V-\nabla(3S'V/S) ---(5)$$

This is the modified general relativistic wave equation for A.

The energy momentum tensor of the electromagnetic field modified by general relativistic (cosmological) corrections:

$$S_{\mu\nu}=(-1/4)F_{\alpha\beta}F^{\alpha\beta}g_{\mu\nu}+F_{\mu\alpha}F^{\alpha}_{\nu}$$

We have

$$F_{\alpha\beta}F^{\alpha\beta}=2F_{0r}F^{0r}+F_{rs}F^{rs}=$$
$$-2F_{0r}F_{0r}/S^2+F_{rs}F_{rs}/S^4$$
$$=-2(A_{r,0}-A_{0,r})^2+(A_{s,r}-A_{r,s})^2/S^4$$
$$=-2((S^2A^r)_{,0}+A^0_{,r})^2+(A^r_{,s}-A^s_{,r})^2/S^2$$
$$=-2E^2/S^2+2B^2/S^4$$

Further,

$$F_{0\alpha}F^{\alpha}_{0}=-F_{0r}F_{0r}/S^2=-E^2/S^2$$
$$F_{0\alpha}F^{\alpha}_{r}=-F_{0s}F_{rs}/S^2=\epsilon(rsk)E_sB_k/S^2=(E\times B)_r/S^2$$
$$F_{r\alpha}F^{\alpha}_{s}=F_{r0}F^{0}_{s}+F_{rk}F^{k}_{s}=F_{0r}F_{0s}-F_{rk}F_{sk}/S^2$$
$$=E_rE_s-\epsilon(rkl)\epsilon(skm)B_lB_m/S^2$$

Consider

$$Q_{rs}=\epsilon(rkl)\epsilon(skm)B_lB_m$$

We get

$$Q_{11}=B_2^2+B_3^2, Q_{22}=B_3^2+B_1^2, Q_{33}=B_1^2+B_2^2,$$
$$Q_{12}=\epsilon(1kl)\epsilon(2km)B_lB_m=-B_1B_2,$$

A.26. Waveguide with inhomogeneous and anisotropic permittivity and permeability placed in a strong gravitational field. In vacuum with only a gravitational field described by the metric tensor $g_{\mu\nu}(x)$, the general relativistic Maxwell equations are

$$(F^{\mu\nu}\sqrt{-g})_{,\nu}=0, F_{\mu\nu}=A_{\nu,\mu}-A_{\mu,\nu}, F^{\mu\nu}=g^{\mu\alpha}g^{\nu\beta}F_{\alpha\beta}$$

We identify

$$F_{0r}=A_{r,0}-A_{0,r}=E_r, r=1,2,3$$

the electric field and

$$F_{rs}=A_{s,r}-A_{r,s}=-\epsilon(rsk)B_k$$

where E is the electric field and B the magnetic field. This identification is due to the fact that in flat space-time, but in the presence of a medium, the Maxwell equations $divB=0$, $curlE=-B_{,t}$ are valid and these imply that $B=curlA, E=-V_{,0}-A_{,t}$ and the above definition of F_{rs} and the corresponding identification with the electric and magnetic fields agree with this definition provided that we choose $A=(-A_r), V=A_0$. In the presence of a medium we have in the frequency domain $D=\epsilon(\omega,r)E(\omega,r), B=\mu(\omega,r).H(\omega,r)$ or equivalently, $H=\mu^{-1}B$ where ϵ,μ are 3×3 matrix valued functions of frequency and spatial position. The other two Maxwell equations are $curlH=D_{,t}$ and $divD=0$ or equivalently in the frequency domain, $curlH=j\omega D, divD=0$ assuming the absence of external sources.

This suggests that to model the anisotropic and inhomogeneous medium in the presence of a static gravitational field, we must introduce the permittivity-permeability tensor $M^{\mu\nu\rho\sigma}(\omega, r)$ and write down the Maxwell equations as

$$(M^{\mu\nu\rho\sigma}F_{\rho\sigma}\sqrt{-g})_{,\nu} = 0$$

where when $\nu = 0$ we replace the time derivative by multiplication with $j\omega$. We rename the quantity $M^{\mu\nu\rho\sigma}(\omega, r)\sqrt{-g(r)}$ as $M^{\mu\nu\rho\sigma}(\omega, r)$ and write down our Maxwell equations as

$$(M^{\mu\nu\rho\sigma}F_{\rho\sigma})_{,\nu} = 0 - - - (1)$$

This suggests that $M^{0r\rho\sigma}F_{\rho\sigma}$ should correspond to D_r and $M^{rs\rho\sigma}F_{\rho\sigma}$ should correspond to $\epsilon(rsk)H_k$, so that (1) is the general relativistic equivalent of the Maxwell equations $divD = 0, curlH = j\omega D$. Note that $M^{\mu\nu\rho\sigma} = -M^{\nu\mu\rho\sigma} = -M^{\mu\nu\sigma\rho}$. Treating the medium effects and the gravitational effects to be small of order δ, we can thus write down the Maxwell curl equations as

$$curlE + j\omega\mu_0 H = \delta(P_1 E + P_2 H) - - - (1)$$

$$curlH - j\omega\epsilon_0 E = \delta(Q_1 E + Q_2 H + Q_3(\nabla \otimes E) + Q_4(\nabla \otimes H)) - - - (2)$$

where $P_1, P_2, Q_1, Q_2, Q_3, Q_4$ are functions of ω, r only.

Remark: The equation $M^{0r\rho\sigma}F_{\rho\sigma} = D_r$ can be expressed in matrix notation as

$$D = S_1 E + S_2 B,$$

and the equation $M^{rs\rho\sigma}F_{\rho\sigma} = \epsilon(rsk)H_k$ can be expressed in matrix notation as

$$H = S_3 E + S_4 B$$

since the components of $F_{\mu\nu}$ are E, B. Thus,

$$B = S_4^{-1}(H - S_3 E) = -S_4^{-1}S_3 E + S_4^{-1}H,$$

$$D = S_1 E + S_2(S_4^{-1}H - S_4^{-1}S_3 E) = (S_1 - S_2 S_4^{-1}S_3)E + S_2 S_4^{-1}H$$

and the $curlE = -j\omega B$ equation becomes

$$curlE = -j\omega(S_4^{-1}H - S_4^{-1}S_3 E)$$

which is of the form (1) while the general relativistic generalization of the equation $curlH = j\omega D$ given by $(M^{rs\rho\sigma}F_{\rho\sigma})_{,s} + (M^{r0\rho\sigma}F_{\rho\sigma})_{,0} = 0$ is of the form (2). Now suppose we expand

$$P_j(\omega, r) = \sum_n P_{jn}(\omega, x, y)exp(-n\beta z),$$

$$Q_j(\omega, r) = \sum_n Q_{jn}(\omega, x, y)exp(-n\beta z),$$

$$E(\omega, r) = \sum_n E_n(\omega, x, y)exp(-(\gamma + n\beta)z)),$$

$$H(\omega, r) = \sum_n H_n(\omega, x, y)exp(-(\gamma + n\beta)z)$$

Note that P_j, Q_j, P_{jn}, Q_{jn} are 3×3 matrices. Substituting into

$$curlE + j\omega\mu_0 H = \delta(P_1 E + P_2 H)$$

$$curlH - j\omega\epsilon_0 E = \delta(Q_1 E + Q_2 H + Q_3(\nabla \otimes E) + Q_4(\nabla \otimes H))$$

gives

$$\nabla \times E_n - (\gamma + n\beta)\hat{z} \times E_n + j\omega\mu_0 H_n = \delta\sum_m(P_{1,n-m}E_m + P_{2,n-m}H_m))$$

$$\nabla \times H_n - (\gamma + n\beta)\hat{z} \times H_n - j\omega\epsilon_0 E_n = delta\sum_m(Q_{1,n-m}E_m + Q_{2,n-m}H_m + Q_{3,n-m}(\hat{x} \otimes E_{m,x} + \hat{y} \otimes E_{m,y}$$

$$-(\gamma + m\beta)\hat{z} \otimes E_m) + Q_{4,n-m}(\hat{x} \otimes H_{m,x} + \hat{y} \otimes H_{m,y} - (\gamma + m\beta)\hat{z} \times H_m))$$

Note: The boundary conditions are that the normal component of B vanishes on the boundary and the tangential components of E also vanish on the boundary. So it may be more convenient to express the Maxwell equations in terms of E, B. So by redefining the matrices Q_j, P_j we write down the Maxwell curl equations as

$$curlE + j\omega B = 0,$$

$$curlB - j\omega\epsilon_0\mu_0 E = \delta(Q_1 E + Q_2 B + Q_3(\nabla \otimes E) + Q_4(\nabla \otimes B))$$

and accordingly,

$$\nabla \times E_n - (\gamma + n\beta)\hat{z} \times E_n + j\omega B_n = 0$$

$$\nabla \times B_n - (\gamma + n\beta)\hat{z} \times B_n - j\omega\epsilon_0\mu_0 E_n = delta\sum_m (Q_{1,n-m}E_m + Q_{2,n-m}B_m + Q_{3,n-m}(\hat{x} \otimes E_{m,x} + \hat{y} \otimes E_{m,y}$$

$$-(\gamma + m\beta)\hat{z} \otimes E_m) + Q_{4,n-m}(\hat{x} \otimes B_{m,x} + \hat{y} \otimes B_{m,y} - (\gamma + m\beta)\hat{z} \times B_m))$$

The boundary conditions imply that $E_{nz}(\omega, x, y)$ vanishes when $x = 0, a, y = 0, b$, $E_{nx}(\omega, x, y)$ vanishes when $y = 0, b$ and $E_{ny}(\omega, x, y)$ vanishes when $x = 0, a$ and further B_{nx} vanishes when $x = 0, a$ while B_{ny} vanishes when $y = 0, b$. These equations can be expressed in the following convenient form:

$$\nabla_\perp E_{nz} \times \hat{z} - (\gamma + n\beta)\hat{z} \times E_{n\perp} + j\omega B_{n\perp} = 0,$$

$$(\hat{z}, \nabla_\perp \times E_{n\perp}) + j\omega B_{nz} = 0$$

$$\nabla_\perp B_{nz} times\hat{z} - (\gamma + n\beta)\hat{z} \times B_{n\perp} - j\omega\epsilon_0\mu_0 E_{n\perp} = \delta.F_{n\perp},$$

$$(\hat{z}, \nabla_\perp \times B_{n\perp}) - j\omega\epsilon_0\mu_0 E_{nz} = \delta F_{nz}$$

where F_n is a linear function of $\{E_m, B_m, E_{m,x}, E_{m,y}, B_{m,x}, B_{m,y}\}_{m\in\mathbb{Z}}$ with coefficients being 3×3 matrix valued functions of ω, x, y. We get from the above equations

$$\nabla_\perp E_{nz} \times \hat{z} - ((\gamma + n\beta)/j\omega\epsilon_0\mu_0)\hat{z} \times (\nabla_\perp B_{nz} \times \hat{z} - (\gamma + n\beta)\hat{z} \times B_{n\perp} - \delta F_{n\perp})$$

$$+j\omega B_{n\perp} = 0$$

or

$$\nabla_\perp E_{nz} \times \hat{z} - ((\gamma + n\beta)/j\omega\epsilon_0\mu_0)(\nabla_\perp B_{nz} + (\gamma + n\beta)B_{n\perp} - \delta\hat{z} \times F_{n\perp})$$

$$+j\omega B_{n\perp} = 0$$

or

$$\nabla_\perp E_{nz} \times \hat{z} - ((\gamma + n\beta)/j\omega\epsilon_0\mu_0)\nabla_\perp B_{nz} - (h_n^2/j\omega\epsilon_0\mu_0)B_{n\perp}$$

$$+\delta((\gamma + n\beta)/j\omega\epsilon_0\mu_0)\hat{z} \times F_{n\perp} = 0$$

or

$$B_{n\perp} = (j\omega\epsilon_0\mu_0/h_n^2)\nabla_\perp E_{nz} \times \hat{z} - ((\gamma + n\beta)/h_n^2)\nabla_\perp B_{nz}$$

$$+\delta((\gamma + n\beta)/h_n^2)\hat{z} \times F_{n\perp}$$

Further, from the above equations,

$$j\omega\epsilon_0\mu_0 E_{n\perp} =$$

$$\nabla_\perp B_{nz} \times \hat{z} - (\gamma + n\beta)\hat{z} \times B_{n\perp} - \delta.F_{n\perp}$$

$$= \nabla_\perp B_{nz} \times \hat{z} - \delta F_{n\perp} - (\gamma + n\beta)[(j\omega\epsilon_0\mu_0/h_n^2)\nabla_\perp E_{nz}$$

$$+((\gamma + n\beta)/h_n^2)\nabla_\perp B_{nz} \times \hat{z} - \delta(\gamma + n\beta)/h_n^2)F_{n\perp}]$$

$$= (\omega^2\mu_0\epsilon_0/h_n^2)\nabla_\perp B_{nz} \times \hat{z} - \delta(\omega^2\mu_0\epsilon_0/h_n^2)F_{n\perp}$$

$$-(j\omega\mu_0\epsilon_0(\gamma + n\beta)/h_n^2)\nabla_\perp E_{nz}$$

or

$$E_{n\perp} = -((\gamma + n\beta)/h_n^2)\nabla_\perp E_{nz} - (j\omega/h_n^2)\nabla_\perp B_{nz} \times \hat{z}$$

$$+\delta(j\omega/h_n^2)F_{n\perp}$$

Substituting these expressions for $E_{n\perp}$ and $B_{n\perp}$ into the equations

$$(\hat{z}, \nabla_\perp \times E_{n\perp}) + j\omega B_{nz} = 0$$

$$(\hat{z}, \nabla_\perp \times B_{n\perp}) - j\omega\epsilon_0\mu_0 E_{nz} = \delta F_{nz}$$

gives us modified Helmholtz equations for E_{nz} and B_{nz}:

$$(j\omega/h_n^2)\nabla_\perp^2 B_{nz} + j\omega B_{nz} + \delta(j\omega/h_n^2)(\hat{z}, \nabla_\perp \times F_{n\perp}) = 0$$

and

$$-(j\omega\epsilon_0\mu_0/h_n^2)\nabla_\perp^2 E_{nz} - j\omega\epsilon_0\mu_0 E_{nz} = \delta[(F_{nz} - (\gamma + n\beta)/h_n^2)\nabla_\perp.F_{n\perp}]$$

or equivalently,

$$(\nabla_\perp^2 + h_n^2)B_{nz} - \delta(\hat{z}, \nabla_\perp \times F_{n\perp}) = 0,$$
$$(\nabla_\perp^2 + h_n^2)E_{nz} - (j\delta h_n^2/\omega\epsilon_0\mu_0))(F_{nz} - (\gamma + n\beta)/h_n^2)\nabla_\perp.F_{n\perp})) = 0$$

Now, we expand upto $O(\delta)$

$$E_n(x, y) = E_n^{(0)}(x, y) + \delta.E_n^{(1)}(x, y) + O(\delta^2),$$
$$B_n(x, y) = B_n^{(0)}(x, y) + \delta.B_n^{(1)}(x, y) + O(\delta^2)$$

We write

$$\gamma = \gamma^{(0)} + \delta\gamma^{(1)} + O(\delta^2)$$

Then, noting that

$$h_n^2 = (\gamma + n\beta)^2 + \omega^2\mu_0\epsilon_0$$

we get

$$h_n^2 = (\gamma^{(0)} + \delta\gamma^{(1)} + n\beta)^2 + \omega^2\mu_0\epsilon_0$$
$$= h_n^{(0)2} + \delta\gamma^{(1)}\alpha_n + O(\delta^2)$$

where

$$h_n^{(0)2} = (\gamma^{(0)} + n\beta)^2 + \omega^2\mu_0\epsilon_0,$$
$$\alpha_n = 2(\gamma^{(0)} + n\beta)$$

We get for the $O(\delta^0) = O(1)$ term,

$$(\nabla_\perp^2 + h_n^{(0)2}B_{nz}^{(0)} = 0,$$
$$(\nabla_\perp^2 + h_n^{(0)2})E_{nz}^{(0)} = 0$$

and for the $O(\delta)$ term,

$$(\nabla_\perp^2 + h_n^{(0)^2})B_{nz}^{(1)} + \gamma^{(1)}\alpha_n B_{nz}^{(0)}$$
$$= (\hat{z}, \nabla_\perp \times F_{n\perp}^{(0)})$$
$$(\nabla_\perp^2 + h_n^{(0)^2})E_{nz}^{(1)} + \gamma^{(1)}\alpha_n E_{nz}^{(0)} = jh_n^{(0)^2}/\omega\epsilon_0\mu_0)(F_{nz}^{(0)} - (\gamma^{(0)} + n\beta)/h_n^{(0)^2})\nabla_\perp.F_{n\perp})^{(0)}$$

Further, using this perturbation expansion, we have for the tangential components of the fields,

$$B_{n\perp}^{(0)} = (j\omega\epsilon_0\mu_0/h_n^{(0)^2})\nabla_\perp E_{nz}^{(0)} \times \hat{z} - ((\gamma + n\beta)/h_n^{(0)^2})\nabla_\perp B_{nz}^{(0)},$$

$$B_{n\perp}^{(1)} = (j\omega\epsilon_0\mu_0/h_n^2)\nabla_\perp E_{nz}^{(1)} \times \hat{z}$$
$$-(2j\omega\epsilon_0\mu_0(\gamma^{(0)} + n\beta)\gamma^{(1)}/h_n^{(0)4})\nabla_\perp E_{nz}^{(0)} \times \hat{z}$$
$$-((\gamma^{(0)} + n\beta)/h_n^{(0)^2})\nabla_\perp B_{nz}^{(1)} - (\gamma^{(1)}/h_n^{(0)2})\nabla_\perp B_{nz}^{(0)}$$
$$+(2\gamma^{(1)}(\gamma^{(0)} + n\beta)/h_n^{(0)4})\nabla_\perp B_{nz}^{(0)}$$
$$(\gamma^{(0)} + n\beta)/h_n^{(0)^2})\hat{z} \times F_{n\perp}^{(0)}$$

and likewise,

$$E_{n\perp}^{(0)} = -((\gamma^{(0)} + n\beta)/h_n^{(0)^2})\nabla_\perp E_{nz}^{(0)} - (j\omega/h_n^{(0)^2})\nabla_\perp B_{nz}^{(0)} \times \hat{z}$$
$$E_{n\perp}^{(1)} = -((\gamma^{(0)} + n\beta)/h_n^{(0)^2})\nabla_\perp E_{nz}^{(1)}$$
$$+\gamma^{(1)}(2(\gamma^{(0)} + n\beta)/h_n^{(0)4} - 1/h_n^{(0)2})\nabla_\perp E_{nz}^{(0)}$$
$$-(j\omega/h_n^{(0)2})\nabla_\perp B_{nz}^{(1)} \times \hat{z} + 2\gamma^{(1)}(j\omega(\gamma^{(0)} + n\beta)/h_n^{(0)4})\nabla_\perp B_{nz}^{(0)} \times \hat{z}$$
$$+(j\omega/h_n^{(0)2})F_{n\perp}^{(0)}$$

We can in accordance with the unperturbed Helmholtz equations and the boundary conditions on the unperturbed fields assume that

$$E^{(0)}_{nz}(\omega,x,y)=\sum_{p,q\geq 1}c_E(\omega,n,p,q)u_{pq}(x,y),u_{pq}(x,y)=(2/\sqrt{ab})sin(p\pi x/a)sin(q\pi y/b),$$

$$B^{(0)}_{nz}(\omega,x,y)=\sum_{p,q\geq 1}c_B(\omega,n,p,q)v_{pq}(x,y),v_{pq}(x,y)=(2/\sqrt{ab})cos(p\pi x/a)cos(q\pi y/b)$$

so that

$$h_n^{(0)2}=\pi^2(p^2/a_2+q^2/b^2)=(\gamma^{(0)}+n\beta)^2+\omega^2\mu_0\epsilon_0$$

This determines

$$\gamma^{0)}=\gamma^{(0)}(n,p,q)=\pm(\pi^2(p^2/a^2+q^2/b^2)-k^2)^{1/2}-n\beta$$

where

$$k^2=\omega^2\mu_0\epsilon_0$$

Then, assuming that $B^{(1)}_{nz}$ and $E^{(1)}_{nz}$ satisfy the same boundary conditions respectively as $B^{(0)}_{nz}$ and $E^{(0)}_{nz}$, it follows from the equations

$$(\nabla_\perp^2+h_n^{(0)^2})B^{(1)}_{nz}+\gamma^{(1)}\alpha_n B^{(0)}_{nz}$$

$$=(\hat{z},\nabla_\perp\times F^{(0)}_{n\perp})$$

$$(\nabla_\perp^2+h_n^{(0)^2})E^{(1)}_{nz}+\gamma^{(1)}\alpha_n E^{(0)}_{nz}=jh_n^{(0)^2}/\omega\epsilon_0\mu_0)(F^{(0)}_{nz}-(\gamma^{(0)}+n\beta)/h_n^{(0)^2})\nabla_\perp.F^{(0)}_{n\perp})$$

after applying Green's theorem that

$$\gamma^{(1)}\alpha_n<v_{pq},B^{(0)}_{nz}>=<v_{pq},(\hat{z},\nabla_\perp\times F^{(0)}_{n\perp})>$$

and

$$\gamma^{(1)}\alpha_n<u_{pq},E^{(0)}_{nz}>=(jh_n^{(0)2}/\omega\mu_0\epsilon_0)<u_{pq},F^{(0)}_{nz}-(\gamma^{(0)}+n\beta)/h_n^{(0)^2})\nabla_\perp.F^{(0)}_{n\perp})>$$

A.27. Green's functions in quantum field theory:
Let $\psi_n(t,r)$ be field operator valued solutions to the Schrodinger equation:

$$i\psi_{n,t}(t,r)=H(r)\psi_n(t,r)=E_n\psi_n(t,r)$$

where

$$H(r)=-\nabla_r^2/2m+V(r)$$

Thus,

$$\psi_n(t,r)=exp(-iE_nt)\psi_n(r)$$

The field Hamiltonian (ie second quantized Hamiltonian) operator is given by

$$H_F=\int\psi_n^*(r)H(r)\psi_n(r)d^3r,\psi_n(r)=\psi_n(0,r)$$

We introduce the number operator

$$N=\sum_n\int\psi_n(r)^*\psi_n(r)d^3r$$

We assume the anticommutation relations

$$\{\psi(t,r),\psi(t,r')^*\}=\delta^3(r-r')$$

for any second quantized solution $\psi(t,r)$ to the Schrodinger equation. We can actually construct the field operators $\psi_n(r)$ as follows. Let $u_n(r)$ be the stationary state solutions to the Schrodinger equation. Assume that the $u_n(r)'s$ are normalized functions in $L^2(\mathbb{R}^3)$. Then, let $a_n,a_n^*,n=1,2,...$ be operators in a Fock space (ie second quantized Hilbert space) satisfying the anticommutation relations

$$\{a_n,a_m^*\}=\delta_{n,m},\{a_n,a_m\}=0,\{a_n^*,a_m^*\}=0$$

We then define

$$\psi(t,r) = \sum_n c_n exp(-iE_n t) a_n u_n(r)$$

We then have

$$i\psi_{,t}(t,r) = \sum_n c_n exp(-iE_n t) a_n E_n u_n(r) = \sum_n c_n exp(-iE_n t) a_n H(r) u_n(r) = H(r)\psi(t,r)$$

Here, the $c_n's$ are complex numbers satisfying $|c_n|^2 = 1$. We have further,

$$\{\psi(t,r), \psi(t,r')^*\} = \sum_{n,m} c_n \bar{c}_m exp(-i(E_n - E_m)t)\{a_n, a_m^*\} u_n(r)\bar{u}_m(r')$$

$$= \sum_{n,m} c_n \bar{c}_m exp(-i(E_n - E_m)t)\delta_{n,m} u_n(r)\bar{u}_m(r') = \sum_n u_n(r)\bar{u}_m(r') = \delta^3(r - r')$$

We have

$$N = \sum_{n,m} \bar{c}_n c_m a_n^* a_m (\int \bar{u}_n(r) u_m(r) d^3r) = \sum_n a_n^* a_n$$

in view of the orthonormality of the eigenfunctions $u_n(r), n = 1, 2, ...$ The field Hamiltonian can also be expressed in terms of the $a_n's$: We define

$$\psi_n(t,r) = c_n.exp(-iE_n t) u_n(r) a_n$$

Then,

$$H_F = \sum_n \int \psi_n(r)^* H(r) \psi_n(r) = \sum_n a_n^* a_n \int \bar{u}_n(r) H(r) u_n(r) d^3r = \sum_n E_n a_n^* a_n$$

since

$$H(r) u_n(r) = E_n u_n(r)$$

Now consider the Green's function

$$G(t,r,t',r') =< T(\psi(t,r).\psi(t',r')^*) >$$

where the expected value is taken in the ground state of H_F, ie vacuum state $|0>$ in which each a_n has eigenvalue zero. T is the time ordering operator. We can write

$$G(t,r,t',r') = \theta(t-t') < \psi(t,r)\psi(t',r')^* > + \theta(t'-t) < \psi(t',r')^*\psi(t,r) >=$$

Now we note that

$$< \psi(t,r)\psi(t',r')^* >= \sum_{n,m} c_n \bar{c}_m exp(-i(E_n t - E_m t')) u_n(r)\bar{u}_m(r') < a_n a_m^* >$$

and since

$$< a_n a_m^* >= \delta_{n,m}$$

(since $a_n a_m^* + a_m^* a_n = \delta_{n,m}$ and $a_n|0>= 0$), it follows that

$$< \psi(t,r)\psi(t',r')^* >= \sum_n exp(-iE_n(t-t') u_n(r)\bar{u}_m(r')$$

We note that

$$< \psi(t',r')^*\psi(t,r) >= 0$$

since $\psi(t,r)|0>= 0$. Thus,

$$G(t,r,t',r') = \theta(t-t')\sum_n exp(-iE_n(t-t')) u_n(r)\bar{u}_n(r') = G_0(t-t',r,r')$$

where

$$G_0(t,r,r') = G(t,r,0,r')$$

We then get

$$\int_{\mathbb{R}} G_0(t,r,r') exp(i(\omega + i\epsilon)t) dt = \sum_n \frac{u_n(r)\bar{u}_n(r')}{i(\omega - E_n + i\epsilon)}$$

We have

$$G_{,t}(t,r,t',r') = \delta(t-t') < \psi(t,r)\psi(t,r')^* > +\theta(t-t') < \psi_{,t}(t,r)\psi(t',r')^* >$$
$$= \delta(t-t')\delta^3(r-r') - i\theta(t-t') < H(r)\psi(t,r)\psi(t',r')^* >$$
$$= \delta(t-t')\delta^3(r-r') - iH(r)G(t,r,t',r')$$

A.28.Design of quantum gates using scattering theory.
Let H_0, H be two self-adjoint operators in the same Hilbert space $\mathcal{H}$. We define

$$W(t) = exp(itH)exp(-itH_0), t \in \mathbb{R}$$

The limits

$$\Omega_+ = slim_{t\to\infty}W(t), \Omega_- = slim_{t\to-\infty}W(t)$$

may exist on different domains. Let D_+ be the domain of Ω_+ and D_- that of Ω_-. Then $\Omega_+^* : \mathcal{H} \to D_+$ is defined and hence

$$S = \Omega_+^*\Omega_- : D_- \to D_+$$

is defined. It is clear that $W(t)$ is a unitary operator on $\mathcal{H}$ and hence

$$< f|\Omega_+^*\Omega_+|g >=< \Omega_+ f|\Omega_+ g >= lim_{t\to\infty} < W(t)f|W(t)g >=< f|g >, f,g \in D_+$$

Thus Ω_+ is unitary when restricted to D_+. Thus, $\Omega_+\Omega_+^* : \mathcal{H} \to \mathcal{H}$ is an orthogonal projection of $\mathcal{H}$ onto D_+. We write $V = H - H_0$ and then find that

$$W(t) - I = \int_0^t W'(s)ds == i\int_0^t exp(isH)V.exp(-isH_0)ds$$

In particular,

$$\Omega_+ = I + i\int_0^\infty exp(isH)V.exp(-isH_0)ds$$

Let E_0 be the spectral measure of H_0 and E that of H. We have

$$\Omega_+^* = I + i\int_0^\infty d/dt(exp(itH_0).exp(-itH))dt = I - i\int_0^\infty exp(itH_0)V.exp(-itH)dt$$

Thus,

$$I = \Omega_+^*\Omega_+ = \Omega_+ - i\int_0^\infty exp(itH_0)V.exp(-itH)\Omega_+ dt$$

Now,

$$exp(-itH)\Omega_+ = lim_{s\to\infty}exp(i(s-t)H)exp(-isH_0) = lim_{s\to\infty}exp(isH).exp(-i(s+t)H_0)$$
$$= \Omega_+ exp(-itH_0)$$

Thus,

$$I = \Omega_+ - i\int_0^\infty exp(itH_0)V\Omega_+ exp(-itH_0)dt$$

or equivalently,

$$\Omega_+ = I + i\int_{[0,\infty)\times\mathbb{R}} exp(it(H_0 - \lambda + i\epsilon))V\Omega_+ dtE_0(d\Lambda)$$
$$= I - \int_{\mathbb{R}}(H_0 - \lambda + i\epsilon)^{-1}V\Omega_+ E_0(d\lambda)$$

The right side integral is to be interpreted as the limit of $\epsilon \to 0+$. This is the rigorous statement of the first Lippman-Schwinger equation in scattering theory. Roughly speaking, if $|\phi_+ >$ is an output free particle state corresponding to an "eigenfunction" of H_0 with "eigenvalue" λ and $|\psi_+ >= \Omega_+|\phi_+ >$ is the corresponding output scattered state, which is an "eigenfunction" of H with the same eigenvalue λ (energy is conserved during the scattering process), then we have

$$|\psi_+ >= |\phi_+ > -(H_0 - \lambda + i\epsilon)^{-1}V|\psi_+ >$$

In a similar way, we get that if $|\phi_- >$ is an input free particle state corresponding to an energy eigenfunction of H_0 with eigenvalue λ and $|\psi_- >= \Omega_-|\phi_- >$ the corresponding input scattered state corresponding to an eigenfunction of H with eigenvalue λ, then we get the second Lippman-Schwinger equation

$$|\psi_- >= |\phi_- > -(H_0 - \lambda - i\epsilon)^{-1}V|\psi_- >$$

This is seen as follows:

$$\Omega_-^* = lim_{t\to -infty} exp(itH_0)exp(-itH) = I + i\int_{-\infty}^{0} exp(itH_0)V.exp(-itH)dt$$

$$= I + i\int_{0}^{\infty} exp(-itH_0)V.exp(itH)dt$$

Thus,

$$I = \Omega_-^*\Omega_- = \Omega_- + i\int_{0}^{\infty} exp(-itH_0)V.exp(itH)\Omega_- dt$$

$$= \Omega_- + i\int_{0}^{\infty} exp(-itH_0)V\Omega_-.exp(itH_0)dt$$

or equivalently,

$$\Omega_- = I - i\int_{0}^{\infty} exp(-it(H_0 - \lambda - i\epsilon))V\Omega_- dtE_0(d\lambda)$$

$$= I - \int_{\mathbb{R}} (H_0 - \lambda - i\epsilon)^{-1}V\Omega_- E_0(d\lambda)$$

which gives the second Lippman-Schwinger equation. Now, we derive an explicit form of the scattering matrix. We define $R_0(\lambda) = (H_0 - \lambda)^{-1}, R(\lambda) = (H - \lambda)^{-1}$ respectively for λ belonging to the resolvent set of H_0 and of H. We have

$$S = \Omega_+^*\Omega_- =$$

$$(I - i\int_{0}^{\infty} exp(itH_0)V.exp(-itH)dt)\Omega_-$$

$$= \Omega_- - i\int_{0}^{\infty} exp(itH_0)V.\Omega_- exp(-itH_0)dt$$

$$= \Omega_- + \int_{\mathbb{R}} (H_0 - \lambda + i\epsilon)^{-1}V.\Omega_- E_0(d\lambda)$$

$$= \Omega_- + \int_{\mathbb{R}} R_0(\lambda - i\epsilon)V\Omega_- E_0(d\lambda)$$

Now,

$$\Omega_- = I - i\int_{-\infty}^{0} exp(itH).V.exp(-itH_0)dt$$

$$= I - i\int_{0}^{\infty} exp(-itH)Vexp(itH_0)dt$$

$$= I - \int_{\times\mathbb{R}} (H - \lambda - i\delta)^{-1}VE_0(d\lambda)$$

Substituting this into the previous expression gives

$$S - I = -\int_{\mathbb{R}} R(\lambda + i\delta)VE_0(d\lambda) + \int_{\mathbb{R}} R_0(\lambda - i\epsilon)VE_0(d\lambda) - \int R_0(\lambda - i\epsilon)VR(\lambda + i\delta)VE_0(d\lambda)$$

A.29. Perturbed Einstein field equations with electromagnetic interactions.
The unperturbed metric is

$$g_{00} = 1, g_{rs} = -S(t)^2\delta_{rs}, g_{0r} = 0,$$

A small change in the coordinate system can be made such that the first order perturbations in the metric $\delta g_{\mu\nu}$ satisfies $\delta g_{0\mu} = 0$. The unperturbed energy-momentum tensor of the matter field is

$$T_{\mu\nu} = (\rho(t) + p(t))V_\mu V_\nu - p(t)g_{\mu\nu}$$

where $V_r = 0, V_0 = 1$ (This is possible because the above unperturbed metric satisfies the comoving condition, ie, particles with $V_r = 0$ satisfy the geodesic equation). Thus,

$$T_{00} = \rho(t), T_{rs} = p(t)S^2(t)\delta_{rs}, g_{0r} = 0$$

We also define the tensor

$$S_{\mu\nu} = T_{\mu\nu} - Tg_{\mu\nu}/2, T = g^{\mu\nu}T_{\mu\nu} = \rho + p - 4p = \rho - 3p$$

We have

$$\delta S_{\mu\nu} = \delta T_{\mu\nu} - T\delta g_{\mu\nu}/2 - g_{\mu\nu}\delta T/2$$

so that since $\delta g_{00} = 0$,

$$\delta S_{00} = \delta T_{00} - \delta T/2 = \delta\rho - \delta\rho/2 + 3\delta p/2 = (3\delta p + \delta\rho)/2$$

Note that $\delta\rho, \delta p, \delta v_\mu, \delta g_{\mu\nu}$ are in general functions of space and time, ie, of $x = (t, r)$.

$$\delta S_{rs} = \delta T_{rs} - T\delta g_{rs}/2 - g_{rs}\delta T/2$$

$$= -\delta p g_{rs} - (\rho - 3p)\delta g_{rs}/2 - g_{rs}(\delta\rho - 3\delta p)/2$$

$$= (3p - \rho)\delta g_{rs}/2 + S^2(\delta\rho - \delta p)\delta_{rs}/2$$

Note that we have the sequence of implications

$$v_\mu v^\mu = 1, V_\mu \delta v^\mu = 0, \delta v^0 = 0$$

More precisely,

$$\delta(g_{\mu\nu}v^\mu v^\nu) = 0$$

gives

$$V^\mu V^\nu \delta g_{\mu\nu} + g_{\mu\nu}(V^\mu \delta v^\nu + V^\nu \delta v^\mu) = 0$$

which implies

$$\delta g_{00} + 2\delta v^0 = 0$$

and since by the coordinate condition, $\delta g_{00} = 0$, it follows that $\delta v^0 = 0$. Hence,

$$\delta v_0 = (\delta g_{0\mu}v^\mu) = \delta g_{0\mu}V^\mu + g_{0\mu}\delta v^\mu = 0$$

which gives

$$\delta v_0 = 0$$

since by the coordinate condition, $\delta g_{0\mu} = 0$ and $g_{0r} = 0$. So

$$\delta T_{00} = (p + \rho)(2V_0\delta v_0) + V_0^2(\delta p + \delta\rho) - p\delta g_{00} - g_{00}\delta p = \delta\rho$$

Further,

$$\delta S_{r0} = \delta S_{0r} = (\rho + p)\delta v_r$$

since $\delta g_{0r} = \delta g_{r0} = 0$. Finally, we need to compute $\delta R_{\mu\nu}$ and equate this to $K.\delta S_{\mu\nu}$. We have

$$\delta R_{\mu\nu} = (\delta\Gamma^\alpha_{\mu\alpha})_{:\nu} - (\delta\Gamma^\alpha_{\mu\nu})_{:\alpha}$$

$$(\delta\Gamma^\alpha_{\mu\alpha})_{:\nu} =$$

$$(\delta\Gamma^\alpha_{\mu\nu})_{:\beta} = (\delta\Gamma^\alpha_{\mu\nu})_{,\beta}$$

$$+\Gamma^\alpha_{\rho\beta}\delta\Gamma^\rho_{\mu\nu} - \Gamma^\rho_{\mu\beta}\delta\Gamma^\alpha_{\rho\nu}$$

$$-\Gamma^\rho_{\nu\beta}\delta\Gamma^\alpha_{\rho\mu}$$

Thus,

$$(\delta\Gamma^\alpha_{\mu\nu})_{:\alpha}$$

$$= (\delta\Gamma^\alpha_{\mu\nu})_{,\alpha}$$

$$+\Gamma^\alpha_{\rho\alpha}\delta\Gamma^\rho_{\mu\nu} - 2\Gamma^\rho_{\mu\alpha}\delta\Gamma^\alpha_{\rho\nu}$$

Thus final form of the first order perturbation to the Ricci tensor is given by

$$\delta R_{\mu\nu} =$$

$$(\delta\Gamma^\alpha_{\mu\alpha})_{,\nu} - \Gamma^\rho_{\mu\nu}\delta\Gamma^\alpha_{\rho\alpha}$$

$$-(\delta\Gamma^\alpha_{\mu\nu})_{,\alpha}$$

$$-\Gamma^\alpha_{\rho\alpha}\delta\Gamma^\rho_{\mu\nu} + 2\Gamma^\rho_{\mu\alpha}\delta\Gamma^\alpha_{\rho\nu}$$

The perturbed Einstein field equations have the general form

$$A_1(\mu\nu, rs, t)\delta g_{rs}(t,r) + A_2(\mu\nu, rsl, t)\delta g_{rs,l}(t,r) + A_3(\mu\nu, rslp, t)\delta g_{rs,lp}(t,r)$$

$$+A_4(\mu\nu, rs, t)\delta g_{rs,0}(t,r) + A_5(\mu\nu, rsl, t)\delta g_{rs,l0}(t,r) + A_6(\mu\nu, rs, t)\delta g_{rs,00}(t,r)$$

$$+A_7(\mu\nu, t)\delta\rho(t,r) + A_8(\mu\nu, r, t)\delta v_r(t,r) = K\delta S_{\mu\nu}(t,r)$$

where $S_{\mu\nu}(t,r)$ is the energy-momentum tensor of the electromagnetic field. We have,

$$S_{\mu\nu} = (-1/4)F_{\alpha\beta}F^{\alpha\beta}g_{\mu\nu} + F_{\mu\alpha}F^\alpha_\nu$$

$$F_{\mu\nu} = A_{\nu,\mu} - A_{\mu,\nu}, (F^{\mu\nu}\sqrt{-g})_{,\nu} = 0$$

$$F_{0r} = A_{r,0} - A_{0,r} = -(S^2(t)A^r)_{,0} - A^0_{,r}$$

$$F_{rs} = A_{s,r} - A_{r,s} = S^2(t)(A^r_{,s} - A^s_{,r})$$

The gauge condition is

$$(A^\mu\sqrt{-g})_{,\mu} = 0$$

which reads

$$(A^0S^3(t))_{,0} + S^3(t)A^r_{,r} = 0$$

The unperturbed Maxwell equations are

$$F^{0r}_{,r} = 0, (F^{r0}S^3)_{,0} + S^3F^{rs}_{,s} = 0$$

$$F^{0r} = g^{00}g^{rr}F_{0r} = (1/S^2)((S^2A^r)_{,0} - A^0_{,r})$$

and so the first Maxwell equation gives

$$(S^2A^r)_{,0r} - \nabla^2A^0 = 0$$

or

$$(S^2)'A^r_{,r} + S^2A^r_{,0r} - \nabla^2A^0 = 0$$

which reads on using the gauge condition,

$$-(S^2)'S^{-3}(A^0S^3)_{,0} - S^2(S^{-3}(A^0S^3)_{,0})_{,0} - \nabla^2A^0 = 0$$

and the second Maxwell equation gives

$$-S(A^s_{,r} - A^r_{,s})_{,s} - ((A^0_{,r} + A^r_{,0})S)_{,0} = 0$$

which simplifies to

$$S\nabla^2A^r - SA^s_{,rs} - (S(A^0_{,r} + A^r_{,0}))_{,0} = 0$$

and this becomes on using the gauge condition,

$$S\nabla^2A^r + S^{-2}(A^0S^3)_{,0r} - (SA^0_{,r})_{,0} - (SA^r_{,0})_{,0} = 0$$

or

$$(\nabla^2A^r - A^r_{,00}) + S^{-3}(A^0S^3)_{,0r} - S^{-1}(SA^0_{,r})_{,0} - S^{-1}S'A^r_{,0} = 0$$

Since the radiation field is homogeneous and isotropic, we assume that the vector potential has mean zero and statistical correlations of the form

$$< A^r(t,r)A^s(t',r') >= P(t,t',r-r')\delta_{rs}$$

We find that since

$$(A^0S^3(t))_{,0} + S^3(t)A^r_{,r} = 0$$

we have

$$A^0_{,0}(t,r)S^3(t) + 3S^2(t)S'(t)A^0(t,r) + S^3(t)A^r_{,r}(t,r) = 0$$

and hence we can assume that $A^0 = 0$ provided that we admit the Coulomb gauge constraint

$$A^r_{,r}(t,r) = 0$$

which gives

$$\frac{\partial P(t,t',r)}{\partial x^s} = 0$$

and so we can assume that $P(t,t',r)$ is independent of the spatial vector $r = (x^s)$. Thus, we can write

$$< A^r(t,r)A^s(t',r') >= P(t,t')\delta_{rs}$$

Now, the Maxwell equation

$$(S^2A^r)_{,0r} - \nabla^2 A^0 = 0$$

is automatically satisfied since $A^0 = 0$ and $A^r_{,r} = 0$. The second Maxwell equation

$$(\nabla^2 A^r - A^r_{,00}) + S^{-3}(A^0S^3)_{,0r} - S^{-1}(SA^0_{,r})_{,0} - S^{-1}S'A^r_{,0} = 0$$

gives using $A^0 = 0$,

$$\nabla^2 A^r - A^r_{,00} - S^{-1}S'A^r_{,0} = 0$$

and hence taking correlations with $A^s(t',r')$, we get

$$\nabla^2 P(t,t') - P_{,tt}(t,t') - S^{-1}(t)S'(t)P_{,t}(t,t') = 0$$

and likewise, with t and t' interchanged. We assume that we have solved this equation to obtain $P(t,t')$. Now, the average energy-momentum tensor of the radiation field before the metric has been perturbed by $\delta g_{\mu\nu}(t,r)$ is given by

$$S_{\mu\nu} = (-1/4) < F_{\alpha\beta}F^{\alpha\beta} > g_{\mu\nu} + < F_{\mu\alpha}F^\alpha_\nu >$$

Now,

$$\begin{aligned} F_{\alpha\beta}F^{\alpha\beta} &= 2F_{0r}F^{0r} + F_{rs}F^{rs} = \\ &-2S^{-2}(A_{r,0} - A_{0,r})^2 + S^{-4}(A_{s,r} - A_{r,s})^2 \\ &= -2S^{-2}((-S^2A^r)_{,0})^2 - A^0_{,r})^2 + (A^s_{,r} - A^r_{,s})^2 \\ &= 2S^{-2}((S^2A^r)_{,0})^2 + (A^s_{,r} - A^r_{,s})^2 \\ &= 2S^2(A^r_{,0})^2 + 2S^{-2}((S^2)')^2(A^r)^2 + (A^s_{,r} - A^r_{,s})^2 \end{aligned}$$

and its average value is

$$\begin{gathered} < F_{\alpha\beta}F^{\alpha\beta} >= \\ 6S^2P_{,tt'}(t,t') + 2S^{-2}((S^2)')^2P(t,t') \end{gathered}$$

since P does not depend on the spatial coordinates. Note that by $P_{,tt'}(t,t)$, we mean that

$$P_{,tt'}(t,t) = lim_{t' \to t}P_{,tt'}(t,t')$$

Thus,

$$< F_{\alpha\beta}F^{\alpha\beta} >= 6S^2(t)P_{,tt'}(t,t) + 8(S')^2P(t,t')$$

We next compute $< F_{\mu\alpha}F^\alpha_\nu >$. For $\mu = \nu = 0$ this is

$$< F_{0r}F^r_0 >= -S^{-2} < (A_{r,0})^2 >= -S^2 < (A^r_{,0})^2 >= -3S^2(t)P_{,tt'}(t,t)$$

For $\mu = r, \nu = 0$, we get

$$< F_{r\alpha} F_0^{\alpha} >=$$

$$< F_{rs} F_0^s >= -S^{-2} < (A_{s,r} - A_{r,s}) A_{s,0} >= -S^2 < (A^s_{,r} - A^r_{,s}) A^s_{,0} >= 0$$

since P is independent of the spatial coordinates. Thus,

$$< S_{00} >= (-3/2) S^2(t) P_{,tt'}(t,t) - 2(S')^2 P(t,t') - 3S^2 P_{,tt'}(t,t) = (-9/2) S^2(t) P_{,tt'}(t,t) - 2S'^2 P(t,t')$$

$$< S_{r0} >=< S_{0r} >= 0$$

Finally,

$$< S_{rs} >= (3/2) S^4(t) P_{,tt'}(t,t) \delta_{rs} + < F_{r0} F_s^0 + F_{rm} F_s^m >$$

$$< F_{r0} F_s^0 >=< A_{r,0} A_{s,0} >= S^4(t) < A^r_{,0} A^s_{,0} >= S^4(t) P_{,tt'}(t,t) \delta_{rs}$$

$$< F_{rm} F_s^m >= (-1/S^2) < F_{rm} F_{sm} >= (-1/S^2) < (A_{m,r} - A_{r,m})(A_{m,s} - A_{s,m}) >= 0$$

since P is independent of the spatial coordinates. Thus,

$$< S_{rs} >= (-1/2) S^4(t) P_{,tt'}(t,t) \delta_{rs}$$

The perturbed Maxwell equations and the perturbed energy-momentum tensor of the electromagnetic field: The perturbed Maxwell equations are

$$(\delta(F^{\mu\nu}\sqrt{-g})_{,\nu} = 0$$

The perturbed gauge condition is

$$(\delta(A^{\mu}\sqrt{-g}))_{,\mu} = 0$$

which gives

$$(S^3(t)\delta A^{\mu} + A^{\mu}\delta\sqrt{-g})_{,\mu} = 0$$

If we assume $\delta A^0 = 0$, then this gauge condition gives

$$S^3(\delta A^r)_{,r} + (A^r \delta\sqrt{-g})_{,r} = 0$$

Now,

$$\delta g = g(-S^{-2})\delta g_{rr}$$

(Note that our coordinate system is chosen so that $\delta g_{0\mu} = 0$. Thus, we get

$$\delta\sqrt{-g} = -\delta g / 2\sqrt{-g} = -S\delta g_{rr}/2$$

Thus the gauge condition with $\delta A^0 = 0$ gives

$$S^2(\delta A^r)_{,r} - A^r \delta g_{ss,r}/2 = 0$$

where we have used $A^r_{,r} = 0$,ie, the Coulomb gauge for the unperturbed em field. This gives us the following gauge condition for the perturbed em field

$$(\delta A^r)_{,r} = S^{-2} A^r \delta g_{ss,r}/2$$

We now write down the first perturbed Maxwell equation using $\delta A^0 = 0$ as

$$S^2 \delta F^{0r}_{,r} - (F^{0r}\delta g_{ss}/2)_{,r} = 0$$

or

$$S^2 \delta A^r_{,0r} - (A^r_{,0}\delta g_{ss}/2)_{,r} = 0$$

which is since $A^r_{,r} = 0$, the same as

$$S^2 \delta A^r_{,0r} - A^r_{,0}\delta g_{ss,r}/2 = 0$$

Substituting for $\delta A^r_{,r}$ from the perturbed gauge condition, we get

$$S^2(S^{-2} A^r \delta g_{ss,r})_{,0} - A^r_{,0}\delta g_{ss,r} = 0$$

or equivalently

$$A^r (S^{-2}\delta g_{ss,r})_{,0} = 0$$

and this condition is impossible to fulfill in general. So we cannot assume the Coulomb condition $\delta A^0 = 0$ for the perturbed situation.

Remark: We have assumed that $A^0 = 0$ and hence that the gauge condition reads $A^r_{,r} = 0$. We have also assumed that $< A^r(t,r)A^s(t',s) >= P(t,t',r-r')\delta_{rs}$ and have hence derived the condition that $P(t,t',r)$ is independent of r. This is rather unrealistic since P being independent of r implies that the vector potentials between all the spatial points have the same correlation. A more realistic assumption would be to take

$$< A^r(t,r)A^s(t',s) >= P(t,t')f^r(r)f^s(r')$$

and hence derive from the gauge condition that

$$div f = f^r_{,r} = 0$$

A more general approach: Assume that the unperturbed metric is $g_{\mu\nu}(x)$ and the unperturbed vector potential is $A^\mu(x)$. Their perturbed versions are $g_{\mu\nu}(x) + \delta g_{\mu\nu}(x)$ and $A^\mu(x) + \delta A^\mu(x)$ respectively. The unperturbed velocity field of matter is $v^\mu(x)$ and it gets perturbed by $\delta v^\mu(x)$. Likewise the unperturbed density and pressure are $\rho(x)$ and $p(x)$ and they get perturbed by $\delta\rho(x)$ and $\delta p(x)$ respectively. The unperturbed four vector potential correlations are

$$L^{\mu\nu}(x,x') =< A^\mu(x)A^\nu(x') >$$

The Einstein field equations are

$$R_{\mu\nu} = K(S_{M\mu\nu} + S_{E\mu\nu}))$$

where $K = -8\pi G$,

$$S_{M\mu\nu} = T_{M\mu\nu} - T_M g_{\mu\nu}/2$$

where

$$T_{M\mu\nu} = (\rho+p)v_\mu v_\nu - pg_{\mu\nu}, T_M = g^{\mu\nu}T_{M\mu\nu} = \rho - 3p$$

so that

$$S_{M\mu\nu} = (\rho+p)v_\mu v_\nu - (p+\rho)g_{\mu\nu}/2$$

and

$$S_{E\mu\nu} = (-1/4)F_{\alpha\beta}F^{\alpha\beta}g_{\mu\nu} + F_{\mu\alpha}F_{\nu\beta}g^{\alpha\beta}$$

$$< S_{E\mu\nu} >= g_{\mu\nu}g^{\alpha\rho}g^{\beta\sigma} < F_{\alpha\beta}F_{\rho\sigma} > +g^{\alpha\beta} < F_{\mu\alpha}F_{\nu\beta} >$$

Now,

$$\begin{aligned}
< F_{\mu\alpha}F_{\nu\beta} > &=< (A_{\alpha,\mu} - A_{\mu,\alpha})(A_{\beta,nu} - A_{\nu,\beta}) > \\
&=< (g_{\alpha\rho}A^\rho)_{,\mu} - (g_{\mu\rho}A^\rho)_{,\alpha})((g_{\beta\sigma}A^\sigma)_{,\nu} - (g_{\nu\sigma}A^\sigma)_{,\beta}) > \\
&= (g_{\alpha\rho,\mu} - g_{\mu\rho,\alpha})(g_{\beta\sigma,\nu} - g_{\nu\sigma,\beta})L^{\rho\sigma}(x,x) \\
&+(g_{\alpha\rho}g_{\beta\sigma} < A^\rho_{,\mu}A^\sigma_{,\nu} > +g_{\alpha\rho}g_{\beta\sigma,\nu} < A^\rho_{,\mu}A^\sigma > \\
&+(g_{\alpha\rho,\mu}g_{\beta\sigma} < A^\rho A^\sigma_{,\nu} > \\
&-g_{\alpha\rho}g_{\nu\sigma,\beta} < A^\rho_{,\mu}A^\sigma > -g_{\alpha\rho,\mu}g_{\nu\sigma} < A^\rho A^\sigma_{,\beta} > \\
&-g_{\alpha\rho}g_{\nu\sigma} < A^\rho_{,\mu}A^\sigma_{,\beta} > -g_{\mu\rho}g_{\beta\sigma} < A^\rho_{,\alpha}A^\sigma_{,\nu} > \\
&-g_{\mu\rho,\alpha}g_{\beta\sigma} < A^\rho A^\sigma_{,\nu} > -g_{\mu\rho}g_{\beta\sigma,\nu} < A^\rho_{,\alpha}A^\sigma > \\
&+g_{\mu\rho}g_{\nu\sigma} < A^\rho_{,\alpha}A^\sigma_{,\beta} > +g_{\mu\rho,\alpha}g_{\nu\sigma} < A^\rho A^\sigma_{,\beta} > +g_{\mu\rho}g_{\nu\sigma,\beta} < A^\rho_{,\alpha}A^\sigma >
\end{aligned}$$

Using this formula, we can express

$$< S_{E\mu\nu}(x) >= C_1(\mu\nu\alpha\beta,x) < A^\alpha(x)A^\beta(x) > +C_2(\mu\nu\alpha\beta\rho,x) < A^\alpha(x)A^\beta_{,\rho}(x) >$$

$$+C_3(\mu\nu\alpha\beta\rho\sigma,x) < A^\alpha_{,\rho}(x)A^\beta_{,\sigma}(x) >$$

where the functions C_k are constructed using the unperturbed metric $g_{\mu\nu}(x)$ and its first order partial derivatives $g_{\mu\nu,\alpha}(x)$.

The perturbation $\delta < S_{E\mu\nu}(x) >$ of the average Maxwell energy momentum tensor is to be evaluated in terms of the metric perturbations. For evaluating this, we have to first express using the Maxwell equations, the perturbation $\delta A^\mu(x)$ of the electromagnetic four vector potential in terms of the unperturbed potentials $A^\mu(x)$ and the metric perturbations $\delta g_{\mu\nu}(x)$. The Maxwell equations are

$$(F^{\mu\nu}\sqrt{-g})_{,\nu} = 0$$

and its perturbation is

$$(\sqrt{-g}\delta F^{\mu\nu} + F^{\mu\nu}\delta\sqrt{-g})_{,\nu} = 0$$

We have

$$\delta\sqrt{-g} = -\delta g/2\sqrt{-g} = -gg^{\mu\nu}\delta g_{\mu\nu}/2\sqrt{-g} = (\sqrt{-g}/2)g^{\mu\nu}\delta g_{\mu\nu}$$

$$\delta F^{\mu\nu} = \delta(g^{\mu\alpha}g^{\nu\beta}F_{\alpha\beta})$$

$$= \delta(g^{\mu\alpha}g^{\nu\beta})F_{\alpha\beta} + g^{\mu\alpha}g^{\nu\beta}\delta F_{\alpha\beta}$$

Now,

$$\delta(g^{\mu\alpha}g^{\nu\beta}) = -g^{\mu\alpha}g^{\nu\rho}g^{\beta\sigma}\delta g_{\rho\sigma}$$

$$-g^{\nu\beta}g^{\mu\rho}g^{\alpha\sigma}\delta g_{\rho\sigma}$$

$$\delta F_{\alpha\beta} = \delta(A_{\beta,\alpha} - A_{\alpha,\beta}) =$$

$$(\delta A_\beta)_{,\alpha} - (\delta A_\alpha)_{,\beta} =$$

$$\delta A_\beta = \delta(g_{\beta mu}A^\mu) = A^\mu\delta g_{\beta\mu} + g_{\beta\mu}\delta A^\mu$$

The perturbation of the gauge condition

$$(A^\mu\sqrt{-g})_{,\mu} = 0$$

is given by

$$(\sqrt{-g}\delta A^\mu)_{,\mu} + (A^\mu\delta\sqrt{-g})_{,\mu} = 0$$

Now

$$\delta\sqrt{-g} = -\delta g/2\sqrt{-g} = -gg^{\mu\nu}\delta g_{\mu\nu}/2\sqrt{-g} = \sqrt{-g}g^{\mu\nu}\delta g_{\mu\nu}/2$$

Combining all these equations, it follows that δA^μ satisfies an equation of the form

$$D_1(\mu\nu\rho\sigma,x)\delta A^\nu_{,\rho\sigma}(x) + D_2(\mu\nu\rho,x)\delta A^\nu_{,\rho}(x) + D_3(\mu\nu\rho\alpha\beta,x)A^{nu}_{,\rho}(x)\delta g_{\alpha\beta}(x)$$

$$+D_4(\mu\nu\alpha\beta,x)A^\nu(x)\delta g_{\alpha\beta}(x)+$$

$$D_5(\mu\nu\rho\alpha\beta\sigma,x)A^\nu_{,\rho}(x)\delta g_{\alpha\beta,\sigma}(x)$$

$$+D_6(\mu\nu\alpha\beta\sigma,x)A^\nu(x)\delta g_{\alpha\beta,\sigma}(x) = 0$$

This equation can formally be solved to give

$$\delta A^\mu(x) = \int M(\mu\nu,\alpha\beta,x,x',x'')A^\nu(x')\delta g_{\alpha\beta}(x'')d^4x'd^4x''$$

The perturbation to $\delta < S_{E\mu\nu}(x) > = < \delta S_{E\mu\nu}(x) >$ to the average energy momentum tensor of the electromagnetic field can be expressed as follows:

$$\delta S_{E\mu\nu}(x) = (-1/4)\delta(F_{\alpha\beta}F^{\alpha\beta}g_{\mu\nu}) + \delta(F_{\mu\alpha}F_{\nu\beta}g^{\alpha\beta})$$

$$= (E_1(\mu\nu\rho\sigma\alpha\beta)(x)A^\rho_{,\sigma}(x) + E_2(\mu\nu\rho\alpha\beta,x)A^\rho(x))\delta A^\alpha_{,\beta}(x)$$

$$+(E_3(\mu\nu\rho\sigma\alpha)(x)A^\rho_{,\sigma}(x) + E_4(\mu\nu\rho\alpha,x)A^\rho(x))\delta A^\alpha(x)$$

$$+(E_5(\mu\nu\rho\delta\alpha\beta\sigma,x)A^\delta(x)A^\rho(x) + E_6(\mu\nu\rho\gamma\delta\alpha\beta\sigma,x)A^\delta(x)A^\rho_{,\gamma}(x))\delta g_{\alpha\beta,\sigma}(x)$$

$$+(E_7(\mu\nu\rho\delta\chi\alpha\beta,x)A^\delta_{,\chi}A^\rho(x) + E_8(\mu\nu\rho\delta\chi\gamma\alpha\beta,x)A^\delta_{,\chi}(x)A^\rho_{,\gamma}(x))\delta g_{\alpha\beta}(x)$$

A.30. BCS theory of superconductivity.
The second quantized Hamiltonian is given by

$$K = \int \psi_a(x)^*(-\nabla^2/2m + \mu)\psi_a(x)d^3x + \int V(x,y)\psi_a(x)^*\psi_a(x)\psi_b(y)^*\psi_b(y)d^3xd^3y$$

with the Fermionic fields ψ_a satisfying the anticommutation relations

$$\{\psi_a(x), \psi_b(y)^*\} = \delta_{ab}\delta^3(x-y),$$

$$\{\psi_a(x), \psi_b(y)\} = 0$$

We define

$$\psi_a(tx) = exp(itK).\psi_a(x).exp(-itK)$$

Then,

$$\psi_{a,t}(tx) = iexp(itK)[K, \psi_a(x)]exp(-itK)$$

The Green's function is defined as

$$G_{ab}(t, x|s, y) =< T(\psi_a(tx)\psi_b(s, y)^*) >= Tr(\rho_G T(\psi_a(tx)\psi_b(sy)^*))$$

where

$$\rho_G = exp(-\beta K)/Tr(exp(-\beta K))$$

We observe that for $t > s$,

$$G_{ab}(tx|sy) = Tr(exp(-\beta K).exp(itK)\psi_a(x)exp(i(s-t)K)\psi_b(y)^*.exp(-isK))$$

$$= Tr(exp((i(t-s)-\beta)K)\psi_a(x)exp(i(s-t)K))\psi_b(y)^*)$$

We make the approximation of replacing

$$\int V(x, y)\psi_a(x)^*\psi_a(x)\psi_b(y)^*\psi_b(y)d^3xd^3y$$

by

$$\int V(x, y)(2 < \psi_a(y)^*\psi_a(y) > \psi_b(x)^*\psi_b(x)d^3xd^3y + \int V(x, y) < \psi_a(x)\psi_b(y)^* > \psi_a(x)^*\psi_b(y)d^3y$$

$$+ \int V(x, x)\psi_a(x)^*\psi_a(x)d^3x + \int V(x, y) < \psi_a(x)\psi_b(y) > \psi_a(x)^*\psi_b(y)^*d^3xd^3y$$

plus other quadratic terms. We can write down the general form of the above approximation to the potential energy integral as

$$\int F_1(x, y) < \psi_b(y)^*\psi_b(y) > \psi_a(x)^*\psi_a(x)d^3xd^3y$$

$$+ \int F_2(x, y) < \psi_b(x)^*\psi_a(y) > \psi_a(y)^*\psi_b(x)d^3xd^3y$$

$$+ \int F_3(x, y) < \psi_b(y)\psi_a(x) > \psi_a(x)^*\psi_b(y)^*d^3xd^3y$$

$$+ \int F_4(x, y) < \psi_b(y)^*\psi_a(x)^* > \psi_a(x)\psi_b(y)d^3xd^3y$$

$$= V_1 + V_2 + V_3 + V_4$$

say. We have

$$[V_1, \psi_c(x')] = \int F_1(x, y) < \psi_b(y)^*\psi_b(y) > [\psi_a(x)^*\psi_a(x), \psi_c(x')]d^3xd^3y'$$

$$= -\int F_1(x, y) < \psi_b(y)^*\psi_b(y) > \delta_{ac}\delta^3(x-x')\psi_a(x)d^3xd^3y$$

$$= -(\int F_1(x', y) < \psi_b(y)^*\psi_b(y) > d^3y)\psi_c(x')$$

$$[V_2, \psi_c(x')] = \int F_2(x, y) < \psi_b(x)^*\psi_a(y) > [\psi_a(y)^*\psi_b(x), \psi_c(x')]d^3xd^3y$$

$$= -\int F_2(x, y) < \psi_b(x)^*\psi_a(y) > \delta_{ac}\delta^3(y-x')\psi_b(x)d^3xd^3y$$

$$= -(\int F_2(x, x') < \psi_b(x)^*\psi_a(x') > \psi_b(x)d^3x$$

$$[V_3, \psi_c(x')] =$$

$$\int F_3(x, y) < \psi_b(y)\psi_a(x) > [\psi_a(x)^*\psi_b(y)^*, \psi_c(x')]d^3xd^3y$$

Now,

$$\psi_a(x)^*\psi_b(y)^*\psi_c(x') = \psi_a(x)^*(\delta_{bc}\delta^3(y-x') - \psi_c(x')\psi_b(y)^*)$$
$$= \delta_{bc}\delta^3(y-x')\psi_a(x)^* - (\delta_{ac}\delta^3(x-x') - \psi_c(x')\psi_a(x)^*)\psi_b(y)^*$$
$$= \delta_{bc}\delta^3(y-x')\psi_a(x)^* - \delta_{ac}\delta^3(x-x')\psi_b(y)^* + \psi_c(x')\psi_a(x)^*\psi_b(y)^*$$

Thus,

$$[V_3, \psi_c(x')] = \int F_3(x,y) < \psi_b(y)\psi_a(x) > (\delta_{bc}\delta^3(y-x')\psi_a(x)^* - \delta_{ac}\delta^3(x-x')\psi_b(y)^*)d^3xd^3y$$

$$= \int F_3(x,x') < \psi_c(x')\psi_a(x) > \psi_a(x)^*d^3x - \int F_3(x',y) < \psi_b(y)\psi_c(x') > \psi_b(y)^*d^3y$$

Finally,

$$[V_4, \psi_c(x')] = \int F_4(x,y) < \psi_b(y)^*\psi_a(x)^* > [\psi_a(x)\psi_b(y), \psi_c(x')]d^3xd^3y = 0$$

A.31. Maxwell's equations in the closed isotropic model of the universe. The metric is

$$d\tau^2 = dt^2 - f^2(r)S^2(t)dr^2 - r^2S^2(t)(d\theta^2 + sin^2(\theta)d\phi^2)$$

where

$$f(r) = (1-kr^2)^{-1/2}$$

Thus,

$$g_{00} = 1, g_{11} = -f^2(r)S^2(t), g_{22} = -r^2S^2(t), g_{33} = -r^2S^2(t)sin^2(\theta)$$
$$\sqrt{-g} = r^2f(r)S^3(t)sin(\theta)$$

Assume $A^0 = 0$. The gauge condition

$$(A^\mu\sqrt{-g})_{,\mu} = 0$$

then gives

$$sin(\theta)(A^1r^2f)_{,1} + r^2f(A^2sin(\theta))_{,2} + r^2fsin(\theta)A^3_{,3} = 0$$

or equivalently,

$$r^{-2}(r^2fA^1)_{,1} + (sin(\theta))^{-1}(A^2sin(\theta))_{,2} + A^3_{,3} = 0$$

The Maxwell equation

$$(F^{0k}\sqrt{-g})_{,k} = 0$$

under the assumption $A^0 = 0$ gives

$$(g^{kk}A_{k,0}\sqrt{-g})_{,k} = 0$$

or equivalently,

$$(g^{kk}\sqrt{-g}(g_{kk}A^k)_{,0})_{,k} = 0$$

or equivalently,

$$(r^2fsin(\theta)(S^2A^k)_{,0})_{,k} = 0$$

The above two equations are not compatible. So we assume that $A^0 \neq 0$ and then the above two equations, ie, the gauge condition and Gauss' law respectively become

$$sin(\theta)r^2f(S^3A^0)_{,0} + sin(\theta)(A^1r^2f)_{,1} + r^2f(A^2sin(\theta))_{,2} + r^2fsin(\theta)A^3_{,3} = 0$$

and

$$0 = (F^{0k}\sqrt{-g})_{,k}$$
$$= (g^{kk}F_{0k}\sqrt{-g})_{,k} =$$
$$(g^{kk}(A_{k,0} - A_{0,k})\sqrt{-g})_{,k} =$$
$$(g^{kk}(g_{kk}A^k)_{,0}\sqrt{-g})_{,k} - (g^{kk}A^0_{,k}\sqrt{-g})_{,k}$$

This further simplifies to

$$((S^2A^k)_{,0}r^2fsin(\theta))_{,k} - S^2(g^{kk}A^0_{,k}r^2fsin(\theta))_{,k} = 0$$

The Maxwell equation

$$(F^{rs}\sqrt{-g})_{,s} + (F^{r0}\sqrt{-g})_{,0} = 0$$

gives taking $r = 1$,

$$(F^{12}\sqrt{-g})_{,2} + (F^{13}\sqrt{-g})_{,3} + (F^{10}\sqrt{-g})_{,0} = 0$$

and likewise for $r = 2, 3$. This equation can be expressed as

$$0 = (F^{12}\sqrt{-g})_{,2} + (F^{13}\sqrt{-g})_{,3} + (F^{10}\sqrt{-g})_{,0}$$
$$= (g^{11}g^{22}(A_{2,1} - A_{1,2})\sqrt{-g})_{,2} + (g^{11}g^{33}(A_{3,1} - A_{1,3})\sqrt{-g})_{,3}$$
$$+(g^{11}(A_{0,1} - A_{1,0})\sqrt{-g})_{,0}$$
$$= (g^{11}g^{22}((g_{22}A^2)_{,1} - g^{11}A^1_{,2})_{,2}+$$
$$(g^{11}g^{33}(g_{33}A^3)_{,1} - (g_{11}A^1)_{,3})\sqrt{-g})_{,3}$$
$$+(g^{11}(A^0_{,1} - (g_{11}A^1)_{,0})\sqrt{-g})_{,0}$$

A.32. Intuitive proof of Cramer's theorem for large deviations for iid random variables. Let $X_i, i = 1, 2, ...$ be iid random variables with

$$\mathbb{E}exp(\lambda X_1) = exp(\Lambda(\lambda))$$

Define

$$I(x) = sup_{\lambda\in\mathbb{R}}(\lambda x - \Lambda(\lambda))$$

Also define

$$S_n = \sum_{i=1}^{n} X_i$$

Then, we have for any Borel set E and any $\lambda \in \mathbb{R}$, the following: Let $x \in E$ be arbitrary. Then,

$$1 = \mathbb{E}[exp(\lambda S_n - n\Lambda(\lambda))] \geq \mathbb{E}[exp(\lambda S_n - n\Lambda(\lambda))\chi_{S_n/n\in E}]$$
$$\geq \mathbb{E}[inf_{x\in E}exp(n\lambda x - n\Lambda(\lambda))\chi_{S_n/n\in E}]$$
$$= inf_{x\in E}exp(n(\lambda x - \Lambda(\lambda)))P(S_n/n \in E)$$

and therefore,

$$n^{-1}.log(P(S_n/n \in E) + inf_{x\in E}(\lambda x - \Lambda(\lambda)) \leq 0$$

or equivalently,

$$n^{-1}.log(P(S_n/n \in E) \leq -inf_{x\in E}(\lambda x - \Lambda(\lambda))$$

If we assume that for each $x \in E$ the function $\lambda \to (\lambda x - \Lambda(\lambda))$ attains it supremum at some $\lambda(x) \in \mathbb{R}$, then we can conclude that

$$n^{-1}.log(P(S_n/n \in E) \leq -inf_{x\in E}I(x)$$

where

$$I(x) = sup_{\lambda\in\mathbb{R}}(\lambda x - \Lambda(\lambda)) = \lambda(x)x - \Lambda(\lambda(x))$$

This is called the Large deviations upper bound. To get the lower bound, we define a new probability measure $\tilde{P}_n$ by the prescription,

$$\tilde{P}_n = exp(\lambda S_n - n\Lambda(\lambda)).P_n$$

where P_n is the distribution of $(X_1, ..., X_n)$, then we have

$$1 = \tilde{P}_n(\mathbb{R}^n) = \int exp(\lambda\xi - n\Lambda(\lambda))dP_n(\xi)$$

and we obtain on differentiating both sides of this equation w.r.t. λ,

$$0 = \int (\xi - n\Lambda'(\lambda))d\tilde{P}_n(\xi)$$

and hence if η denotes the mean of X_1 under $\tilde{P}_1$, then we have

$$0 = \eta - \Lambda'(\lambda)$$

ie,

$$\eta = \Lambda'(\lambda)$$

Note that $X_1, ..., X_n$ under $\tilde{P}_n$ are iid random variables with distribution $\tilde{d}P_1(x) = exp(x\lambda - \Lambda(\lambda))dP_1(x)$. Thus, we get for $\lambda > 0$ and $\eta > 0$,

$$P_n(E) = \int_E exp(-\lambda\xi + n\Lambda(\lambda))d\tilde{P}_n(\xi) =$$

$$\mathbb{E}_{\tilde{P}}[exp(-n(\lambda S_n/n - \Lambda(\lambda)))\chi_{S_n/n \in E}]$$

$$\geq \mathbb{E}_{\tilde{P}}[exp(-n(\lambda S_n/n - \Lambda(\lambda)))\chi_{S_n/n \in E}\chi_{|S_n/n - \eta| \leq \epsilon}]$$

We are assuming that λ is a function of η determined by the condition $\Lambda'(\lambda) = \eta$. It follows that

$$P_n(E) \geq sup_{\eta+\epsilon \in E} exp(-n(\lambda(\eta+\epsilon) - \Lambda(\lambda)))\tilde{P}_n(|\{S_n/n \in E\} \bigcap \{|S_n/n - \eta| \leq \epsilon\})$$

Now assume that $E = (-\infty, a]$. By the law of large numbers for iid random variables, we have

$$lim_{n\to\infty}\tilde{P}_n(|S_n/n - \eta| \leq \epsilon) = 1$$

$\forall \epsilon > 0$. Note that $\tilde{P}$ is the probability measure on $\mathbb{R}^{\mathbb{Z}_+}$ having marginals $\tilde{P}_n, n = 1, 2, ...$ If we further assume that $\eta < a$, the it follows by the law of large numbers that

$$\tilde{P}_n(\{S_n/n \in E\} \cap \{|S_n/n - \eta| < \epsilon) \to 1$$

and hence

$$liminf_{n\to\infty} n^{-1}.log(P_n(E)) \geq sup_{\eta \in E}(-\lambda\eta + \Lambda(\lambda))$$

We can write the above as

$$liminf_{n\to\infty} n^{-1}.log(P_n(E)) \geq sup_{\eta \in E}(-\lambda\eta + \Lambda(\lambda)) = -inf_{\eta \in E} I(\eta)$$

A similar argument works for $\eta > a$.

A.33. Current in the BCS theory of superconductivity:
$\psi_\alpha(x)$ are the Fermionic field operators. They satisfy the anticommutation rules

$$\{\psi_\alpha(x), \psi_\beta(y)^*\} = \delta_{\alpha\beta}\delta^3(x-y),$$

$$\{\psi_\alpha(x), \psi_\beta(y)\} = 0, \{\psi_\alpha(x)^*, \psi_\beta(y)^*\} = 0$$

The BCS Hamiltonian has the form

$$H = (-1/2m)\int \psi_\alpha(x)^*(\nabla + ieA(x))^2\psi_\alpha(x)d^3x + \mu\int \psi_\alpha(x)^*\psi_\alpha(x)dx$$

$$+\int V_1(x,y) < \psi_\alpha(x)\psi_\beta(y) > \psi_\alpha(x)^*\psi_\beta(y)^*dxdy$$

$$+\int \bar{V}_1(x,y) < \psi_\beta(y)^*\psi_\alpha(x)^* > \psi_\beta(y)\psi_\alpha(x)dxdy$$

$$+\int V_2(x,y) < \psi_\alpha(x)\psi_\beta(x)^* > \psi_\alpha(y)^*\psi_\beta(y)dxdy$$

$$+\int V_3(x,y) < \psi_\alpha(x)\psi_\beta(y)^* > \psi_\alpha(x)^*\psi_\beta(y)dxdy$$

where for Hermitianity of H, we require that

$$\bar{V}_2(x,y) = V_2(x,y), \bar{V}_3(x,y) = V_3(y,x)$$

The Gibbs density operator for this Hamiltonian is

$$\rho_G = exp(-\beta H)/Tr(exp(-\beta H))$$

and averages are defined w.r.t this density, ie for any operator X constructed out of the quantum fields $\psi_\alpha(x), \psi_\alpha(x)^*, \alpha = 1, 2, x \in \mathbb{R}^3$, we define

$$< X >= Tr(\rho_G X)$$

We write

$$T = (-1/2m)\int \psi_\alpha(x)^*(\nabla + ieA(x))^2\psi_\alpha(x)dx$$

Then,

$$[T, \psi_\alpha(x)] = (1/2m)(\nabla + ieA(x))^2\psi_\alpha(x)$$

Also

$$[\int psi_\beta(y)^* psi_\beta(y)d^3y, \psi_\alpha(x)] =$$

$$-\psi_\alpha(x)$$

$$[V, \psi_\gamma(z)] =$$

$$\int V_1(x, y) < \psi_\alpha(x)\psi_\beta(y) > [\psi_\alpha(x)^*\psi_\beta(y)^*, \psi_\gamma(z)]dxdy$$

$$+\int \bar{V}_1(x, y) < \psi_\beta(y)^*\psi_\alpha(x)^* > [\psi_\beta(y)\psi_\alpha(x), \psi_\gamma(z)]dxdy$$

$$+\int V_2(x, y) < \psi_\alpha(x)\psi_\beta(x)^* > [\psi_\alpha(y)^*\psi_\beta(y), \psi_\gamma(z)]dxdy$$

$$+\int V_3(x, y) < \psi_\alpha(x)\psi_\beta(y)^* > [\psi_\alpha(x)^*\psi_\beta(y), \psi_\gamma(z)]dxdy$$

$$= \int \delta_{1\alpha\gamma}(x, z)V_1(x, z)\psi_\alpha(x)^*dx - \int \Delta_{1\gamma\alpha}(z, x)V_1(z, x)\psi_\alpha(x)^*dx$$

$$+\int V_2(x, z)\Delta_{2\gamma\alpha}(x, x)\psi_\alpha(z)dx$$

A.34. Some aspects of nonlinear filtering theory. Let X_t be a Markov process with transition generator kernel $K_t(x, y)$, ie,

$$\mathbb{E}(f(X_{t+dt})|X_t = x) - f(x) = dt\int K_t(x, y)f(y)dy + o(dt) = (K_t f)(x)dt + o(dt)$$

Assume that the measurement process is z_t defined by

$$dz_t = h_t(X_t)dt + dv_t$$

where v_t is a Levy process, ie, an independent increment process with moment generating function given by

$$\mathbb{E}exp(sv_t) = exp(t\psi(s))$$

Let $Z_t = \sigma(z_s, s \leq t)$, ie, the measurement process upto time t and for any function $\phi(x)$ defined on the state space of the Markov process X_t, we define

$$\pi_t(\phi) = \mathbb{E}(\phi(X_t)|Z_t)$$

We can write a stochastic differential equation for the process $\pi_t(\phi)$ as

$$d\pi_t(\phi) = F_t(\phi)dt + \sum_{k=1}^{\infty} G_{kt}(\phi)(dz_t)^k$$

where $F_t(\phi), G_{kt}(\pi)$ are Z_t measurable functions. The processes $F_t(\phi)$ and $G_{kt}(\phi)$ need to be calculated. We define a process y_t via the sde

$$dy_t = \sum_{k\geq 1} f_k(t)(dz_t)^k y_t, y_0 = 1$$

where the $f_k's$ are arbitrary non-random functions. Then y_t is a process adapted to Z_t and we have the relation

$$\mathbb{E}[(\phi(X_t) - \pi_t(\phi))y_t] = 0$$

by the definition of conditional expectation. We write $\phi_t = \pi_t(\phi)$. It follows that

$$\mathbb{E}[(d\phi_t - d\pi_t(\phi))y_t] + \mathbb{E}[(\phi_t - pi_t(\phi))dy_t] + \mathbb{E}[(d\phi_t - d\pi_t(\phi))dy_t] = 0$$

by Ito's formula. This equation can be expressed as

$$\mathbb{E}[(d\phi_t - d\pi_t(\phi))y_t] + \sum_{k\geq 1} f_k(t)\mathbb{E}[(\phi_t - \pi_t(\phi))(dz_t)^k y_t] + \sum_k \mathbb{E}[(d\phi_t - d\pi_t(\phi))(dz_t)^k y_t] = 0$$

and from the arbitrariness of the functions f_k, we get

$$\mathbb{E}[(d\phi_t - d\pi_t(\phi))|Z_t] = 0 - - - (1)$$

$$\mathbb{E}[(\phi_t - \pi_t(\phi))(dz_t)^k|Z_t] + \mathbb{E}[(d\phi_t - d\pi_t(\phi))(dz_t)^k|Z_t] = 0, k \geq 1 - - - (2)$$

We note that

$$(dz_t)^k = (dv_t)^k, k \geq 2$$

and hence writing

$$\mathbb{E}[(dv_t)^k] = \mu_k dt$$

we get from (1) and (2), assuming $\mu_1 = 0$,

$$F_t(\phi) + \sum_{k\geq 1} \mu_k G_{kt}(\phi) = \pi_t(K_t\phi) - - - (3),$$

$$(\pi_t(h_t\phi) - \pi_t(\phi)\pi_t(h_t)) + \pi_t((K_t\phi)h_t) - F_t(\phi)\pi_t(h_t) - \sum_{r\geq 1} \mu_{r+1}G_{rt}(\phi)\pi_t(h_t) = 0,$$

$$\mu_k\pi_t(K_t(\phi)) - \sum_{r\geq 1} \mu_{r+k}G_{rt}(\phi) = 0, k \geq 2$$

A.35. Some aspects of supersymmetry. Assume that $x^\mu, \mu = 0, 1, 2, 3$ are the Bosonic spatial variables and $\theta^\mu, \mu = 0, 1, 2, 3$ are the Fermionic anticommuting variables. A super field $S(x, \theta)$ can be expanded as

$$S(x, \theta) = S_0(x) + S_{1\mu}(x)\theta^\mu + S_{2\mu\nu}(x)\theta^\mu\theta^\nu + S_{3\mu\nu\rho}(x)\theta^\mu\theta^\nu\theta^\rho$$
$$+S_4(x)\theta^1\theta^2\theta^3\theta^4$$

Note that a product of more than four $\theta's$ is zero. The sum in each term is over the repeated variables and we may assume that the summation is over $\mu < \nu < \rho < \sigma$. We now need to introduce an infinitesimal supersymmetry transformation as a super vector field and describe the transformation laws of the component fields $S_0, S_{1\mu}, S_{2\mu\nu}, S_{3\mu\nu\rho}, S_4$ under such a supersymmetry transformation. The infinitesimal super-symmetry transformation may be derived by assuming the following group conservation law for Bosonic and Fermionic variables $(x, \theta), x = (x^\mu), \theta = (\theta^\mu)$:

$$(x^\mu, \theta^\alpha).(x'\mu, \theta'^\alpha) = (x^\mu + x'^\mu + \theta^T\gamma^\mu\theta', \theta^\alpha + \theta'^{alpha})$$

We express this equation as

$$(x, \theta).(x', \theta') = m((x, \theta), (x', \theta')) = (m_1((x, \theta), (x', \theta')), m_2((x, \theta), (x', \theta')))$$

where

$$m^1((x, \theta), (x', \theta')) = (x^\mu + x'^\mu + \theta^T\gamma^\mu\theta)$$

$$m_2((x, \theta), (x', \theta')) = (\theta^\alpha + \theta'^\alpha)$$

Then to get left and right-invariant vector fields on the supermanifold described by Bosonic and Fermionic coordinates (x, θ), we define

$$D_{x\mu} = \frac{\partial m_1^\nu}{\partial x'^\mu}\frac{\partial}{\partial x^\nu} + \frac{\partial m_2^\nu}{\partial x'^\mu}\frac{\partial}{\partial \theta^\nu}$$

$$= \frac{\partial}{\partial x^\mu} = \partial_{x\mu}$$

$$D_{\theta\alpha} = \frac{\partial m_1^\nu}{\partial \theta'^\alpha} \frac{\partial}{\partial x^\nu} + \frac{\partial m_2^\nu}{\partial \theta'^\alpha} \frac{\partial}{\partial \theta^\nu}$$

$$= (\theta^T \gamma^\nu)_\alpha \partial_{x\mu} + \partial_{\theta\alpha}$$

$$= (\theta^T \gamma^\mu)_\alpha \frac{\partial}{\partial x^\mu} + \frac{\partial}{\partial \theta^\alpha}$$

A.36. Comparison between the classical and quantum motions of a simple pendulum perturbed by white Gaussian noise.

Classical case:

$$\theta''(t) = -a.sin(\theta(t)) + \sigma w(t)$$

$$w(t) = B'(t)$$

$B(t)$ is standard Brownian motion. We solve this by perturbation theory: Let

$$\theta(t) = \theta_0(t) + \sigma\theta_1(t) + \sigma^2\theta_2(t) + ...$$

Then substituting this expression into the equation of motion and equating coefficients of $\sigma^m, m = 0, 1, 2$ successively gives us

$$\theta_0''(t) = -a.sin(\theta_0(t)), \theta_1''(t) = -a.cos(\theta_0(t))\theta_1(t) + B'(t),$$

$$\theta_2''(t) = -a.cos(\theta_0(t))\theta_2(t) + (a/2)sin(\theta_0(t))\theta_1(t)^2$$

Define the matrix

$$A(t) = \begin{pmatrix} 0 & 1 \\ -a.cos(\theta_0(t)) & 0 \end{pmatrix},$$

and the 2×2 state transition matrix $\Phi(t, \tau)$ by

$$\frac{\partial \Phi(t,\tau)}{\partial t} = A(t)\Phi(t,\tau), t \geq \tau, \Phi(\tau,\tau) = I_2$$

Then,

$$\theta_1(t) = \int_0^t \Phi_{12}(t,\tau)dB(\tau),$$

$$\theta_2(t) = (a/2) \int \Phi_{12}(t,\tau)sin(\theta_0(\tau))\theta_1(\tau)^2 d\tau$$

Using these formulas, the statistics of $\theta(t)$ can be computed upto $O(\sigma^2)$. We leave the problem of calculating the mean and autorcorrelation of $\theta(t)$ upto $O(\sigma^2)$ as an exercise to the reader.

The quantum case: Unitarity of the Schrodinger evolution operator $U(t)$ is guaranteed if we take into account an Ito correction term in the Hamiltonian. Thus, the Schrodinger evolution is given by

$$dU(t) = [-(iH_0 + \sigma^2 V^2/2)dt - i\sigma V dB(t)]U(t)$$

where

$$H_0 = -\partial_\theta^2/2ml^2 - mgl.cos(\theta)$$

We note that in this formalism, $U(t)$ should be regarded as an integral kernel $U(t, \theta, \theta')$ so that

$$(U(t)\psi)(\theta) = \int U(t,\theta,\theta')\psi(\theta')d\theta'$$

Heisenberg matrix mechanics: Let X be an observable and define

$$X(t) = U(t)^* X U(t)$$

Then, assuming that $V = V(\theta)$ is a multiplication operator,

$$dX(t) = dU(t)^* XU(t) + U(t)^* XdU(t) + dU(t)^* XdU(t)$$

$$= U(t)^*(i[H_0, X]dt - (\sigma^2/2)(V^2X + XV^2 - VXV)dt + i\sigma[V, X]dB(t))U(t)$$

For example, taking $X = \theta, p_\theta$ respectively gives

$$d\theta(t) = U(t)^*(p_\theta/ml^2 - (\sigma^2/2)(V^2\theta + \theta V^2 - V\theta V) + i\sigma[V, \theta]dB(t))U(t)$$

$$= p_\theta(t)/ml^2,$$

$$dp_\theta(t) = U(t)^*(i[-mglcos(\theta), p_\theta] - (\sigma^2/2)(V^2p_\theta + p_\theta V^2 - Vp_\theta V) + i\sigma[V, p_\theta]dB(t))U(t)$$

$$= U(t)^*(-mglsin(\theta) - (\sigma^2/2)(V[V, p_\theta] + [p_\theta, V]V) + i\sigma[V, p_\theta]dB(t))U(t)$$

$$= -mglsin(\theta(t)) - \sigma V'(\theta(t))dB(t)$$

Thus, we get the quantum analogue of the classical sde. However, to be more precise and obtain further generalizations, we must take our noise processes as the creation, annihilation and conservation processes of the Hudson-Parthasarathy quantum stochastic calculus.

HP calculus generalizations: Consider the qsde

$$dU(t) = (-(iH_0 + P)dt + L_1dA(t) - L_2dA(t)^* + Sd\Lambda(t))U(t)$$

where L_1, L_2, P, S are system operators chosen to make $U(t)$ unitary for all t. Here,

$$H_0 = -\nabla^2/2m + U(r)$$

acts in the system Hilbert space $\mathfrak{h} = L^2(\mathbb{R}^3)$. Let X be a system observable. At time t it evolves under HP noisy Heisenberg dynamics to

$$X(t) = U(t)^*XU(t)$$

$X(t)$ is defined on $\mathfrak{h} \otimes \Gamma_s(L^2(\mathbb{R}_+))$ and $A(t), A^*(t), \Lambda(t)$ are the HP noise operator processes satisfying the quantum Ito formula

$$dA(t)dA(t)^* = dt, dA^*dA = 0, (dA)^2 = 0, (dA^*)^2 = 0, (d\Lambda)^2 = d\Lambda,$$

$$dA.d\Lambda = dA, d\Lambda.dA^* = dA^*, d\Lambda.dA = 0, dA^*d\Lambda = 0$$

Remark: Formally, we can write

$$d\Lambda = dA^*.dA/dt$$

and hence using $dA.dA^* = dt$, we get

$$d\Lambda.dA^* = dA^*, dA.d\Lambda = dA$$

We get

$$dX(t) = dU(t)^*XU(t) + U(t)^*XdU(t) + dU(t)^*XdU(t) =$$

$$U(t)^*(i[H_0, X]dt - (PX + XP)dt + (L_1^*X - XL_2 + S^*XL_1^*)dA^*(-L_2^*X + XL_1^* - L_2^*XS)dA + (S^*X + XS^* + S^*XS)d\Lambda)U(t)$$

Exercise: Evaluate the Heisenberg equations of motion taking $X = q_k, X = p_k, k = 1, 2, 3$ where $r = (q_1, q_2, q_3), p = (p_1, p_2, p_3)$. Assume that the system operators L_1, L_2, S, P are arbitrary functions of q, p subject to the constraint that makes $U(t)$ unitary for all t. Specialize to the case when L_1, L_2, S, P are functions of q alone.

A.37. The magnetic field produced by a transmission line current when hysteresis effects are taken into account. The line equations are

$$V'(\omega, z) + Z(\omega, z)I(\omega, z) + \delta \int H_1(\omega_1, \omega - \omega_1, z)I(\omega_1, z)I(\omega - \omega_1, z)d\omega_1 = 0$$

$$I'(\omega, z) + Y(\omega, z)V(\omega, z) + \delta \int H_2(\omega_1, \omega - \omega_1, z)V(\omega_1, z)V(\omega - \omega_1, z)d\omega_1 = 0$$

Here,

$$Z(\omega, z) = R(z) + j\omega L(z), Y(\omega, z) = G(z) + j\omega C(z)$$

We can expand in a Fourier series

$$Z(\omega, z) = \sum_{n \in \mathbb{Z}} Z_n(\omega)exp(jn\beta z), \beta = 2\pi/d$$

$$Y(\omega, z) = \sum_n Y_n(\omega)exp(jn\beta z)$$

$$V^{(0)}(\omega,z)=\sum_n V_n^{(0)}(\omega)exp((\gamma+jn\beta)z),$$

$$I^{(0)}(\omega,z)=\sum_n I_n^{(0)}(\omega)exp((\gamma+jn\beta)z)$$

$$V(\omega,z)=V^{(0)}(\omega,z)+\delta V^{(1)}(\omega,z)+O(\delta^2)$$
$$I(\omega,z)=I^{(0)}(\omega,z)+\delta I^{(1)}(\omega,z)+O(\delta^2)$$

Then, substituting these expressions into the above line differential equations and equating coefficients of δ^0 and δ^1 respectively gives us

$$(\gamma+jn\beta)V_n^{(0)}(\omega)+\sum_m Z_{n-m}(\omega)I_m^{(0)}(\omega)=0,$$

$$(\gamma+jn\beta)I_n^{(0)}(\omega)+\sum_m Y_{n-m}(\omega)V_m^{(0)}(\omega)=0$$

$$\frac{\partial V^{(1)}(\omega,z)}{\partial z}+Z(\omega,z)I^{(1)}(\omega,z)+H_1.(I^{(0}\otimes I^{(0)})(\omega,z)=0,$$
$$\frac{\partial I^{(1)}(\omega,z)}{\partial z}+Y(\omega,z)I^{(1)}(\omega,z)+H_2(V^{(0)}\otimes V^{(0)})(\omega,z)=0$$

Let $\Phi(\omega,z,z')\in\mathbb{C}^{2\times 2}$ denote the state transition matrix corresponding to the forcing matrix

$$A(\omega,z)=\begin{pmatrix}0 & -Z(\omega,z)\\ -Y(\omega,z) & 0\end{pmatrix}$$

Then, the solutions to the first order line voltage and current perturbations are given by

$$V^{(1)}(\omega,z)=-\int_0^z \Phi_{11}(\omega,z,z')H_1.(I^{(0}\otimes I^{(0)})(\omega,z')dz'$$

$$-\int_0^z \Phi_{12}(\omega,z,z')H_21.(V^{(0}\otimes V^{(0)})(\omega,z')dz'$$

$$I^{(1)}(\omega,z)=-\int_0^z \Phi_{21}(\omega,z,z')H_1.(I^{(0)}\otimes I^{(0)})(\omega,z')dz'$$

$$-\int_0^z \Phi_{22}(\omega,z,z')H_2.(V^{(0)}\otimes V^{(0)})(\omega,z')dz'$$

Denoting the infinite dimensional matrix $((Z_{n-m}(\omega)))$ by $\mathbf{Z}(\omega)$, $((Y_{n-m}(\omega)))$ by $\mathbf{Y}(\omega)$, the diagonal matrix $diag[n:n\in\mathbb{Z}]$ by $\mathbf{D}$, the unperturbed voltage and current Fourier coefficient vectors $((V_n^{(0)}(\omega)))$ and $((I_n^{(0)}(\omega)))$ by $\mathbf{V}^{(0)}(\omega)$ and $\mathbf{I}^{(0)}(\omega)$ respectively, the above unperturbed eigenvalue equations can be expressed as

$$\mathbf{T}(\omega)\mathbf{u}(\omega)=-\gamma(\omega)\mathbf{u}(\omega)---(1)$$

where

$$\mathbf{T}(\omega)=\begin{pmatrix}j\beta\mathbf{D} & \mathbf{Z}(\omega)\\ \mathbf{Y}(\omega) & j\beta\mathbf{D}\end{pmatrix},$$

$$\mathbf{u}(\omega)=\begin{pmatrix}\mathbf{V}^{(0)}(\omega)\\ \mathbf{I}^{(0)}(\omega)\end{pmatrix}$$

Let $-\gamma_n(\omega),\mathbf{u}_n(\omega),n=1,2,...$ be the complete set of eigenvalues and corresponding eigenvectors of $\mathbf{T}(\omega)$ as in (1). Then we can write

$$\mathbf{u}(\omega)=\sum_{n\in\mathbb{Z}} c_n(\omega)\mathbf{u}_n(\omega)$$

and we have on defining the infinite dimensional column vector

$$\mathbf{e}_n(\omega,z)=((exp((\gamma_n(\omega)+jk\beta)z)))_{k\in\mathbb{Z}},$$

the following expression for the unperturbed line voltage and current:

$$V^{(0)}(\omega,z)=\sum_n c_n(omega)\mathbf{e}_n(\omega,z)^T\mathbf{v}_n(\omega),$$

$$I^{(0)}(\omega, z) = \sum_n c_n(\omega)\mathbf{e}_n(\omega, z)^T \mathbf{w}_n(\omega)$$

where

$$\mathbf{u}_n(\omega) = \begin{pmatrix} \mathbf{v}_n(\omega) \\ \mathbf{w}_n(\omega) \end{pmatrix}$$

The far field magnetic vector potential produced by this unperturbed line current is given by

$$A_z(\omega, r) = (\mu.exp(-jKr)/r) \int_0^d I^{(0)}(\omega, \xi) exp(jK\xi.cos(\theta)) d\xi$$

$$= (\mu.exp(-jKr)/r) P(\omega, \theta)$$

where $K = \omega/c$ and

$$P(\omega, \theta) = \int_0^d I^{(0)}(\omega, \xi) exp(jk\xi.cos(\theta)) d\xi$$

A.38. Problems in quantum mechanics with solutions:

[1] Obtain the supersymmetry generators in the form of super vector fields for four Boson and four Fermionic variables $(x^\mu), (\theta^\mu)$.

hint: Consider the generators of the form

$$D_\alpha = (A^\mu \theta)_\alpha \partial/partialx^\mu + B_\alpha^\beta \partial/\partial\theta^\beta$$

and

$$\bar{D}_\alpha = (C^\mu \theta)_\alpha \partial/partialx^\mu + F_\alpha^\beta \partial/\partial\theta^\alpha$$

The matrices A^μ, C^μ and the complex numbers B_α^β and F_α^{beta} are to be chosen so that the supersymmetric commutation relations hold in the form:

$$[D_\alpha, \bar{D}_\beta] = K_{\alpha\beta}^\mu \partial/\partial x^\mu$$

[2] Consider a periodic potential $V(r), r \in \mathbb{R}^3$ with three linearly independent period vectors, $a_1, a_2, a_3 \in \mathbb{R}^3$ so that

$$V(r + n_1 a_1 + n_2 a_2 + n_3 a_3) = V(r), n_1, n_2, n_3 \in \mathbb{Z}$$

Expand V as a 3-D Fourier series using the reciprocal lattice vectors and hence formulate the stationary state Schrodinger equation

$$-\nabla^2 \psi(r)/2m + V(r)\psi(r) = E\psi(r), r \in \mathbb{R}^3$$

so that in view of the periodicity of V, we have

$$\psi(r + n_1 a_1 + n_2 a_2 + n_3 a_3) = C_1^{n_1} C_2^{n_2} C_3^{n_3} \psi(r), n_1, n_2, n_3 \in \mathbb{Z}$$

where $|C_k| = 1$. If there are N_k atoms along a_k $k - 1, 2, 3$, then we may apply the periodic boundary conditions on ψ without loss of generality in the form

$$\psi(r + N_k a_k) = \psi(r), k = 1, 2, 3$$

This leads to

$$C_k^{N_k} = 1, k = -1, 2, 3$$

so that

$$C_k = exp(2\pi i l_k/N_k), k = 1, 2, 3$$

for some $l_k \in \{0, 1, ..., N_k - 1\}$. Now define the Bloch wave functions $u_{l_1 l_2 l_3}(r)$ by

$$\psi(r) = exp(2\pi i(l_1 b_1/N_1 + l_2 b_2/N_2 + l_3 b_3/N_3, r)) u_{l_1 l_2 l_3}(r)$$

where $\{b_1, b_2, b_3\}$ are the reciprocal lattice vectors corresponding to $\{a_1, a_2, a_3\}$. In other words,

$$(b_i, a_j) = \delta_{ij}$$

Then, we have

$$u_{l_1 l_2 l_3}(r + m_1 a_1 + m_2 a_2 + m_3 a_3) = u_{l_1 l_2 l_3}(r), m_1 m_2, m_3 \in \mathbb{Z}$$

ie, $u_{l_1l_2l_3}$ is also periodic with periods a_1, a_2, a_3 like $V(r)$ and hence it can also be developed into a Fourier series:

$$u_{l_1l_2l_3}(r) = \sum_{n_1n_2n_3\in\mathbb{Z}} U_{l_1l_2l_3}[n_1n_21n_3]exp(2\pi i(n_1b_1 + n_2b_2 + n_3b_3, r))$$

$$V(r) = \sum_{n_1n_2n_3\in\mathbb{Z}} V[n_1n_2n_3]exp(2\pi i(n_1b_1 + n_2b_2 + n_3b_3, r))$$

The problem is to subsitute these two Fourier series expansions for the wave function and the potential into the stationary state Schrodinger wave equation and derive a difference equation for $U_{l_1l_2l_3}[n_1n_2n_3]$.

[3] Quantum control of the Hudson-Parthasarathy equation based on the Belavkin filter observer.
HP equn:

$$dU(t) = (-(iH + P)dt + L_1dA + L_2dA^* + Sd\Lambda)U(t)$$

For unitarity of $U(t)$, we require

$$0 = d(U^*U) = dU^*.U + U^*.dU + dU^*.dU$$

This gives using the quantum Ito formula

$$dAdA^* = dt, d\Lambda.dA^* = dA^*, dA.d\Lambda = dA, (d\Lambda)^2 = d\Lambda,$$

(all the other products of differentials are zero),

$$P = L_2^*L_2/2,$$

$$L_2^* + L_1 + L_2^*S = 0,$$

$$L_1^* + L_2 + S^*L_2 = 0,$$

$$S + S^* + S^*S = 0$$

L_1, L_2, S, H, P are system operators with H, P self-adjoint. We have

$$(I + S)^*(I + S) = I + S^* + S + S^*S = I$$

so we can write

$$I + S = exp(i\lambda(t)Z)$$

where Z is selfadjoint and $\lambda(t)$ is a real valued controllable function of time. Then, we have

$$L_1 = -L_2^*(I + S) = -L_2^*exp(i\lambda(t)Z)$$

We write

$$L_2 = L = \sum_{k=1}^{p} c_k(t)N_k = L(\{c_k(t)\})$$

Then

$$L_1 = -(\sum_k \bar{c}_k(t)N_k^*))exp(i\lambda(t)Z) = L_1(\{c_k(t)\}, \lambda(t)),$$

$$S = exp(i\lambda(t)Z) - 1 = S(\lambda(t))$$

$c_k(t)$ are complex valued controllable functions of time. The real time control algorithms for controlling the $c_k(t)'s$ and $\lambda_k(t)$ are based on minimizing the expected value of the Belavkin observer error

$$E(t) = X_d(t) - \pi_t(X)$$

where $X_d(t)$ is the desired system state at time t. This is a system observable and $\pi_t(X)$ is a measurable function of the noise algebra $\sigma(Y)_t$ upto time t. The noise is taken as

$$Y(t) = U(t)^*Y_i(t)U(t), Y_i(t) = a_1B(t) + a_2\Lambda(t), B(t) = A(t) + A(t)^*, a_1, a_2 \in \mathbb{R}$$

It is easily verified that Y statisfies the non-demolition abelian property, ie,

$$[Y(t), Y(s)] = 0\forall t, s \geq 0$$

and

$$[Y(s), j_t(X)] = 0, t \geq s, j_t(X) = U(t)^* XU(t) = U(t)^*(X \otimes I)U(t)$$

We have

$$Y(t) = j_t(Y_i(t)) = U(t)^* Y_i(t)U(t),$$

$$dY(t) = dY_i(t) + dU(t)^* dY_i(t).U(t) + U(t)^* dY_i(t)dU(t) =$$

$$dY_i(t) + U(t)^*(L_1^* dA^* + L_2^* dA + S^* d\Lambda)dY_i + dY_i(L_1 dA + L_2 dA^* + Sd\Lambda))U(t)$$

Now,

$$dY_i(L_1 dA + L_2 dA^* + Sd\Lambda) =$$

$$(a_1(dA + dA^*) + a_2 d\Lambda)(L_1 dA + L_2 dA^* + Sd\Lambda) =$$

$$a_1 L_2 dt + a_2 Sd\Lambda + a_2 L_2 dA^* + a_1 SdA$$

We thus get

$$dY(t) = dY_i(t) + j_t(a_1(L_2 + L_2^*))dt + j_t(a_2(S + S^*))d\Lambda + j_t(a_2 L_2 + a_1 S^*)dA^* + j_t(a_2 L_2^* + a_1 S)dA$$

write

$$\pi_t(X) = \mathbb{E}(j_t(X)|\sigma(Y)_t)$$

where the expectation is taken in the state $|f\phi(u)>$ with $f \in \mathfrak{h}$, the system Hilbert space and $|\phi(u)>= exp(-\parallel u \parallel^2 /2)|e(u)>$ the normalized exponential vector in the Boson Fock space $\Gamma_s(L_2(\mathbb{R}_+))$. We assume that the optimal filter is described by the following qsde:

$$d\pi_t(X) = F_t(X)dt + G_t(X)dY(t)$$

where $F_t(X), G_t(X) \in \sigma(Y)_t$. Then applying the orthogonality principle

$$\mathbb{E}[(j_t(X) - \pi_t(X))C(t)] = 0$$

for $C(t)$ defined via the qsde

$$dC(t) = f(t)C(t)dY(t), C(0) = 1$$

and using the arbitrariness of $f(t)$ gives us

$$\mathbb{E}[(dj_t(X) - d\pi_t(X))|\sigma(Y)_t] = 0,$$

$$\mathbb{E}[(dj_t(X) - d\pi_t(X))dY(t)|\sigma(Y)_t] + \mathbb{E}[(j_t(X) - \pi_t(X))dY(t)|\sigma(Y)_t] = 0$$

These two equations give us two equations for the observables $F_t(X)$ and $G_t(X)$, solving which the Belavkin filter is obtained. To derive these two equations, we compute using quantum Ito's formula,

$$dj_t(X) = dU^* XU + U^* XdU + dU^* XdU =$$

$$= U^*(i[H, X] - PX - XP + L_2^* XL_2)dt + (L_2^* X + XL_1 + L_2^* XS)dA + (L_1^* X + XL_2 + S^* XL_2)dA^* + (S^* X + XS + S^* XS)d\Lambda)U$$

$$= j_t(\theta_0(X))dt + j_t(\theta_1(X))dA_t + j_t(\theta_2(X))dA_t^* + j_t(\theta_3(X))d\Lambda_t$$

where

$$\theta_0(X) = i[H, X] - L^* LX/2 - XL^* L/2 + L^* XL^*$$

$$\theta_1(X) = L^* X - XL^*(I + S) + L^* XS,$$

$$\theta_2(X) = -(1 + S)^* LX + XL - S^* XL,$$

$$\theta_3(X) = S^* X + XS + S^* XS$$

Note that

$$L_2 = L = L(\{c_k(t)\}), S = S(\lambda(t))$$

and

$$L_1 = -L^*(1 + S) = L_1(\{c_k(t), \lambda(t))$$

The equation

$$\mathbb{E}[(dj_t(X) - d\pi_t(X))|\sigma(Y)_t] = 0,$$

thus gives

$$\pi_t(\theta_0(x))+\pi_t(\theta_1(X))u(t)+\pi_t(\theta_2(x))\bar{u}(t)+\pi_t(\theta_3(X))|u(t)|^2$$

$$-G_t(X)[a_1(u(t)+\bar{u}(t))+a_2|u(t)|^2+\pi_t(a_1(L_2+L_2^*))+\pi_t(a_2(S+S^*))|u(t)|^2+pi_t(a_2L_2+a_1S^*)\bar{u}(t)+\pi_t(a_2L_2^*+a_1S)u(t)]=0$$

or equivalently,

$$F_t(X)=$$

$$\pi_t(\theta_0(x))+\pi_t(\theta_1(X))u(t)+\pi_t(\theta_2(x))\bar{u}(t)+\pi_t(\theta_3(X))|u(t)|^2$$

$$-G_t(X)[a_1(u(t)+\bar{u}(t))+\pi_t(a_1(L_2+L_2^*))+\pi_t(a_2(S+S^*))|u(t)|^2+pi_t(a_2L_2+a_1S^*)\bar{u}(t)+\pi_t(a_2L_2^*+a_1S)u(t)]$$

The equation

$$\mathbb{E}[(dj_t(X)-d\pi_t(X))dY(t)|\sigma(Y)_t]+\mathbb{E}[(j_t(X)-\pi_t(X))dY(t)|\sigma(Y)_t]=0$$

gives

$$\pi_t(\theta_0(X))+\pi_t(\theta_1(X))u(t)+\pi_t(\theta_2(X))\bar{u}(t)+\pi_t(\theta_3(X))|u(t)|^2$$

$$-F_t(X)-G_t(X)\mathbb{E}[(dY(t))^2|\sigma(Y)_t]+\pi_t(X)(a_1(u(t)+\bar{u}(t))+a_2|u(t)|^2+\pi_t(a_1(L_2+L_2^*))+\pi_t(a_2(S+S^*))|u(t)|^2$$

$$+\pi_t(a_2L_2+a_1S^*)\bar{u}(t)+\pi_t(a_2L_2+a_1S)u(t)=0$$

Now, from the equation

$$dY(t)=dY_i(t)+j_t(a_1(L_2+L_2^*))dt+j_t(a_2(S+S^*))d\Lambda+j_t(a_2L_2+a_1S^*)dA^*+j_t(a_2L_2^*+a_1S)dA$$

$$=j_t(a_1(L_2+L_2^*))dt+j_t(a_2(S+S^*+1))d\Lambda+j_t(a_2L_2+a_1(S^*+1))dA^*+j_t(a_2L_2^*+a_1(S+1))dA$$

we get using quantum Ito's formula and the homomorphism property of j_t,

$$\mathbb{E}[(dY(t))^2|\sigma(Y)_t]=$$

$$\pi_t(a_2^2(S+S^*+1)^2)|u(t)|^2+\pi_t((a_2L_2^*+a_1(S+1))(a_2L_2+a_1(S^*+1)))+\pi_t((a_2(S+S^*+1)(a_2L_2+a_1(S^*+1)))\bar{u}(t)$$

$$+\pi_t(a_2(a_2L_2^*+a_1(S+1))(S+S^*+1))u(t)$$

[4](Reference: S.Wienberg, The quantum theory of fields, vol.II, Cambridge University Press)Evaluate approximately the path integeral

$$Z(J)=\int exp(iI(\phi)-i\int J(x)\phi(x)d^4x)D\phi$$

where $J(x)$ is an external current source and

$$I(\phi)=\int(\partial_\mu\phi)(\partial^\mu\phi)/2-m^2\phi^2/2-\epsilon V(\phi)d^4x$$

ie, the Klein-Gordon action functional with a small perturbative correction. Calculate using this expression the propagators

$$\int exp(iI(\phi)\phi(x_1)...\phi(x_n)D\phi/Z(0)$$

by using the formula

$$\frac{\partial^n bZ(J)}{\partial J(x_1)...\partial J(x_n)}|_{J=0}$$

$$=i^n\int exp(iI(\phi))\phi(x_1)...\phi(x_n)d\phi$$

Now derive the equation for J at which $log(Z(J))-\int J(x)\phi_0(x)d^4x$ becomes stationary for a fixed field ϕ_0. Denoting this stationary solution by $J_0(x)$, show that

$$-\phi_0(x)+\delta logZ(J_0)/\delta J(x)=0$$

or equivalently,

$$\phi_0(x)+iZ(J_0)^{-1}\int exp(i\int(I(\phi)-J_0(x)\phi(x))d^4x)\phi(x).D\phi=0$$

We write

$$J_0(x) = J_0(\phi_0)(x)$$

or inverting this,

$$\phi_0(x) = \phi_0(J_0)(x)$$

It follows that

$$(\delta\phi_0(x)/\delta J_0(y) = \delta^2 log(Z(J_0))/\delta J(x)\delta J(y)$$

We define

$$\Gamma(\phi_0) = log(Z(J_0)) - \int J_0(x)\phi_0(x)d^4x$$

Thus,

$$\delta\Gamma(\phi_0)/\delta\phi_0(y) = \int \phi_0(x)(\delta J_0(x)/\delta\phi_0(y))d^4x - \int (\delta J_0(x)/\delta\phi_0(y))\phi_0(x)d^4x - J_0(y)$$
$$= -J_0(y)$$

$\Gamma(\phi_0)$ is called the quantum effective action and the above equation is called the equation of motion for the quantum effective ation.

[5] Calculation of path integrals for gauge invariant theories: Let $\phi(x)$ be the set of fields and $f[\phi]$ a gauge fixing functional. We consider a path integral of the form

$$X = \int G[\phi]B[f[\phi]]F[\phi]D\phi$$

where $F[\phi]$ is the Jacobian determinant:

$$F[\phi] = det(df[\phi_\Lambda]/d\Lambda)|_{\Lambda=id}$$

where $\phi \to \phi_\Lambda$ is the gauge transformed field. Λ is the gauge transorming function. We wish to show that in a certain sense, X does not depend on the gauge fixing functional f. We assume that the combined action functional $G[\phi]$ along with the measure $D\phi$, ie, $G[\phi]D\phi$ is invariant under the gauge transformation, ie, for all gauge transformations Λ, we have

$$G[\phi_\Lambda]D\phi_\Lambda = G[\phi]D\phi$$

Then, it follows that

$$X = \int G[\phi_\Lambda]B[f[\phi_\Lambda]]F[\phi_\Lambda]D\phi_\Lambda$$
$$= \int G[\phi]B[f[\phi_\Lambda]]F[\phi_\Lambda]D\phi$$

Now, if Λ and λ are two gauge transformations, then

$$det(df[\phi_{\Lambda o\lambda}]/d\lambda)|_{\lambda=id} =$$
$$F[\phi_\Lambda]/\rho(\Lambda)$$

where

$$\rho(\Lambda)^{-1} = det(d(\Lambda o\lambda)/d\lambda)_{\lambda=id}$$

So,

$$X = \int G[\phi]B[f[\phi_\Lambda]]\rho(\Lambda)det(df[\phi_\Lambda]/d\Lambda)D\phi$$

Integrating this equation w.r.t Λ over the gauge group gives us after appropriately normalizing the Haar measure,

$$X = \int G[\phi].B[f]]df.D\phi = C\int G[\phi]D\phi$$

where C is a constant defined by

$$C = \int B[f]Df = \int B[f[\phi_\Lambda]]det(df[\phi_\Lambda]/d\Lambda)\rho(\Lambda)d\Lambda$$

where we use the fact that $\rho(\Lambda)$ as defined above is the left invariant Haar density on the gauge group and also the assumption that the gauge group acts transitively on the matter fields ϕ.

[6] Quantum teleportation: The simplest version of this idea involves transmitting one qubit of information by transmitting just two classical bits of information. Alice and Bob share an entangled state $|00>+|11>$ apart from normalization factor of $2^{-1/2}$. Alice prepares another state

$$|\psi>=c_1|1>+c_2|0>$$

where $c_1, c_2 \in \mathbb{C}$ and $|c_1|^2+|c_2|^2=1$ which she wishes to transmit to Bob by making use of the entangled state that she shares with Bob. The overall state of Alice and Bob is thus

$$|\psi>(|00>+|11>)=(c_1|1>+c_2|0>)(|11>+|00>)=$$

$$c_1|111>+c_1|100>+c_2|011>+c_2|000>$$

The first two qubits of this state can be controlled only by Alice and the last only by Bob. Alice applies a phase gate to her two qubits thus obtaining the overall state as

$$|\phi>=c_1|111>+c_1|100>+c_2|011>-c_2|000>$$

We express this state in terms of the orthogonal one qubit state

$$|+>=|1>+|0>, |->=|1>-|0>$$

or equivalently,

$$|1>=|+>+|->, |0>=|+>-|->$$

Thus the overall state of Alice and Bob is given by

$$|\phi>=c_1(|+>+|->)(|+>+|->)|1>+c_1(|+>+|->)(|+>-|->)|0>$$

$$+c_2(|+>-|->)(|+>+|->)|1>-c_2(|+>-|->)(|+>-|->)|0>$$

Alice now performs a measurement on her two qubits in this shared state using the orthonormal basis $|++>, |+->, |-+>, |-->$. If she measures $|++>$, then clearly from the above expression for $|\phi>$, Bob's state collapses to

$$c_1|1>+c_1|0>+c_2|1>-c_2|0>=c_1(|1>+|0>)+c_2(|1>-|0>)$$

If Alice measures $|+->$, then Bob's state collapses to

$$c_1|1>-c_1|0>+c_2|1>+c_2|0>=c_1(|1>-|0>)+c_2(|1>+|0>)$$

If Alice measures $|-+>$, then Bob's state collapses to

$$c_1|1>+c_1|0>-c_2|1>+c_2|0>=c_1(|1>+|0>)+c_2(|0>-|1>)$$

Finally, if Alice measures $|-->$, then Bob's state collapses to

$$c_1|1>-c_1|0>-c_2|1>-c_2|0>=c_1(|1>-|0>)-c_2(|1>+|0>)$$

Using two classical bits, Alice reports to Bob the outcome of her measurements. If she reports $|++>$, then Bob applies the unitarty gate U_1 to his state defined by

$$U_1(|1>+|0>)\sqrt{2}=|1>, U_1(|1>-|0>)/\sqrt{2}=|0>,$$

to recover $|\psi>$. If Alice reports $|+->$, then Bob applies the unitary gate U_2 to his state defined by

$$U_2(|1>-|0>)/\sqrt{2}=|1>, U_2(|1>+|0>)/\sqrt{2}=|0>$$

to recover $|\psi>$. If Alice reports $|-+>$, then Bob applies the unitary gate U_3 to his state defined by

$$U_3(|1>+|0>)/\sqrt{2}=|1>, U_3(|1>-|0>)/\sqrt{2}=-|0>$$

to recover $|\psi>$. Finally, if Alice reports $|-->$, then Bob applies the unitary gate U_4 to his state defined by

$$U_4(|1>-|0>)/\sqrt{2}=|1>, U_4(|1>+|0>)/\sqrt{2}=-|0>$$

to recover the state $|\psi>$.

[7] Quantum Boltzmann equation. N identical particles. $\rho(t)$ is the joint density matrix of the particles. The kth particle acts in the Hilbert space $\mathcal{H}_k$. The Hilbert space of the whole system is

$$\mathcal{H} = \bigotimes_{k=1}^{N} \mathcal{H}_k$$

Each $\mathcal{H}_k$ is an identical copy of a fixed Hilbert space $\mathcal{H}_0$. ρ satisfies the Schrodinger-Von-Neumann-Liouville equation

$$i\rho'(t) = [H, \rho(t)]$$

where

$$H = \sum_{k=1}^{N} H_k + \sum_{1\leq k<j\leq N} V_{kj}$$

where H_k acts in the Hilbert space $\mathcal{H}_k$ and V_{kj} acts in the Hilbert space $\mathcal{H}_k \otimes \mathcal{H}_j$. Each V_{kj} is to be an identical copy of V_{12}.

[8] Supersymmetry generators. Let $\theta^k, k = 1, 2, 3, 4$ be the anticommuting Fermionic variables and $x^\mu, \mu = 1, 2, 3, 4$ the space-time Bosonic variables. Define super vector fields by

$$L_\alpha = (A^\mu\theta)_\alpha \partial/\partial x^\mu + B_\alpha^\beta \partial/\partial\theta^\beta$$

where

$$(A^\mu\theta)_\alpha = A^\mu_{\alpha\beta}\theta^\beta$$

Also define otheta super vector fields by

$$M_\alpha = (C^\mu\theta)_\alpha \partial/\partial x^\mu + D_\alpha^\beta \partial/\partial\theta^\beta$$

We wish to choose the matrices A, B, C, D so that the supersymmetry anticommutation relations are satisfied:

$$\{L_\alpha, M_\beta\} = F^\mu_{\alpha\beta}\partial/\partial x^\mu$$

We recall the Fermionic anticommutation relations:

$$\{\partial/\partial\theta^k, \theta^l\} = \delta^l_k$$

in contrast to the Bosonic commutation relations

$$[\partial/\partial x^\mu, x^\nu] = \delta^\nu_\mu$$

Then, we find that

$$\{A^\mu_{\alpha\rho}\theta^\rho\partial/\partial x^\mu, C^\nu_{\beta\sigma}\theta^\sigma\partial/\partial x^\nu\} = 0$$

since $\partial/\partial x^\mu$ commute amongst each other and with θ^k while the $\theta^{k's}$ anticommute amongst each other. Also, we have

$$\{B^\rho_\alpha\partial/\partial\theta^\rho, D^\sigma_\beta\partial/\partial\theta^\sigma\} = 0$$

and hence we get finally,

$$\begin{gathered}\{L_\alpha, M_\beta\} = \\ \{(A^\mu\theta)_\alpha\partial/\partial x^\mu + B^\rho_\alpha\partial/\partial\theta^\rho, (C^\nu\theta)_\beta\partial/\partial x^\nu + D^\sigma_\beta\partial/\partial\theta^\sigma\} \\ \{(A^\mu\theta)_\alpha\partial/\partial x^\mu, D^\sigma_\beta\partial/\partial\theta^\sigma\} \\ \{B^\rho_\alpha\partial/\partial\theta^\rho, C^\nu\theta)_\beta\partial/\partial x^\nu\} \\ = A^\mu_{\alpha\rho}D^\sigma_\beta\{\theta^\rho, \partial/\partial\theta^\sigma\}\partial/\partial x^\mu \\ +B^\rho_\alpha C^\mu_{\beta\sigma}\{\partial/\partial\theta^\rho, \theta^\sigma\}\partial/\partial^\mu \\ = A^\mu_{\alpha\rho}D^\sigma_\beta\delta^\rho_\sigma\partial/\partial x^\mu \\ +B^\rho_\alpha C^\mu_{\beta\sigma}\delta^\sigma_\rho\partial/\partial x^\mu\end{gathered}$$

$$= (A^{\mu}_{\alpha\rho}D^{\rho}_{\beta} + B^{\rho}_{\alpha}C^{\mu}_{\beta\rho})\partial/\partial x^{\mu}$$

[9] Cartan's equations of structure. The basic identity used in proving these is

$$d\omega(X,Y) = X(\omega(Y)) - Y(\omega(X)) - \omega([X,Y])$$

where X, Y are vector fields and ω is a one form. We choose a local basis $\{e_\alpha\}$ for the tangent space at each point of an open neighbourhood of the differentiable manifold. Then let $\{e^\alpha\}$ be its dual basis so that $e^\alpha(e_\beta) = \delta^\alpha_\beta$. Then, the torsion is given by

$$T(X,Y) = \nabla_X Y - \nabla_Y X - [X,Y]$$
$$= \nabla_X(Y^a e_a) - \nabla_Y(X^a e_a) - [X^a e_a, Y^b e_b]$$
$$= X(Y^a)e_a + Y^a\nabla_X(e_a) - Y(X^a)e_a - X^a\nabla_Y(e_a) - X(Y^b)e_b - Y^bX(e_b) + Y(X^a)e_a + X^aY(e_a)$$
$$= Y^a\nabla_X(e_a) - X^a\nabla_Y(e_a) - Y^aX(e_a) + X^aY(e_a)$$
$$= e^a(Y)\omega^b_a(X)e_b - e^a(X)\omega^b_a(Y)e_b + e^a(X)Y(e_a) - e^a(Y)X(e_a)$$

Note that $X(e_a)$ and $Y(e_a)$ are second order differential operators. Now,

$$de^a(X,Y) = X(e^a(Y)) - Y(e^a(X)) - e^a([X,Y])$$
$$= X(Y^a) - Y(X^a) - e^a(X(Y^b e_b) - Y(X^b e_b))$$
$$= X(Y^a) - Y(X^a) - e^a(X(Y^b)e_b + Y^bX(e_b) - Y(X^b)e_b - X^bY(e_b))$$
$$= X(Y^a) - Y(X^a) - X(Y^a) + Y(X^a) + e^a(Y^bX(e_b) - X^bY(e_b))$$
$$= e^a(Y^bX(e_b) - X^bY(e_b))$$

Hence, we can write the above equation as

$$T^a(X,Y) = e^a(T(X,Y)) = \omega^b_a \wedge e^a(X,Y) - de^a(X,Y)$$

or equivalently,

$$T^a = \omega^b_a \wedge e^a - de^a$$

This is Cartan's first equation of structure.

[10] An application of the Riesz-Thorin theorem. Let $f \in L^q(\mathbb{R}^n).q \geq 1$ and let p be defined by $1/q + 1/p = 1$. Let Tf denote the Fourier transform of f. Then,

$$\| Tf \|_\infty \leq C_1. \| f \|_1$$

and

$$\| Tf \|_2 = C_2. \| f \|_2$$

The latter is the Parseval theorem. If follows from the Riesz-Thorin interpolation theorem which can be derived using the Hadamard theorem that for $2 \leq p < \infty$ so that $1 \leq q < 2$, we have

$$F(q) \leq F(1)^x F(2)^{1-x}$$

where

$$F(q) = \| Tf \|_p / \| f \|_q, p = q/(q-1)$$

and $x \in [0,1]$ is defined by

$$x + 2(1-x) = q$$

ie

$$x = 2 - q$$

Now,

$$F(1) = \| Tf \|_\infty / \| f \|_1 \leq C_1,$$
$$F(2) = C_2$$

Thus,

$$\| Tf \|_p \leq C_1^x C_2^{1-x} \| f \|_q$$

This is called the Hadamard-Young inequality (Reference: W.O.Amrein, Hilbert space methods in quantum mechanics).

[11] Induced representations for the semidirect product: Let N be an Abelian group, H another group and N, H are both subgroups of a group G, so that $G = N \otimes_s H$. In other words, N is normalized by H, ie, $hNh^{-1} = N, h \in H$. Every element $g \in G$ can be uniquely expressed as $g = nh, n \in N, h \in H$. So we can write $(n,h) = nh$. We have

$$n_1 h_1 n_2 h_2 = n_2 \tau_{h_1}(n_2) h_1 h_2 = (n_1 \tau_{h_1}(n_2), h_1 h_2)$$

where

$$\tau_h \in Aut(N), \tau_h(n) = hnh^{-1}$$

and $h \to \tau_h$ is a homomorphism from the group H into the group $Aut(N)$. Let $\chi \in \hat{N}$ and for $h \in H$ define $\beta_h : \hat{N} \to \hat{N}$ by

$$\beta_h \chi(n) = \chi(h^{-1} n h) = \chi(\tau_h^{-1}(n))$$

Then $h \to \beta_h$ is a homomorphism from H into the group $Aut(\hat{N})$. Let O_χ be the orbit of χ under H, with β as the group action, ie,

$$O_\chi = \{\beta_h \chi : h \in H\}$$

Choose and fix a $\chi_0 \in \hat{N}$ and let

$$H_0 = \{h \in H : \beta_h \chi_0 = \chi_0\}$$

so that H_0 is a subgroup of H called the stability group of χ_0 or sometimes as the little group at χ_0. Then, it is clear that the manifold O_{χ_0} is isomorphic to the manifold H/H_0 under the action of the group H, ie, $hH_0 \to \beta_h \chi_0, h \in H$. Suppose L is an irreducible representation of H_0 in a vector space V. Then, $nh \to \chi_0(n) L(h)$ is an irreducible representation of $G_0 = N \otimes_s H_0$ in V. We denote this representation by $\chi_0 \otimes L$. The representation property follows from

$$\chi_0(n) L(h) \chi_0(n') L(h') = \chi_0(nn') L(hh')$$

on the one hand and on the other

$$(\chi_0 \otimes L)(nhn'h') = (\chi_0 \otimes L)(n \tau_h(n') hh') = \chi_0(n \tau_h(n')) L(hh')$$

$$= \chi_0(n) \chi_0(\tau_h(n')) L(hh') = \chi_0(n) \chi_0(n') L(hh') = \chi_0(nn') L(hh')$$

since $\chi_0 o \tau_h = \beta_h^{-1} \chi_0 = \chi_0$. Now let $U = Ind_{G_0}^G (\chi_0 \otimes L)$. Then, we claim that U is an irreducible representation of G. We note that U acts in the vector space X of all functions $f : G \to V$ for which $f(gg_0^{-1}) = (\chi_0 \otimes L)(g_0) f(g), g_0 \in G_0, G \in G$ and is defined on such functions by $(U(g_1)f)(g) = f(g_1^{-1} g), g, g_1 \in G$. Clearly, Clearly, if we choose one element from each coset gG_0 of G_0 in G, then f on G is completely determined from its values at these elements. Specifically, writing $G = \bigcup_{\alpha \in I} g_\alpha G_0$ with $g_\alpha G_0 \bigcap g_\beta G_0 = \phi$ for all $\alpha \neq \beta$, we have $f(g_\alpha g_0) = (\chi_0 \otimes L)(g_0^{-1}) f(g_\alpha), g_0 \in G_0, \alpha \in I$ and this proves our claim. Thus, the representation space of U can equivalently be chosen as $G/G_0 = (N \otimes_s H)/(N \otimes_s H_0) = H/H_0 = O_{\chi_0}$ all equalities being manifold isomorphisms under appropriate group actions. Thus, the representation space X of U can equivalently be taken as the set of all functions from O_{χ_0} into V. Specifically, choose $f_1 : O_{\chi_0} \to V$ and define for $g = nh, n \in N, h \in H$, $f_0(g) = f(nh) = \chi_0(n) f_1(h \chi_0)$ $(h \chi_0 = \beta_h(\chi_0))$. Then, if $n_0 \in N, h_0 \in H_0$, we have setting $g_0 = n_0 h_0$ that $g_0^{-1} = h_0^{-1} n_0^{-1} = h_0^{-1} n_0^{-1} h_0 h_0^{-1} = \tau_{h_0}^{-1}(n_0^{-1}) h_0^{-1}$ and hence

$$f_0(gg_0^{-1}) = \chi_0(n \tau_{h_0}^{-1}(n_0^{-1})) f_1(hh_0^{-1} \chi_0)$$

$$= \chi_0(n) f_1(h \chi_0) = f_0(g)$$

This shows that f_0 is a function on G/G_0 or equivalently on H/H_0. This means that we have an isomorphism between the space X and the set of all functions on O_{χ_0}. Further, the action of U on f_1 os obtained as follows. Let $g = nh, g_1 = n_1 h_1 \in G$. Then under $U(g)$ $f_0(g_1)$ goes to $f_0(g^{-1} g_1)$ and

$$g^{-1} g_1 = h^{-1} n^{-1} n_1 h_1 = \tau_h^{-1}(n^{-1} n_1) h^{-1} h_1,$$

$$f_0(g^{-1} g_1) = \chi_0(\tau_h^{-1}(n^{-1} n_1)) f_1(h^{-1} h_1 \chi_0)$$

$$= h.\chi_0(n^{-1} n_1) f_1(h^{-1} h_1 \chi_0)$$

$$= (hh_1.\chi)(n^{-1}n_1)f(h^{-1}\chi), \chi = h_1\chi_0$$

Instead, if we write

$$f_0(g) = A(n,h)f_1(h\chi_0), g = nh, n \in N, h \in H$$

where $A(n,h)$ is a linear operator in V, then

$$f_0(g^{-1}g_1) = A(\tau_h^{-1}(n^{-1}n_1), h^{-1}h_1)f_1(h^{-1}h_1\chi_0)$$

If we denote this by $A(n_1,h_1)\tilde{f}_1(h_1\chi_0)$, then we see that f_1 transforms to $\tilde{f}_1$ under $U(g) = U(nh)$, where

$$\tilde{f}_1(h_1\chi_0) = A(n_1,h_1)^{-1}A(\tau_h^{-1}(n^{-1}n_1), h^{-1}h_1)f_1(h^{-1}h_1\chi_0)$$

or equivalently writing $\chi = h_1\chi_0$, we get

$$\tilde{f}_1(\chi) = A(n_1,h_1)^{-1}A(\tau_h^{-1}(n^{-1}n_1), h^{-1}h_1)f_1(h^{-1}\chi)$$

For this to be a valid group action, the matrix $A(n_1,h_1)^{-1}A(\tau_h^{-1}(n^{-1}n_1), h^{-1}h_1)$ must be independent of n_1. For example, taking

$$A(n_1,h_1) = \chi(n_1)L_{h_1^{-1}\gamma(\chi)}, \chi = h_1\chi_0$$
$$= h_1\chi_0(n_1)L_{h_1^{-1}\gamma(h_1\chi_0)}$$

we get

$$A(n_1,h_1)^{-1}A(\tau_h^{-1}(n^{-1}n_1), h^{-1}h_1)$$
$$= h_1.\chi_0(n_1^{-1})L^{-1}_{\gamma(h_1\chi_0)}h_1.\chi_0(n^{-1}n_1)L_{h_1^{-1}h\gamma(h^{-1}h_1\chi_0)}$$
$$= h_1\chi_0(n^{-1})L_{\gamma(h_1\chi_0)^{-1}h\gamma(h^{-1}h_1\chi_0)}$$
$$= \chi(n^{-1})L_{\gamma(\chi)^{-1}h\gamma(h^{-1}\chi)}$$

Thus, we can write

$$U(nh)f_1(\chi) = \bar{\chi}(n)L_{\gamma(\chi)^{-1}h\gamma(h^{-1}\chi}f_1(h^{-1}\chi)$$

To check the representation property, we observe that

$$U(h_2)(U(h_1)f_1)(\chi) =$$
$$L_{\gamma(\chi)^{-1}h_2\gamma(h_2^{-1}\chi)}(U(h_1)f_1)(h_2^{-1}\chi)$$
$$= L_{\gamma(\chi)^{-1}h_2\gamma(h_2^{-1}\chi)}L_{\gamma(h_2^{-1}\chi)^{-1}h_1\gamma(h_1^{-1}h_2^{-1}\chi)}f_1(h_1^{-1}h_2^{-1}\chi)$$
$$= L_{\gamma(\chi)^{-1}h_2h_1\gamma((h_2h_1)^{-1}\chi)}f_1((h_2h_1)^{-1}\chi)$$
$$= U(h_2h_1)f_1(\chi)$$

verifying the representation property.

Now, suppose U is a unitary representation of $G = N \otimes_s H$. Then, we have for $n \in N, h \in H$ that $U(nh) = V(n)W(h)$, where $V(n) = U(n)$ is a unitary representation of N and $W(h) = U(h)$ is a unitary representation of H. We have

$$W(h)^{-1}V(n)L(h) = U(h^{-1}nh) = V(h^{-1}nh) = V(\tau_h^{-1}(n))$$

Now $V(n), n \in N$ is a commuting family of unitary operators and hence these operators are simultaneously diagonable. Thus, in an appropriate basis, we can write

$$V(n) = diag[\chi_k(n), k = 1, 2, ..., N]$$

where χ_k are characters of N.

[12] Weyl operator and quantum Gaussian states.
Let a, a^* be annihilation and creation operators in $L^2(\mathbb{R}) = \Gamma_s(\mathbb{C})$ so that $[a, a^*] = 1$. Then for $z \in \mathbb{C}$,

$$W(z) = exp(\bar{z}a - za^*)$$

is a unitary operator in $L^2(\mathbb{R})$ since $\bar{z}a - za^*$ is skew Hermitian. We write

$$W(tz) = F(t) = G(t)exp(t\bar{z}a)$$

Then,

$$F'(t) = (G'(t) + G(t)\bar{z}a)exp(t\bar{z}a) = F(t)(\bar{z}a - za^*) = G(t)exp(t\bar{z}a)(\bar{z}a - za^*)$$

from which it follows that

$$G'(t) = -G(t)exp(t\bar{z}ad(a))(za^*) = -G(t)(za^* + t|z|^2)$$

and hence

$$G(t) = exp(-tza^* - t^2|z|^2/2)$$

Thus,

$$W(z) = exp(-|z|^2/2)exp(\bar{z}a).exp(-za^*)$$

Likewise, for $z, u \in \mathbb{C}$, we write

$$F(t) = W(t(z+u)) = G(t)W(tu)$$

Then,

$$F'(t) = G(t)W(tu)((\bar{z} + \bar{u})a - (z+u)a^*) = (G'(t) + G(t)(\bar{u}a - ua^*))W(tu)$$

and hence

$$G'(t) = G(t)W(tu)(\bar{z}a - za^*)W(-tu) = G(t)exp(t.ad(\bar{u}a - ua^*))(\bar{z}a - za^*)$$
$$= G(t)(\bar{z}a - za^* + t(-z\bar{u} + u\bar{z}))$$

Thus,

$$G(1) = exp((u\bar{z} - z\bar{u})/2).W(z)$$

or equivalently,

$$W(z+u) = exp(iIm(\bar{u}z))W(z)W(u)$$

More generally, considering the isomorphism $L^2(\mathbb{R}^N) = \Gamma_s(\mathbb{C}^N)$ and annihilation and creation operators $a_1, ..., a_N, a_1^*, ..., a_N^*$ with the canonical commutation relations $[a_p, a_q^*] = \delta_{p,q}, [a_p, a_q] = 0, [a_p^*, a_q^*] = 0$, we get

$$W(z+u) = exp(-iIm(< z, u >)W(z)W(u)$$

where

$$W(z) = exp(\sum_{j=1}^{N}(\bar{z}_j a_j - z_j a_j^*)), z = [z_1, ..., z_N]^T$$

Let ρ be a state in the Hilbert space $L^2(\mathbb{R}^N) = \Gamma_s(\mathbb{C})$. Then its quantum Fourier transform is given by

$$\hat{\rho}(z) = Tr(\rho W(z))$$

We have

$$\hat{\rho}(z_a - z_b) = Tr(\rho W(z_a - z_b)) =$$
$$Tr(\rho W(z_a)W(z_b)^*)exp(iIm(< z_a, z_b >))$$

and hence the matrix $((\hat{\rho}(z_a - z_b)exp(-iIm(< z_a, z_b >)))_{1\leq a,b\leq p}$ is positive definite for any p complex vectors $z_a, a = 1, 2, ..., p$ in $\mathbb{C}^N$. Let ρ be a Gaussian state, ie, its quantum Fourier transform has the form

$$\hat{\rho}(x + iy) = exp(l^T x + m^T y - [x^T, y^T]S[x^T, y^T]^T)$$

where l, m are complex vectors and S is a symmetric complex matrix. We write $R(z) = [x^T, y^T]^T$ and hence, we can write

$$\hat{\rho}(z) = exp(p^T R(z) - R(z)^T SR(z)/2)$$

We have for $z, u \in \mathbb{C}^N$,

$$Im(< z, u >) = R(z)^T JR(u)$$

where

$$J = \begin{pmatrix} 0_N & I_N \\ -I_N & 0 \end{pmatrix} \in \mathbb{R}^{2N\times 2N}$$

Thus, the requirement of the the above positive definitivity amounts to requiring that

$$2S - iJ \geq 0$$

When this condition is satisfied, we call ρ a quantum Gaussian state.

[13] Problem: Consider Dirac's equation in a noisy electromagnetic field described by the qsde

$$dU(t) = [(\alpha, pdt + e(L^\alpha_\beta d\Lambda^\beta_\alpha(t))) + \beta mdt) + (V(q)dt + eM^\alpha_\beta d\Lambda^\beta_\alpha(t))U(t)$$

where the system Hilbert space is $\mathfrak{h} = L^2(\mathbb{R}^3)$ and the noise Boson-Fock space is $\Gamma_s(L^2(\mathbb{R}_+)\otimes\mathbb{C}^d)$. The system operators $L^\alpha_\beta and M^\alpha_\beta$ are assumed to be functions of q, p only. The electron charge e is a perturbation parameter. Formally, the noisy magnetic vector potential is

$$A_t = L^\alpha_\beta d\Lambda^\beta_\alpha(t)/dt$$

If we assume that L^α_β are functions of the position vector observable q only, then A_t becomes a a multiplication operator when restricted to the system Hilbert space $\mathfrak{h}$ only. The noisy electric scalar potential is

$$\Phi_t = M^\alpha_\beta d\Lambda^\beta_\alpha(t)/dt$$

and if we assume that M^α_β are functions of q only, then the electric scalar potential becomes a multiplication operator when restricted to the system Hilbert space only. To get an approximate solution for $U(t)$, we write

$$U(t) = U_0(t) + eU_1(t) + e^2U_2(t) + O(e^3)$$

Problem: Calculate the matrix elements of $U(t)$ between the states $|f\phi(u) >$ and $|g\phi(v) >$ upto $O(e^2)$ where $|f >, |g > \in \mathfrak{h}$ and $u, v \in L^2(\mathbb{R}_+) \otimes \mathbb{C}^d$, $|\phi(u) >= exp(- \parallel u \parallel^2 /2)|e(u) >$ and likewise for $|\phi(v) >$. Interpret this result in terms of transition probabilities for the system between two energy levels by taking an average over the noise space assuming that the noise space remains in the coherent state $|\phi(u) >$ before and after the transition.

[14] Alternative description of the induced representation of a semidirect product: $G = N\otimes_s H, hNh^{-1} = N, \forall h \in H, N$ is Abelian. Choose $\chi_0 \in \hat{N}$. Let

$$H_0 = \{h \in H : h.\chi_0 = \chi_0\}$$

where

$$h.\chi(n) = \chi(h^{-1}nh), h \in H, n \in N, \chi \in \hat{N}$$

Let O_{χ_0} denote the orbit of χ_0 under H, ie,

$$O_{\chi_0} = \{h\chi_0 : h \in H\} \subset \hat{N}$$

The manifolds H/H_0 and O_{χ_0} are isomorphic under the H-group action. This follows by defining the map $\phi : H/H_0 \to O_{\chi_0}$ by $\phi(hH_0) = h\chi_0, h \in H$. The map is well defined since $hH_0 = h'H_0$ implies $h'^{-1}h \in H_0$ implies $h'^{-1}h\chi_0 = \chi_0$ implies $h\chi_0 = h'\chi_0$. The map ϕ is a bijection since $\phi(hH_0) = \phi(h'H_0)$ implies $h\chi_0 = h'\chi_0$ implies $h'^{-1}h\chi_0 = \chi_0$ implies $h'^{-1}h \in H_0$ implies $hH_0 = h'H_0$. This proves the injectivity of ϕ. Surjectivity is obvious. Let L be a representation of H_0 in a vector space V. Now Let Y denote the vector space of all functions $f : H \to V$ such that $f(hh_0^{-1}) = L(h_0)f(h), h_0 \in H_0, h \in H$. Define $U^L : H \to Y$ by $U^L(h_1)f(h) = f(h_1^{-1}h), h_1, h \in H$. Then, U^L is a representation of H in Y and is called the representation of H induced by the representation L of H_0. Now we give an alternate description of the induced representation. Let $\tilde{Y}$ denote the vector space of all functions $f : O_{\chi_0} \to V$. Define $T : \tilde{Y} \to Y$ by the formula

$$Tf(h) = A(h)f(h\chi_0), h \in H$$

where $A(h)$ is an appropriate matrix, ie $A : H \to GL(V)$. In order that $Tf \in Y$, we require that $Tf(hh_0^{-1}) = L(h_0)Tf(h), h_0 \in H_0, h \in H$, ie,

$$A(hh_0^{-1})f(hh_0^{-1}\chi_0) = L(h_0)A(h)f(h\chi_0), h \in H, h_0 \in H_0$$

or equivalently, since $h_0^{-1}\chi_0 = \chi_0$,

$$A(hh_0^{-1}) = L(h_0)A(h), h_0 \in H_0, h \in H$$

This can be satisfied for example by choosing

$$A(h) = L(\gamma(h\chi_0)^{-1}h)^{-1} = L(h^{-1}\gamma(h\chi_0))$$

We then check that

$$A(hh_0^{-1})A(h)^{-1} = L(h_0h^{-1}\gamma(hh_0^{-1}\chi_0)L(\gamma(h\chi_0)^{-1}h) = L(h_0)$$

Thus,

$$(Tf)(h) = L(h^{-1}\gamma(h\chi_0))f(h\chi_0), h \in H$$

We have for $h, h_1 \in H$,

$$(U^L(h_1)Tf)(h) = Tf(h_1^{-1}h) = L(h^{-1}h_1\gamma(h_1^{-1}h\chi_0))f(h_1^{-1}h\chi_0)$$

If we equate this to $TV^L(h_1)f(h) = L(h^{-1}\gamma(h\chi_0))(V^L(h_1)f)(h\chi_0)$, then we must have

$$(V^L(h_1)f)(h\chi_0) =$$

$$L(\gamma(h\chi_0)^{-1}hh^{-1}h_1\gamma(h_1^{-1}h\chi_0))$$

$$= L(\gamma(h\chi_0)^{-1}h_1\gamma(h_1^{-1}h\chi_0))$$

or equivalently, writing $\chi = h.\chi_0$, we get

$$V^L(h_1)f(\chi) = L(\gamma(\chi)^{-1}h_1\gamma(h_1^{-1}\chi))$$

V^L therefore defines a representation of H in $\tilde{Y}$ that is equivalent to the induced representation U^L of H in Y.

[15] EM wave propagation in an inhomogeneous waveguide: The permittivity and permeability fields are represented as

$$\epsilon(x, y, z) = \sum_n \epsilon[n, x, y]exp(-n\beta z),$$

$$\mu(x, y, z) = \sum_n \mu[n, x, y]exp(-n\beta z)$$

For example, if the length of the guide is d, we can take $\beta = j2\pi/d$, ie, develop ϵ and μ as Fourier series in the z variable. We also expand the electric and magnetic fields as Fourier series in the z variable with an additional complex propagation constant γ:

$$E(x, y, z) = \sum_n E[n, x, y]exp(-(\gamma + n\beta)z),$$

$$H(x, y, z) = \sum_n H[n, x, y]exp(-(\gamma + n\beta)z)$$

We also expand the current density $J(x, y, z)$ within the guide as

$$J(x, y, z) = \sum_n J[n, x, y]exp(-jn\beta z)$$

Substituting these into the Maxwell curl equations

$$curlE(x, y, z) = -j\omega\mu(x, y, z)H(x, y, z),$$

$$curlH(x, y, z) = J(x, y, z) + j\omega\epsilon(x, y,)E(x, y, z)$$

and equating coefficients of $exp(-jn\beta z)$ for each integer n gives us

$$\nabla_\perp E_z[n, x, y] \times \hat{z} - (\gamma + n\beta)\hat{z} \times E_\perp[n, x, y] = -j\omega \sum_m \mu[m, x, y]H_\perp[n - m, x, y],$$

$$\nabla_\perp H_z[n, x, y] \times \hat{z} - (\gamma + n\beta)\hat{z} \times H_\perp[n, x, y] = j\omega \sum_m \epsilon[m, x, y]E_\perp[n - m, x, y]$$

$$+J_\perp[n, x, y]$$

for the $x - y$ components. For the z component, we have

$$(\hat{z}, \nabla_\perp \times H_\perp) = J_z + j\omega\epsilon E_z,$$

$$(\hat{z}, \nabla_\perp \times E_\perp) = -j\omega\mu H_z$$

or in the transform domain,

$$(\hat{z}, \nabla_\perp \times E_\perp[n,x,y]) = -j\omega \sum_m \mu[m,x,y]H_z[n-m,x,y],$$

$$(\hat{z}, \nabla_\perp \times H_\perp[n,x,y]) = J_z[n,x,y] + j\omega \sum_m \epsilon[m,x,y]E_z[n-m,x,y]$$

To cast these equations in the format of perturbation theory, we assume that $\epsilon[n,x,y]$ and $\mu[n,x,y]$ for $n \neq 0$ are small compared to $\epsilon[0,x,y]$ and $\mu[0,x,y]$ and hence attach a small pertrubation parameter δ to the former terms. We then get

$$\nabla_\perp E_z[n,x,y] \times \hat{z} - (\gamma+n\beta)\hat{z} \times E_\perp[n,x,y] + j\omega\mu[0,x,y]H_\perp[n,x,y] = -j\delta\omega \sum_{m\neq 0} \mu[m,x,y]H_\perp[n-m,x,y],$$

$$\nabla_\perp H_z[n,x,y] \times \hat{z} - (\gamma+n\beta)\hat{z} \times H_\perp[n,x,y] - j\omega\epsilon[0,x,y]E_\perp[n,x,y] - J_\perp[n,x,y] = j\delta\omega \sum_{m\neq 0} \epsilon[m,x,y]E_\perp[n-m,x,y]$$

or in the transform domain,

$$(\hat{z}, \nabla_\perp \times E_\perp[n,x,y]) + j\omega\mu[0,x,y]H_z[n,x,y] = -j\delta\omega \sum_{m\neq 0} \mu[m,x,y]H_z[n-m,x,y],$$

$$(\hat{z}, \nabla_\perp \times H_\perp[n,x,y]) - J_z[n,x,y] - j\omega\epsilon[0,x,y]E_z[n,x,y] = j\omega\delta \sum_{m\neq 0} \epsilon[m,x,y]E_z[n-m,x,y]$$

[16] Symmetry breaking: Let $\psi(x) \in \mathbb{C}^N$ be the matter field. It transforms under gauge group $G \subset U(N)$. Under a local gauge transformation, this is given by $\psi(x) \to g(x)\psi(x)$, where $g(x) \in G$. Now, assume that we can write $\psi(x) = \gamma(x)\tilde{\psi}(x)$ where $\gamma(x)$ is a representative element of a coset in G/H with H a subgroup of G. Thus, $\tilde{\psi}(x)$ transforms according to the "broken subgroup" H and $g(x) \in G$ transforms $\psi(x)$ to $g(x)\psi(x) = g(x)\gamma(x)\tilde{\psi}(x)$ and we can write $g(x)\gamma(x) = \gamma'(x)h(x)$ where $h(x) \in H$ and either $\gamma'(x) = e$ or else $\gamma'(x) \notin H$. Assume that $A_\mu(x)$ are the gauge fields taking values in the Lie algebra of G, ie in $\mathfrak{g}$. $\psi(x)$ transforms according to G and the corresponding transformation law of $A_\mu(x)$ under G is obtained by using the fact that

$$g(x)(\partial_\mu + ieA_\mu(x))\psi(x) = (\partial_\mu + A'_\mu(x))g(x)\psi(x)$$

for this ensures that if the Lagrangian density $L(\psi(x), \nabla_\mu\psi(x))$ is globally G-invariant, ie

$$L(g\psi(x), g\nabla_\mu\psi(x)) = L(\psi(x), \nabla_\mu\psi(x))$$

for all $g \in G$, where

$$\nabla_\mu = \partial_\mu + ieA_\mu(x)$$

is the gauge covariant derivative, then it would follow that L is also locally G-invariant, ie,

$$L(\psi'(x), \nabla'_\mu\psi'(x)) = L(\psi(x), \nabla_\mu\psi(x))$$

where

$$\nabla'_\mu = \partial_\mu + ieA'_\mu(x)$$

It follows that the transformation law of the matter field $\psi(x)$ under the gauge transformation is $\psi'(x) = g(x)\psi(x)$ while the corresponding transformation law of the gauge field $A_\mu(x)$ under the gauge transformation is given by

$$(\partial_\mu + ieA'_\mu(x)) = g(x)(\partial_\mu + ieA_\mu(x))g(x)^{-1}$$

(as an operator equation), or equivalently,

$$ieA'_\mu(x) = ieg(x)A_\mu(x)g(x)^{-1} + g(x)\partial_\mu(g(x)^{-1})$$

or

$$A'_\mu(x) = g(x)A_\mu(x)g(x)^{-1} + (i/e)(\partial_\mu g(x))g(x)^{-1}$$

This equation is the non-Abelian generalization of the Abelian U(1) gauge transformation of the Dirac field in an electromagnetic field where the wave function $\psi(x) \in \mathbb{C}^4$ transforms to $\psi'(x) = exp(i\phi(x))\psi(x)$ where $\phi(x) \in \mathbb{R}$ and correspondingly the electromagnetic four potential $A_\mu(x)$ changes to $A'_\mu(x)$ so that

$$(\partial_\mu + ieA'_\mu)exp(i\phi) = exp(i\phi)(\partial_\mu + ieA_\mu)$$

or equivalently,

$$ieA'_\mu = ieA_\mu - i\partial_\mu\phi$$

or equivalently,

$$A'_\mu(x) = A_\mu(x) - e^{-1}\partial_\mu\phi(x)$$

which is the standard Lorentz gauge transformation. Coming back to the general non-Abelian case, we assume that if $\{t_a\}$ are the generators of the Lie algebra of G, then

$$gt_a g^{-1} = \sum_b D_{ba}(g)t_b$$

In other words, $D(g)$ is the adjoint representation of G acting in $\mathfrak{g}$. Then,

$$gA_\mu(x)g^{-1} = g(\sum_a A^a_\mu t_a)g^{-1}$$

$$= \sum_a A^a_\mu g t_a g^{-1} = \sum_a A^a_\mu D_{ba}(g)t_b$$

This means that

$$A'_\mu(x) = \sum_b D_{ba}(g)A^a_\mu(x)t_b + (i/e)(\partial_\mu g(x))g(x)^{-1}$$

Now suppose that the matter Lagrangian density in the absence of the gauge field has the form

$$L = \frac{1}{2}(\partial_\mu\psi(x))^*(\partial^\mu\psi(x)) - m^2\psi(x)^*\psi(x)$$

This Lagrangian density is invariant under global G-transformations since $g \in U(N)$. Then since

$$\psi(x) = \gamma(x)\tilde{\psi}(x)$$

we get

$$\partial_\mu\psi(x) = (\partial_\mu\gamma)\tilde{\psi} + \gamma\partial_\mu\tilde{\psi}$$

so that

$$L = (\partial_\mu\tilde{\psi})(\partial^\mu\psi)^* + \tilde{\psi}^*(\partial_\mu\gamma)^*(\partial^\mu\gamma)\tilde{\psi}$$

$$+2Re(\tilde{\psi}^*\partial_\mu\gamma)^*\gamma\partial^\mu\tilde{\psi}) - m^2\tilde{\psi}(x)^*\tilde{\psi}(x)$$

It is clear that This Lagrangian density of the matter part ie those terms invoving only $\tilde{\psi}$ and not γ is invariant under $\tilde{\psi}(x) \to h\tilde{\psi}(x), h \in H$ and further that the above Lagrangian does not contain terms that are quadratic in $\gamma(x)$ not involving their derivatives. This means that $\tilde{\psi}(x)$ has mass while the other part $\gamma(x)$ does not have mass. $\gamma(x)$ is called the Goldstone-massless part of the field and $\tilde{\psi}$ is called the massive part. Thus, broken symmetries are associated with the productions of massless Goldstone Bosons. Just as ψ was separated into a massive part $\tilde{\psi}(x)$ and a massless part $\gamma(x)$, we can also separate the gauge field $A_\mu(x)$ into two parts:

$$\tilde{\psi}(x) = \gamma(x)^{-1}\psi(x),$$

$$\tilde{A}_\mu(x) = \gamma(x)^{-1}A_\mu(x)\gamma(x)$$

or equivalently,

$$\tilde{A}^a_\mu(x) = \sum D_{ab}(\gamma(x)^{-1})A^b_\mu(x)$$

Then we find that

$$\psi'(x) = g(x)\psi(x) = g(x)\gamma(x)\tilde{\psi}(x),$$

$$g(x)\gamma(x) = \gamma'(x)h(x),$$

$$(\partial_\mu g)\gamma + g\partial_\mu\gamma = (\partial_\mu\gamma')h + \gamma'\partial_\mu h$$

and hence

$$\gamma^{-1}g^{-1}(\partial_\mu g)\gamma + \gamma^{-1}\partial_\mu\gamma =$$
$$h^{-1}\gamma' - 1(\partial_\mu\gamma')h + h^{-1}partial_\mu h$$

Thus,

$$A'_\mu(x) = g(x)A_\mu(x)g(x)^{-1} + (i/e)(\partial_\mu g(x))g(x)^{-1} - - - (1)$$

implies

$$\tilde{A}'_\mu = \gamma'^{-1}A'_\mu\gamma' =$$
$$\gamma'^{-1}g(x)A_\mu(x)g(x)^{-1}\gamma' + (i/e)\gamma'^{-1}(\partial_\mu g(x))g(x)^{-1}\gamma'$$

Noting that

$$\gamma'^{-1}g = h\gamma^{-1}$$

it follows that this equals

$$\tilde{A}'_\mu =$$
$$h\gamma^{-1}A_\mu\gamma h^{-1} + (i/e)\gamma'^{-1}((-g(\partial_\mu\gamma)\gamma^{-1}g^{-1} + (\partial_\mu\gamma')\gamma'^{-1}$$
$$+\gamma'(\partial_\mu h)h^{-1}\gamma'^{-1})\gamma'$$
$$= h\tilde{A}_\mu h^{-1} + (i/e)(-h\gamma^{-1}(\partial_\mu\gamma)h^{-1} + \gamma'^{-1}\partial_\mu\gamma' + (\partial_\mu h)h^{-1})$$

or equivalently,

$$\tilde{A}'_\mu - (i/e)\gamma'^{-1}\partial_\mu\gamma' =$$
$$h(\tilde{A}_\mu(-i/e)\gamma^{-1}\partial_\mu\gamma)h^1 + (i/e)(\partial_\mu h)h^{-1} - - - - - (2)$$

where we have used the fact that the equation

$$\gamma^{-1}g^{-1}(\partial_\mu g)\gamma + \gamma^{-1}\partial_\mu\gamma =$$
$$h^{-1}\gamma' - 1(\partial_\mu\gamma')h + h^{-1}partial_\mu h$$

implies

$$(\partial_\mu g)\gamma + g\partial_\mu\gamma = (\partial_\mu\gamma')h + \gamma'(\partial_\mu h)$$

which again implies

$$(\partial_\mu g)g^{-1} + g(\partial_\mu\gamma)\gamma^{-1}g^{-1} = (\partial_\mu\gamma')h\gamma^{-1}g^{-1}$$
$$+\gamma'(\partial_\mu h)\gamma^{-1}g^{-1}$$

which implies

$$(\partial_\mu g)g^{-1} + g(\partial_\mu\gamma)\gamma^{-1}g^{-1} = (\partial_\mu\gamma')\gamma'^{-1}$$
$$+\gamma'(\partial_\mu h)h^{-1}\gamma'^{-1}$$

Equation (2) should be compared to equation (1). (1) tells us how the gauge fields A_μ transform under a local gauge transformation $g(x)$ coming from the big group G. (2) tells us how the massless components of the gauge fields $\tilde{A}_\mu$ transform locally under the broken subgroup H. It tells us that when symmetry breaking takes place, then the massless gauge field should be taken as $\tilde{A}_\mu - (i/e)\gamma^{-1}\partial_\mu\gamma$ rather than as $\tilde{A}_\mu$. In other words, the massless Goldstone components γ of the matter field ψ also becomes a part of the massless component of the gauge field $\tilde{A}_\mu$ and the sum total of these two determines the effective gauge field which transforms under the broken subgroup H just as the original gauge field A_μ transformed under the larger group G. It is possible to express all these formulae in terms of functions of the space-time coordinates by first choosing a set of generators $t_i, i = 1, 2, ..., p$ of the Lie algebra of H and then extending this to a basis of the Lie algebra of G, ie, $\{t_1, ..., t_p, x_1, ..., x_q\}, p + q = n = dimG$. Then, we can write

$$\gamma(x) = exp(\sum_{i=1}^{q}(\xi_i(x)x_i)$$

$\xi_i(x)$ can be taken as the Goldstone Boson fields and quantities like $(\partial_\mu\gamma(x))\gamma(x)^{-1}$ can be expressed as nonlinear functions of the $\xi_i(x)$. In this way, the Lagrangian density can be expressed in terms of the massless part $\tilde{A}_\mu$ of the Gauge fields, the massive part $\tilde{\psi}$ of the matter field ψ and the massless Goldstone Boson fields $\xi_i(x)$.

[17] Supersymmetric field theories: Let $\theta^a, a = 0,1,2,3$ be anticommuting Majorana Fermions: Writing $\theta = (\theta^a)$, we have

$$\theta^* = \begin{pmatrix} 0 & e \\ -e & 0 \end{pmatrix}\theta$$

where

$$e = \begin{pmatrix} i\sigma^2 & 0 \\ 0 & i\sigma^2 \end{pmatrix}$$

with

$$\sigma^\mu, \mu = 0,1,2,3$$

begin the Pauli spin matrices:

$$\sigma^0 = I_2, \sigma^1 = \begin{pmatrix} 0 & 1 \\ 1 & 0 \end{pmatrix},$$

$$\sigma^2 = \begin{pmatrix} 0 & -i \\ i & 0 \end{pmatrix}, \sigma^3 = \begin{pmatrix} 1 & 0 \\ 0 & -1 \end{pmatrix}$$

We write

$$\sigma_\mu = \eta_{\mu\nu}\sigma^\nu$$

so that

$$\sigma_0 = \sigma^0, \sigma_r = -\sigma^r, r = 1,2,3$$

The Dirac gamma matrices are defined as

$$\gamma^\mu = \begin{pmatrix} 0 & \sigma^\mu \\ sigma_\mu & 0 \end{pmatrix}$$

They satisfy

$$\{\gamma^\mu, \gamma^\nu\} = 2\eta^{\mu\nu} I_4$$

Define the left Chiral and right Chiral fermionic variables

$$\theta_L = ((1+\gamma_5)/2)\theta, \theta_R = ((1-\gamma_5)/2)\theta$$

where

$$\gamma_5 = \begin{pmatrix} I_2 & 0 \\ 0 & -I_2 \end{pmatrix}$$

Clearly,

$$\gamma^0\gamma^1\gamma^2\gamma^3 = i\gamma_5$$

and hence from the anticommutativity of the γ^μ, we deduce that

$$\{\gamma^\mu, \gamma_5\} = 0$$

We also have

$$\theta = \theta_L + \theta_R,$$

$$\partial/\partial\theta = ((1+\gamma_5)/2)(\partial/\partial\theta_L) + ((1-\gamma_5)/2)(\partial/\partial\theta_R)$$

Define the supersymmetry generators as

$$L = \gamma_5\epsilon\partial/\partial\theta + \gamma^\mu\theta\partial_\mu$$

where

$$\partial_\mu = \partial/\partial x^\mu$$

Also define

$$\bar{L} = -\gamma_5\epsilon L$$

where

$$\epsilon = \begin{pmatrix} e & 0 \\ 0 & e \end{pmatrix}$$

Clearly,

$$e^2 = -I_2, \epsilon^2 = -I_4, e^T = -e, \epsilon^T = -\epsilon, \gamma_5^T = \gamma_5, [\gamma_5, \epsilon] = 0$$

and hence

$$(\gamma_5\epsilon)^T = -\gamma_5\epsilon, (\gamma_5\epsilon)^2 = -I_4$$

so we can also write

$$\bar{L}^T = L^T\gamma_5\epsilon$$

We get

$$\bar{L} =$$

$$\partial/\partial\theta - \gamma_5\epsilon\gamma^\mu\theta\partial_\mu$$

Then using

$$\{\partial/\partial\theta^a, \theta^b\} = \delta^a_b$$

we get

$$\{L_a, \bar{L}_b\} = \{(\gamma^\mu\theta)_a\partial_\mu, \partial/\partial\theta^b\}$$
$$+\{(\gamma_5\epsilon)_{ac}\partial/\partial\theta^c, -(\gamma_5\epsilon\gamma^\mu\theta)_b\partial_\mu\}$$
$$= (\gamma^\mu)_{ab}\partial_\mu - (\gamma_5\epsilon)_{ac}(\gamma_5\epsilon\gamma^\mu)_{bd}\delta_{cd}\partial_\mu$$
$$= (\gamma^\mu)_{ab}\partial_\mu - (\gamma_5\epsilon)_{ac}(\gamma_5\epsilon\gamma^\mu)_{bc}\partial_\mu$$
$$= (\gamma^\mu)_{ab}\partial_\mu - [(\gamma_5\epsilon)\gamma^{\mu T}(\gamma_5\epsilon)^T]_{ab}\partial_\mu$$
$$= (\gamma^\mu_{ab} + \gamma_5\epsilon\gamma^T_\mu\gamma_5\epsilon)\partial_\mu$$
$$= 2\gamma^\mu_{ab}\partial_\mu$$

since

$$\gamma_5\epsilon\gamma^{\mu T}\gamma_5\epsilon = \begin{pmatrix} e & 0 \\ 0 & -e \end{pmatrix}(\begin{pmatrix} 0 & \sigma^T_\mu \\ \sigma^{\mu T} & 0 \end{pmatrix}\begin{pmatrix} e & 0 \\ 0 & -e \end{pmatrix} = \gamma^\mu$$

Formally, we can write this anticommutation relation as

$$\{L, \bar{L}\} = 2\gamma^\mu\partial_\mu$$

It is easy to see that the six matrices $\epsilon, \gamma_5\epsilon, \epsilon\gamma^\mu, \mu = 0, 1, 2, 3$ form a basis for the space of all skew-symmetric 4×4 matrices. We can thus expand any such matrix X as

$$X = a_0\epsilon + a_1\gamma_5\epsilon + b_\mu\epsilon\gamma^\mu$$

where a_0, a_1, b_μ are complex constants. To evaluate these constants, we note that

$$Tr(\epsilon^2) = -4, Tr(\epsilon\gamma_5\epsilon) = 0, Tr(\gamma_\mu\epsilon^2) = 0,$$

$$Tr((\gamma_5\epsilon)(\gamma_5\epsilon)) = -4,$$
$$Tr((\gamma_5\epsilon)\gamma_\mu\epsilon) = 0,$$
$$Tr(\gamma_\nu)\epsilon^2\gamma^\mu) = -4\delta^\mu_\nu$$

It follows that

$$a_0 = -Tr(\epsilon X)/4, a_1 = -Tr(\gamma_5\epsilon X)/4, b_\mu = -Tr(\gamma_\mu\epsilon X)/4$$

In particular, we find that since $\theta\theta^T$ is a 4×4 skew-symmetric matrix,

$$\theta\theta^T = (-1/4)((\theta^T\epsilon\theta)\epsilon + (\theta^T\gamma_5\epsilon\theta)\gamma_5\epsilon + (\theta^T\gamma_\mu\epsilon\theta)\epsilon\gamma^\mu)$$

Now let $S(x,\theta)$ be any superfield. It can be expanded in powers of θ upto fourth degree with the coefficients being functions of x only. It follows that we can expand it as

$$S(x,\theta) = C(x) + \theta^T\epsilon\omega(x) + \theta^T\epsilon\theta M(x) + \theta^T\gamma_5\epsilon\theta N(x) + \theta^T\epsilon\gamma^\mu\theta V_\mu(x)+$$

$$(\theta^T\epsilon\theta)\theta^T\gamma_5\epsilon\beta(x) + (\theta^T\epsilon\theta)^2 K(x)$$

This is a consequence of the fact that any first degree polynomial in θ can be expressed as $\theta^T\epsilon\omega$ where ω has four components, any second degree polynomial in θ is a linear combination of $\theta^T\epsilon\theta$, $\theta^T\gamma_5\epsilon\theta$ and $\theta^T\epsilon\gamma^\mu\theta, \mu = 0, 1, 2, 3$ since if M is any symmetric matrix, then $\theta^T M\theta = 0$ and further that as noted above, $\epsilon, \gamma_5\epsilon$ and $\epsilon\gamma^\mu, \mu = 0, 1, 2, 3$ is a basis for

4×4 skew symmetric matrices. We now evaluate the change in the superfield S under a supersymmetry transformation $\alpha^T L$ where α is a Fermionic 4-vector. We have

$$\alpha^T L(C(x)) =$$

$$\alpha^T(\gamma_5\epsilon\partial/\partial\theta + \gamma^\mu\theta\partial_\mu)(C(x))$$

$$= \alpha^T\gamma^\mu\theta C_{,\mu}(x)$$

$$\alpha^T L(\theta^T\epsilon\omega(x)) =$$

$$\alpha^T(-\gamma_5\omega(x) + \gamma^\mu\theta\theta^T\epsilon\omega_{,\mu}(x))$$

$$\alpha^T L(\theta^T\epsilon\theta M(x)) =$$

$$= -2\alpha^T\gamma_5\theta M(x) + \theta^T\epsilon\theta\alpha^T\gamma^\mu\theta M_{,\mu}(x))$$

Note that we have used the fact that $\theta^T\epsilon\theta$ commutes with θ (two anticommutations make one commutation).

$$\alpha^T L(\theta^T\gamma_5\epsilon\theta N(x)) =$$

$$-2\alpha^T\theta N(x) + \theta^T\gamma_5\epsilon\theta\alpha^T\gamma^\mu\theta N_{,\mu}(x)$$

$$\alpha^T L(\theta^T\epsilon\gamma^\mu\theta V_\mu(x))$$

$$= \alpha^T(-2\gamma_5\gamma^\mu\theta V_\mu(x) + \theta^T\epsilon\gamma^\mu\theta\gamma^\nu\theta V_{\mu,\nu}(x))$$

$$= -2\alpha^T\gamma_5\gamma^\mu\theta V_\mu + \theta^T\epsilon\gamma^\mu\theta\alpha^T\gamma^\nu\theta V_{\mu,\nu}$$

$$\alpha^T L(\theta^T\epsilon\theta\theta^T\gamma_5\epsilon\beta(x))$$

$$= -2\alpha^T\gamma_5\theta\theta^T\gamma_5\epsilon\beta(x))$$

$$-\theta^T\epsilon\theta\alpha^T\beta(x)$$

$$+\theta^T\epsilon\theta\gamma^\mu\theta\theta^T\gamma_5\epsilon\beta_{,\mu}(x)$$

We now recall that

$$\theta\theta^T = (-1/4)(\theta^T\epsilon\theta\epsilon + \theta^T\gamma_5\epsilon\theta\gamma_5\epsilon + \theta^T\epsilon\gamma^\mu\theta\gamma_\mu\epsilon)$$

and hence

$$\alpha^T L(\theta^T\epsilon\theta\theta^T\gamma_5\epsilon\beta(x))$$

$$= -2\alpha^T\gamma_5\theta\theta^T\gamma_5\epsilon\beta(x))$$

$$-\theta^T\epsilon\theta\alpha^T\beta(x)$$

$$-(1/4)\theta^T\epsilon\theta\alpha^T\gamma^\mu(\theta^T\epsilon\theta\epsilon + \theta^T\gamma_5\epsilon\theta\gamma_5\epsilon + \theta^T\epsilon\gamma^\mu\theta\gamma_\mu\epsilon)\gamma_5\epsilon\beta_{,\mu}(x)$$

$$= -2\alpha^T\gamma_5\theta\theta^T\gamma_5\epsilon\beta(x))$$

$$-\theta^T\epsilon\theta\alpha^T\beta(x)$$

$$-(1/4)\theta^T\epsilon\theta(\theta^T\epsilon\theta\alpha^T\gamma^\mu\epsilon + \theta^T\gamma_5\epsilon\theta\alpha^T\gamma^\mu\gamma_5\epsilon + \theta^T\epsilon\gamma^\nu\theta\alpha^T\gamma^\mu\gamma_\nu\epsilon)\gamma_5\epsilon\beta_{,\mu}(x)$$

Now using

$$\theta^T\epsilon\theta.\theta^T\gamma_5\epsilon\theta = 0,$$

$$\theta^T\epsilon\theta.\theta^T\epsilon\gamma^\nu\theta = 0$$

we get

$$\alpha^T L(\theta^T\epsilon\theta\theta^T\gamma_5\epsilon\beta(x)) =$$

$$-2\alpha^T\gamma_5\theta\theta^T\gamma_5\epsilon\beta(x))$$

$$-\theta^T\epsilon\theta\alpha^T\beta(x)$$

$$+(1/4)(\theta^T\epsilon\theta)^2\alpha^T\gamma^\mu\gamma_5\beta_{,\mu}(x)$$

Finally, we find that

$$\alpha^T L((\theta^T\epsilon\theta)^2 K(x))$$

$$= \alpha^T(-4\gamma_5\theta\theta^T\epsilon\theta K(x))$$

Now denoting the inifnitesimal supersymmetry transformation by

$$\delta = \alpha^T L$$

we obtain on comparing the coefficients of different powers of θ, the following transformation rules for the components of the superfield under a supersymmetry transformation:

$$\delta C(x) = -\alpha^T \gamma_5 \omega(x)$$

$$\theta^T \epsilon \delta\omega(x) =$$

$$\alpha^T \gamma^\mu \theta C_{,\mu}(x)$$

$$-2\alpha^T \gamma_5 \theta M(x) - 2\alpha^T \theta N(x)$$

$$-2\alpha^T \gamma_5 \gamma^\mu \theta V_\mu(x)$$

or equivalently,

$$\delta\omega(x) =$$

$$-\gamma^\mu \epsilon \alpha C_{,\mu}(x)$$

$$+2\gamma_5 \epsilon \alpha M(x)$$

$$+2\epsilon\alpha N$$

$$+2\gamma^\mu \epsilon \gamma_5 \alpha V_\mu(x)$$

where we have used the fact that $\gamma^\mu \epsilon$ and ϵ are skew symmetric. Further,

$$\theta^T \epsilon \theta \delta M(x) + \theta^T \gamma_5 \epsilon \theta \delta N(x) + \theta^T \epsilon \gamma^\mu \theta \delta V_\mu(x)$$

$$= \alpha^T \gamma^\mu \theta \theta^T \epsilon \omega_{,\mu}(x))$$

$$-2\alpha^T \gamma_5 \theta \theta^T \gamma_5 \epsilon \beta(x))$$

$$-\theta^T \epsilon \theta \alpha^T \beta(x)$$

from which we deduce by expanding $\theta\theta^T$ as earlier,

$$\delta M(x) =$$

$$(1/4)\alpha^T \gamma^\mu \omega_{,\mu}(x)$$

$$-(3/2)\alpha^T \beta(x),$$

$$\delta N(x) = (1/4)\theta^T \gamma^\mu \gamma_5 \omega_{,\mu}(x)$$

$$+(1/2)\alpha^T \gamma_5 \beta(x)$$

$$\delta V_\mu(x)$$

$$= (1/4)\alpha^T \gamma^\nu \gamma_\mu \omega_{,\nu}(x))$$

$$(1/2)\alpha^T \gamma_\mu \beta(x)$$

where we have used the fact that γ_5 commutes with ϵ and anticommutes with γ^μ. Note that if we have an equation of the form $a(x)\theta\theta^T$=a quadratic polynomial in θ, then we can expand $\theta\theta^T$ as a linear combination of $\theta^T\epsilon\theta$, $\theta^T\gamma_5\epsilon\theta$ and $\theta^T\epsilon\gamma^\mu\theta$ and equate the corresponding coefficients. We also have the useful fact that if any two of these six distinct quadratic polynomials are multiplied, we get zero. Next, equating cubic terms in θ,

$$(\theta^T \epsilon \theta)\theta^T \gamma_5 \epsilon \delta\beta(x) =$$

$$\theta^T \epsilon \theta \alpha^T \gamma^\mu \theta M_{,\mu}(x))$$

$$+\theta^T \gamma_5 \epsilon \theta \alpha^T \gamma^\mu \theta N_{,\mu}(x)$$

$$+\theta^T \epsilon \gamma^\mu \theta \alpha^T \gamma^\nu \theta V_{\mu,\nu}$$

$$-4\alpha^T \gamma_5 \theta \theta^T \epsilon \theta K(x))$$

from which we deduce again by the above mentioned principles, (multiply both sides by θ, expand $\theta\theta^T$ as above and use the fact that the product of this quantity with $\theta^T\gamma_5\epsilon\theta$ is zero and also its product with $\theta^T\epsilon\gamma^\mu\theta$ is zero:

$$(1/4)(\theta^T\epsilon\theta)^2\gamma_5\delta\beta(x) =$$

$$-(1/4)\theta^T\epsilon\theta)^2\epsilon\gamma^{\mu T}\alpha M_{,\mu}(x)$$

$$-(1/4)(\theta^T\gamma_5\epsilon\theta)^2\gamma_5\epsilon\gamma^{\mu T}\alpha N_{,\mu}(x)$$

$$(-1/4)(\theta^T\epsilon\gamma^\mu)\theta)(\theta^T\epsilon\gamma^\rho\theta)\gamma_\rho\epsilon\gamma^{\nu T}\alpha V_{\mu,\nu}$$

$$+(\theta^T\epsilon\theta)^2\epsilon\gamma_5\alpha K(x)$$

Equivalently,

$$\gamma_5\delta\beta(x) =$$

$$-\epsilon\gamma^{\mu T}\alpha M_{,\mu}(x) + 4\epsilon\gamma_5\alpha K(x)$$

$$+\gamma_5\epsilon\gamma^{\mu T}\alpha N_{,\mu}(x)$$

$$-\eta^{\mu\rho}\gamma_\rho\epsilon\gamma^{\nu T}\alpha V_{\mu,\nu}$$

Here, we make use of the following easily verifiable identities:

$$(\theta^T\epsilon\theta)^2 = 8\theta^0\theta^1\theta^2\theta^3$$

$$(\theta^T\gamma_5\epsilon\theta)^2 = -8\theta^0\theta^1\theta^2\theta^3$$

$$(\theta^T\epsilon\gamma^\mu\theta)(\theta^T\epsilon\gamma^\nu\theta) =$$

$$= 8\theta^0\theta^1\theta^2\theta^3\eta^{\mu\nu}$$

Multiplying both sides of the above identity by γ_5 gives us

$$\delta\beta(x) =$$

$$-\gamma_5\epsilon\gamma^{\mu T}\alpha M_{,\mu}(x) + 4\epsilon\alpha K(x)$$

$$+\epsilon\gamma^{\mu T}\alpha N_{,\mu}(x)$$

$$-\gamma_5\gamma^\mu\epsilon\gamma^{\nu T}\alpha V_{\mu,\nu}$$

Finally, equating the fourth degree term in θ gives us

$$(\theta^T\epsilon\theta)^2\delta K(x) =$$

$$(1/4)(\theta^T\epsilon\theta)^2\alpha^T\gamma^\mu\gamma_5\beta_{,\mu}(x)$$

or equivalently,

$$\delta K(x) = (1/4)\alpha^T\gamma^\mu\gamma_5\beta_{,\mu}(x)$$

Now writing

$$\beta(x) = \lambda(x) + a.\gamma^\mu\omega_{,\mu}(x)$$

and

$$K(x) = D(x) + b\partial^\mu\partial_\mu C(x)$$

and using $\{\gamma_5, \gamma^\mu\} = 0$ and $\{\gamma^\mu, \gamma^\nu\} = \eta^{\mu\nu}I_4$, we get

$$\delta D(x) + b.\partial^\mu\partial_\mu\delta C(x) =$$

$$(1/4)\alpha^T\gamma^\nu\gamma_5(\lambda_{,\nu} + a.\gamma^\mu\omega_{,\mu\nu})$$

$$= (-1/4)\alpha^T\gamma_5\gamma^\nu\lambda_{,\nu} - (a/8)\gamma_5\partial^\mu\partial_\mu\omega$$

So we would expect that

$$\delta D(x) = (-1/4)\alpha^T\gamma_5\gamma^\nu\lambda_{,\nu},$$

$$b\delta C(x) = (-a/8)\alpha^T\gamma_5\omega(x)$$

On the other hand, we have already seen that

$$\delta C(x) = -\alpha^T \gamma_5 \omega(x)$$

So we must choose

$$a = 8b$$

We note that $\delta D(x)$ is a perfect four divergence and hence its integral over the whole of four dimensional space-time is zero. Thus, $\int D(x)d^4x$ is invariant under the supersymmetry transformation $\delta = \alpha^T L$ and hence is a candidate Lagrangian density for supersymmetric theories. We also have

$$\delta\beta(x) = \delta\lambda(x) + a.\gamma^\mu \delta\omega_{,\mu}(x)$$
$$-\gamma_5 \epsilon \gamma^{\mu T} \alpha M_{,\mu}(x)$$
$$+\epsilon\gamma^{\mu T} \alpha N_{,\mu}(x)$$
$$-\gamma_5 \gamma^\mu \epsilon \gamma^{\nu T} \alpha V_{\mu,\nu}$$
$$+4\epsilon\alpha(D(x) + b\partial^\mu \partial_\mu C(x))$$

So,

$$\delta\lambda(x) =$$
$$-\gamma_5 \epsilon \gamma^{\mu T} \alpha M_{,\mu}(x)$$
$$+\epsilon\gamma^{\mu T} \alpha N_{,\mu}(x)$$
$$-\gamma_5 \gamma^\mu \epsilon \gamma^{\nu T} \alpha V_{\mu,\nu}$$
$$+4\epsilon\alpha(D(x) + b\partial^\mu \partial_\mu C(x))$$
$$-a.\gamma^\mu(-\gamma^\nu \epsilon\alpha C_{,\nu}(x)$$
$$+2\gamma_5 \epsilon\alpha M(x) + 2\epsilon\alpha N$$
$$+2\gamma^\nu \epsilon\gamma_5 \alpha V_\nu(x))_{,\mu}$$

or equivalently using

$$\gamma_5 \gamma^\mu = -\gamma^\mu \gamma_5,$$

and

$$\epsilon\gamma^{\mu T} = \gamma^\mu \epsilon,$$

we get

$$\delta\lambda =$$
$$(1-2a)(\gamma^\mu \gamma_5 \epsilon\alpha M_{,\mu}$$
$$+(1-2a)\gamma^\mu \epsilon\alpha N_{,\mu}$$
$$+(\gamma^\mu \gamma^\nu - 2a\gamma^\nu \gamma^\mu)\epsilon\gamma_5 \alpha V_{\mu,\nu}$$
$$+4\epsilon\alpha(D + b\partial_\mu \partial^\mu C) + a\gamma^\mu \gamma^\nu C_{,\mu\nu}$$

and hence, it follows by taking $a = 1/2$ and noting that $(\gamma^\mu \gamma^\nu C_{,\mu\nu} = (1/2)\partial_\mu \partial^\mu C$, that

$$\delta\lambda = [\gamma^\mu, \gamma^\nu]\epsilon\gamma_5 \alpha V_{\mu,\nu} + 4\epsilon\alpha D$$

Suppose we choose $D = 0, \lambda = 0$. Then,

$$\delta D = 0, \delta\lambda = [\gamma^\mu, \gamma^\nu]\epsilon\gamma_5 \alpha V_{\mu,\nu}$$

It follows then that if $V_{\mu,\nu} - V_{\nu,\mu} = 0$, then

$$\delta\lambda = 0$$

This means that a superfield $S[x, \theta]$ with the constraints $D = 0, \lambda = 0$ after a supersymmetry transformation, again maintains the same constraints provided that $V_{\mu,\nu} - V_{\nu,\mu} = 0$, ie, provided that

$$V_\mu(x) = \partial_\mu Z(x)$$

for some scalar field $Z(x)$. A superfield S with these constraints, ie, $D = 0, \lambda = 0$ is called a Chiral field. We denote such a superfield by $\Phi(x, \theta)$.

Remark: Apparently, some errors in the signs of the variables occur above. These are easily rectified if we take into account that α is a Fermionic parameter vector which anticommutes with θ. We would then have $\alpha^T\theta = -\theta^T\alpha$ and more generally for a bosonic matrix A, we would have $\alpha^T A\theta = -\theta^T A\alpha$.

[18] Quantum Control in the sense of Belavkin and Luc Bouten:

The HP equation describing unitary evolution in the tensor product of the system Hilbert space and the Boson Fock space of the noisy bath is given by

$$dU(t) = (-i(H + L^*L/2)dt + LdA^*(t) - L^*SdA(t) + (S - I)d\Lambda(t))U(t)$$

where $S^*S = I, H^* = H$. Using quantum Ito's formula

$$dA.dA^* = dt, dAd\Lambda = dA, d\Lambda.dA^* = dA^*, (d\Lambda)^2 = d\Lambda^* = d\Lambda$$

with all the other product differentials zero (ie, of $o(dt)$). The corresponding quantum process equation describing noisy evolution of a system observable X coupled to the bath is given by

$$j_t(X) = U(t)^*XU(t) = U(t)^*(X \otimes I)U(t)$$

Application of quantum Ito's formula gives

$$dj_t(X) = j_t(\theta_0(X))dt + j_t(\theta_1(X))dA(t) + j_t(\theta_2(X))dA(t)^* + j_t(\theta_3(X))d\Lambda(t)$$

where the structure maps $\theta_k, k = 0, 1, 2, 3$ of this Evans-Hudson flow are given by

$$\theta_0(X) = i[H, X] - (1/2)(L^*LX + XL^*L - 2L^*XL)$$

$$\theta_1(X) = [L^*, X]S, \theta_2(X) = S^*[X, L], \theta_3(X) = S^*XS - X$$

The measurement process is assumed to be

$$Y_{out}(t) = U(t)^*Y_{in}(t)U(t), Y_{in}(t) = A(t) + A(t)^*$$

This is a non-demolition measurement in the sense of Belavakin:

$$[Y_{out}(t), Y_{out}(s)] = [Y_{out}(t), j_s(X)] = 0, s \geq t$$

We have by quantum Ito's formula,

$$dY_{out}(t) = dY_{in}(t) + j_t(S - I)dA + j_t(S^* - I)dA^* + j_t(L + L^*)dt$$

$$= j_t(S)dA + j_t(S^*)dA^* + j_t(L + L^*)dt$$

Let

$$\eta_{out}(t) = \{Y_{out}(s) : s \leq t\}$$

and

$$\pi_t(X) = \mathbb{E}(j_t(X)|\eta_{out}(t))$$

where expectations are taken w.r.t. the state $|f \otimes \psi(\beta) >$ where $|f > \in \mathfrak{h}, < f, f >= 1$ with $\mathfrak{h}$ being the system Hilbert space in which the operators H, S, L, X are defined and $|\psi(\beta) >= exp(- \parallel \beta \parallel^2 /2)|e(\beta) >$ with $|e(\beta) >$ being the exponential vector

$$|e(\beta) >= \bigoplus_{n=0}^{\infty} \beta^{\otimes n}/\sqrt{n!}$$

where $\beta \in L^2(\mathbb{R}_+)$. If $\beta = 0$, then the bath is in the vacuum state. Belavkin's filtering equation is

$$d\pi_t(X) = \pi_t(\mathcal{L}_t X) + (\pi_t(XM_t + M_t^*X) - \pi_t(M_t + M_t^*)\pi_t(X))dW(t)$$

where

$$dW(t) = dY_{out}(t) - (\pi_t(M_t + M_t^*) + \beta(t) + \bar{\beta}(t))dt$$

and

$$M_t = L + (S - I)\beta(t),$$
$$\mathcal{L}_t X = \theta_0(X) + \beta(t)\theta_1(X) + \bar{\beta}(t)\theta_2(X) + |\beta(t)|^2\theta_3(X)$$

In the special case when $\beta = 0$ (vacuum state of the bath), we get

$$M_t = L, \mathcal{L}_t X = \theta_0(X) = i[H, X] - (1/2)(L^*LX + XL^*L - 2L^*XL)$$

Writing

$$\pi_t(X) = Tr(\rho(t)X)$$

where $\rho(t)$ can be regarded as a classical random process with values in the space of density operators in the system Hilbert space $\mathfrak{h}$, we can express the Belavkin equation as

$$d\rho(t) = \mathfrak{L}_t^*(\rho(t))dt + ((M_t\rho(t) + \rho(t)M_t^*) - Tr(\rho(t)(M_t + M_t^*))\rho(t))(dY_{out}(t) - Tr(\rho(t)(M_t + M_t^*)) + 2Re(\beta(t)))dt)$$

It is easy to see that

$$L_t^*(\rho) = \theta_0^*(\rho) + \beta(t)\theta_1^*(\rho) + \bar{\beta}(t)\theta_2^*(\rho) + |\beta(t)|^2\theta_3^*(\rho)$$

where

$$\theta_0^*(\rho) = -i[H, \rho] - (1/2)(L^*L\rho + \rho L^*L - 2L\rho L^*)$$
$$\theta_1^*(\rho) = S\rho L^* - L^*S\rho = [S\rho, L^*],$$
$$\theta_3^*(\rho) = S\rho S^* - \rho$$

Note that for any map θ in $\mathcal{B}(\mathfrak{h})$, the Banach space of bounded operators in $\mathfrak{h}$, we define its dual map θ^* by the prescription

$$Tr(Y\theta(X))Tr(\theta^*(Y)X)$$

Sometimes it is more convenient to define the dual map by the prescription

$$Tr(Y^*\theta(X)) = Tr(\theta^*(Y)^*X)$$

This is true especially if the operators in question are non-Hermitian. This latter definition can be expressed as

$$< Y, \theta(X) >=< \theta^*(Y), X >$$

where $< .,. >$ is the inner product on $\mathcal{B}(\mathfrak{h})$ defined by

$$< X, Y >= Tr(X^*Y)$$

We note that

$$dW(t) = dY_{out}(t) - (\pi_t(M_t + M_t^*) + 2Re(\beta(t)))dt$$
$$= j_t(S)dA(t) + j_t(S^*)dA(t)^* + j_t(L + L^*)dt - (\pi_t(M_t + M_t^*) + 2Re(\beta(t))dt$$

and hence

$$\mathbb{E}(dW(t)|\eta_{out}(t)) =$$
$$\pi_t(S)\beta(t)dt + \pi_t(S^*)\bar{\beta}(t)dt - \pi_t(L + L^*)dt - \pi_t(M_t + M_t^*)dt - 2Re(\beta(t))dt$$
$$= 0$$

and further,

$$(dW(t))^2 = j_t(SS^*)dt = dt$$

by quantum Ito's formula. Hence, the innovations process $\{W(t) : t \geq 0\}$ is a classical Brownian motion w.r.t. the filtration $\{\eta_{out}(t), t \geq 0\}$. We now assume that $t = 0$ so that $t + dt = dt$. We choose a Hermitian system observable Z and apply the control unitary

$$U_{dt}^c = exp(iZdY(t))$$

where $Y = Y_{out}$. Since $t = 0$, $j_t = j_0 = I$ and hence

$$dY(t) = SdA(t) + S^*dA(t)^* + (L + L^*)dt - ((M_t + M_t^*) + 2Re(\beta(t))dt$$
$$= S(dA(t) - \beta(t)dt) + S^*(dA(t)^* - \bar{\beta}(t)dt)$$

Note that in the general case, ie, $t > 0$, we have

$$dW(t) = j_t(S)dA(t) + j_t(S^*)dA(t)^* + j_t(L + L^*)dt - (\pi_t(M_t + M_t^*) + 2Re(\beta(t))dt$$
$$= (j_t(S)dA - \pi_t(S)\beta(t)dt) + (j_t(S^*)dA^* - \pi_t(S^*)\bar{\beta}(t)dt)$$
$$+(j_t(L + L^*) - \pi_t(L + L^*))dt$$

We also note that

$$dY_{out}(t) = dW(t)$$

in the special case when $\beta = 0$ and $L^* = -L$. The resulting state at time $t + dt = dt$ after applying the control under the condition $\beta = 0$ (so that $M_t = L$) is given by

$$\rho_c(dt) = U_c(dt)\rho(dt)dU_c(dt)^*$$

$$(I + iZdY - Z^2dt/2)(\rho_c(0) + \mathcal{L}^*(\rho_c(0))dt + ((L\rho_c(0) + \rho_c(0)L^*) - Tr(\rho_c(0)(L + L^*))\rho_c(0))dW(t))(I - iZdY - Z^2dt/2)$$

[19] Proof of the Shannon Cq coding theorem. A is a finite alphabet and for each $x \in A$, $\rho(x)$ is a density matrix in the finite dimensional Hilbert space $\mathcal{H} = \mathbb{C}^N$. Let P be a probability distribution on A and define the set of all δ-typical sequences of length n:

$$T(P, n, \delta) = \{u \in A^n : |N(x|u) - nP(x)| \leq \delta\sqrt{nP(x)(1 - P(x))} \forall x \in A\}$$

Here, $N(x|u)$ is the number of times x occurs in the sequence u. It is clear that $u \in T(P, n, \delta)$ implies

$$2^{-nS(P)-K\delta\sqrt{n}} \leq P^{\otimes n}(u) = \Pi_{x\in A}P(x)^{N(x|u)} \leq 2^{-nS(P)+K\delta\sqrt{n}}$$

where

$$S(P) = -\sum_{x\in A} P(x)log(P(x)), K = \sum_{x\in A} \sqrt{P(x)(1 - P(x))}log(P(x))$$

The greedy algorithm: Choose the maximal integer M such that for a given probability distribution P on A and given $\epsilon > 0$ there exist sequences $u_1, ..., u_M \in T(P, n, \delta)$ and operators $D_1, ..., D_M \geq 0$ in $\mathcal{H}$ such that $\sum_{i=1}^M D_i \leq I, Tr(\rho(u_i)D_i) \geq 1 - \epsilon \forall i$ and $D_i \leq E(n, u_i, \delta) \forall i$. Here,

$$E(n, u, \delta) = \bigotimes_{x\in A} E(\rho(x)^{\otimes N(x|u)}, \delta)$$

where if ρ is any density matrix and n any positive integer, then if

$$\rho = \sum_i |i > p(i) < i|$$

is the spectral representation of ρ, then

$$E(\rho^{\otimes n}, \delta) = \sum_{N(i|i_1,...,i_n)-np(i)|\leq\delta\sqrt{np(i)(1-p(i))}\forall i} |i_1, ..., i_n >< i_1, ..., i_n|$$

It is easily seen that

$$2^{-nS(\rho)-\delta K_1\sqrt{n}} \leq \rho^{\otimes n}E(\rho^{\otimes n}, \delta) \leq 2^{-nS(\rho)+\delta K_1\sqrt{n}}$$

where

$$S(\rho) = -\sum_i p(i)log(p(i)), K_1 = \sum_i \sqrt{p(i)(1 - p(i))}log(p(i))$$

This simple fact is based on the identity

$$p(i_1)...p(i_n) = \Pi_i p(i)^{N(i|i_1,...,i_n)} = 2^{\sum_i N(i|i_1,...,i_n)log(p(i))}$$

and

$$\rho^{\otimes n}|i_1, ..., i_n >= p(i_1)...p(i_n)|i_1, ..., i_n >$$

It follows that on setting

$$\rho^{\otimes n}(u) = \bigotimes_{x \in A} \rho(x)^{\otimes N(x|u)}$$

we get

$$2^{-n\sum_{x\in A} P_u(x)S(\rho(x)) - \delta\sum_{x\in A} K_1(x)\sqrt{N(x|u)}} \le \rho^{\otimes n}(u)E(n,u,\delta) \le 2^{-n\sum_{x\in A} P_u(x)S(\rho(x)) + \delta\sum_{x\in A} K_1(x)\sqrt{N(x|u)}}$$

where

$$P_u(x) = N(x|u)/n$$

and

$$K_1(x) = \sum_i \sqrt{P_{\rho(x)}(i)(1 - P_{\rho(x)}(i))log(P_{\rho(x)}(i))}$$

where

$$\rho(x) = \sum_i |i > P_{\rho(x)}(i) < i|$$

is the spectral representation of $\rho(x)$. Writing

$$K_0 = \sum_{x \in A} K_1(x)$$

it follows that

$$2^{-n\sum_{x\in A} P_u(x)S(\rho(x)) - \delta K_0\sqrt{n}} \le \rho^{\otimes n}(u)E(n,u,\delta) \le 2^{-n\sum_{x\in A} P_u(x)S(\rho(x)) + \delta K_0\sqrt{n}}$$

Remark: If $u \in T(P,n,\delta)$, then

$$|N(x|u) - nP(x)| \le \delta\sqrt{nP(x)(1-P(x))}\forall x \in A$$

and hence

$$P(x) - \delta\sqrt{P(x)(1-P(x))/n} \le P_u(x) = N(x|u)/n \le P(x) + \delta\sqrt{P(x)(1-P(x))/n}, x \in A$$

and so

$$P(x) - \delta/\sqrt{2n} \le P_u(x) \le P(x) + \delta/\sqrt{2n}, x \in A$$

In other words, for any $\epsilon > 0$, there exists n_0 such that $n \ge n_0$ implies

$$|P_u(x) - P(x)| \le \epsilon, x \in A, u \in T(P,n,\delta)$$

where $\delta > 0$ is given. It follows from this observation and the above inequality that if $u \in T(P,n,\delta)$, then there is a K_2 independent of n such that

$$2^{-n\sum_{x\in A} P(x)S(\rho(x)) - \delta K_2\sqrt{n}} \le \rho^{\otimes n}(u)E(n,u,\delta) \le 2^{-n\sum_{x\in A} P(x)S(\rho(x)) + \delta K_2\sqrt{n}}$$

For example, we can take

$$K_2 = K_0 + \sum_{x\in A}(P(x)(1-P(x))^{1/2}S(\rho(x))$$

A lower bound on M in the greedy algorithm:

Lemma: In the greedy algorithm, define $\sum_{i=1}^{M} D_i = D$. Then, $Tr(\rho(u)D) \ge \gamma\forall u \in T(P,n,\delta)$. For suppose $Tr(\rho(u)D) \le \gamma$ for some $u \in T(P,n,\delta)$. Then since $\gamma \le 1-\epsilon$, it follows that $u \notin \{u_1, ..., u_M\}$. Now define

$$D' = (1-D)^{1/2}E(n,u,\delta)(1-D)^{1/2}$$

Then, (writing $\rho(u)$ for $\rho^{\otimes n}(u) = \otimes_{x\in A}\rho(x)^{\otimes N(x|u)}$), we have

$$Tr(\rho(u)D') = Tr(\rho(u)E(n,u,\delta)) - Tr(\rho(u)(E(n,u,\delta) - (1-D)^{1/2}E(n,u,\delta)(1-D)^{1/2}))$$

$$\ge Tr(\rho(u)E(n,u,\delta)) - \| \rho(u) - (1-D)^{1/2}\rho(u)(1-D)^{1/2} \|_1$$

$$\ge Tr(\rho(u)E(n,u,\delta)) - \alpha.Tr(\rho(u)D) \ge Tr(\rho(u)E(n,u,\delta)) - \alpha\gamma$$

Now observe that for any density ρ,

$$Tr(\rho^{\otimes n}E(\rho^{\otimes n},\delta)) = P_{\rho^{\otimes n}}(\bigcap_i\{|N(i|i_1,...,i_n) - nP_\rho(i)| \le \delta\sqrt{nP_\rho(i)(1-P_\rho(i))})$$

$$\geq 1 - \sum_i P(|N(i|i_1, ..., i_n) - nP_\rho(i)| \geq \delta\sqrt{nP_\rho(i)(1 - P_\rho(i))}$$

$$\geq 1 - N/\delta^2 - - - (1)$$

where we have used the union bound, the Chebyshev inequality and the fact that for each i, $N(i|i_1, ..., i_n)$ is a random variable having mean $nP_\rho(i)$ and variance $\sqrt{nP_\rho(i)(1 - P_\rho(i))}$. Note that we also have

$$P(T(P, n, \delta)^c) = P(u \in A^n : |N(x|u) - nP(x)| > \delta\sqrt{nP(x)(1 - P(x))} for some x \in A)$$

$$\leq \sum_{x \in A} P(u \in A^n : |N(x|u) - nP(x)| > \delta\sqrt{nP(x)(1 - P(x))})$$

$$\leq a/\delta^2$$

by the union bound and Chebyshev's inequality. Thus,

$$P(T(P, n, \delta)) \geq 1 - a/\delta^2$$

From the above arguments (ie equn(1)),

$$Tr(\rho(u)E(n, u, \delta)) = \Pi_{x \in A} Tr(\rho(x)^{\otimes N(x|u)} E(\rho(x)^{N(x|u)}, \delta))$$

$$\geq (1 - N/\delta^2)^a$$

Note:If $\delta = O(n^{s/2})$ where $0 < s < 1$, then $(1 - N/\delta^2)^a$, then it follows that $Tr(\rho(u)E(n, u, \delta))$ can be made as close to unity as possible by choosing n sufficiently large and quantities like $2^{-nS(\rho) \pm K'\delta\sqrt{n}}$ are of the order $2^{-nS(\rho) \pm K" n^\theta}$ where $0 < \theta < 1$. Equivalently, the logarithm of these quantities divided by n would converge to $-S(\rho)$ as $n \to \infty$ with the probabilities $Tr(\rho(u)E(n, u, \delta))$ converging to zero.

We have now derived the inequality

$$Tr(\rho(u)D') \geq Tr(\rho(u)E(n, u, \delta)) - \alpha\gamma \geq (1 - N/\delta^2)^a - \alpha\gamma \geq 1 - \epsilon$$

if for sufficiently large n provided δ is allowed to vary with n as in the above note and $\gamma < \epsilon/\alpha$ is assumed. We thus get a contradiction to the maximality of M on using the fact that

$$0 \leq D' \leq 1 - D$$

ie

$$D + D' \leq 1$$

This completes the proof of the lemma.

A lower bound on M:

Lemma: Let $0 \leq Z, T \leq$ be operators and ρ a density operator in a Hilbert space such that for some $\theta > 0$, we have $\rho T \leq \theta T$. Then,

$$Tr(Z) \geq \theta^{-1}(Tr(\rho Z) - Tr(\rho(1 - T)))$$

An intuitive interpretation of this inequality: Z, T are viewed as events. If Z occurs, then T may or may not occur. If Z and T occur, then the probability of this is the trace of $\rho T \cap Z$ which is smaller than the trace of $\theta T \cap Z$ which in turn is smaller than the trace of θZ. If Z occurs and T does not occur then the probability of this is the trace of $\rho Z \cap (1 - T)$ which is smaller than the trace of $\rho(1 - T)$. Combining these two facts, we get

$$Tr(\rho Z) \leq \theta Tr(Z) + Tr(\rho(1 - T))$$

and the proof of the Lemma is complete.

Now define

$$\bar{\rho} = \sum_{x \in A} P(x)\rho(x)$$

Then, we have

$$\bar{\rho}^{\otimes n} E(\bar{\rho}^{\otimes n}, \delta) \leq 2^{-nS(\bar{\rho}) + K_3\delta\sqrt{n}} E(\bar{\rho}^{\otimes n}, \delta)$$

where

$$K_3 = \sum_j \sqrt{P_\rho(i)(1 - P_\rho(i))} log(P_\rho(i))$$

Hence by the above lemma,

$$Tr(D) \geq 2^{nS(\bar{\rho})-K_3\delta\sqrt{n}}(Tr(\bar{\rho}^{\otimes n}D) - Tr(\bar{\rho}(\bar{\rho}^{\otimes n}(1 - E(\bar{\rho}^{\otimes n}, \delta)))$$

Now, since

$$\bar{\rho}^{\otimes n} = \sum_{u \in A^n} P^{\otimes n}(u)\rho(u)$$

it follows that

$$Tr(\bar{\rho}^{\otimes n}D) \geq \sum_{u \in T(P,n,\delta)} P^{\otimes n}(u)Tr(\rho(u)D)$$

$$\geq \gamma P^{\otimes n}(T(P,n,\delta)) \geq \gamma(1 - N/\delta^2)$$

and further,

$$Tr(\bar{\rho}^{\otimes n}E(\bar{\rho}^{\otimes n}, \delta)) = P_{\bar{\rho}}^{\otimes n}(T(P_{\bar{\rho}}, n, \delta))$$

$$\geq 1 - N/\delta^2$$

Thus,

$$Tr(D) \geq 2^{nS(\bar{\rho})-K_3\delta\sqrt{n}}(1 - 2N/\delta^2)$$

On the other hand,

$$Tr(D) = \sum_{i=1}^{M} Tr(D_i) \leq \sum_{i=1}^{M} Tr(E(n, u_i, \delta))$$

$$\leq M.2^{n\sum_{x \in A} P(x)S(\rho(x))+K_5\delta\sqrt{n}}$$

Combining the above two inequalities gives us

$$M \geq (1 - 2N/\delta^2)2^{n(S(\bar{\rho})-\sum_{x \in A} P_u(x)S(\rho(x)))-K_5\delta\sqrt{n}} - - - (2)$$

which is the desired lower bound on M.

Remark: Here we have made used of the following fact:

$$\rho(u)E(n,u,\delta) = \bigotimes_{x \in A} \rho(x)^{\otimes N(x|u)}E(\rho(x)^{\otimes N(x|u)}, \delta)$$

$$\geq (\Pi_{x \in A} 2^{-N(x|u)S(\rho(x))-\delta\sqrt{N(x|u)})K_1(x)}E(n,u,\delta)$$

where as defined earlier

$$K_1(x) = \sum_i \sqrt{P_{\rho(x)}(i)(1 - P_{\rho(x)}(i))log(P_{\rho(x)}(i)}$$

It follows that

$$Tr(E(n,u,\delta)) \leq 2^{n\sum_{x \in A} P_u(x)S(\rho(x))+K_5\delta\sqrt{n}}$$

and hence (2) is obtained. Note that $P_{u_i}(x)$ differs from $P(x)$ by an amount of order $\sqrt{n}$ since $u_i \in T(P,n,\delta)$. Here,

$$K_5 = \sum_{x \in A} K_1(x)$$

An upper bound on M: Consider the spectral representation of $\bar{\rho}$:

$$\bar{\rho} = \sum_i |i > P_{\bar{\rho}}(i) < i|$$

and with respect to the onb $\{|i>: 1 \leq i \leq N\}$ of this representation, define the density operators

$$\tilde{\rho}(x) = \sum_i |i >< i|\rho(x)|i >< i|, x \in A$$

Then, $\tilde{\rho}(x), x \in A$ form a commuting family of operators in $\mathcal{H} = \mathbb{C}^N$. Define the typical projections $\tilde{E}(n,u,\delta)$ in the usual way corresponding to this family $\tilde{\rho}(x), x \in A$. In other words, if $|j_1, ..., j_{N(x|u)} >$ are the eigenvectors of $\tilde{\rho}(x)^{\otimes N(x|u)}$, then

$$\tilde{\rho}(u) = \bigotimes_{x \in A} \tilde{\rho}(x)^{\otimes N(x|u)}$$

$$\tilde{E}(n,u,\delta) = \bigotimes_{x\in A} E(\tilde{\rho}(x)^{\otimes N(x|u)},\delta)$$

$$= \bigotimes_{x\in A} \sum_{(j_1,...,j_{N(x|u)})\in T(j,x)\forall j} |j_1,...,j_{N(x|u)} >< j_1,...,j_{N(x|u)}|$$

where

$$T(j,x) = \{(j_1,...,j_{N(x|u)}) : N(j|j_1,...,j_{N(x|u)}) - N(x|u)p(j|x)| \leq \delta\sqrt{N(x|u)p(j|x)(1-p(j|x))}\}$$

with

$$p(j|x) =< j|\rho(x)|j >$$

ie the probability of the jth eigenvector of $\bar{\rho}$ occurring given that the system is in the state $\rho(x)$. Note that we can write

$$\tilde{E}(n,u,\delta) = \sum |v >< v|$$

where the summation is over all n-length sequences v of the form

$$v = \bigcup_{x\in A}(j_1,...,j_{N(x|u)})$$

with

$$(j_1,...,j_{N(x|u)}) \in T(j,x)\forall j = 1,2,...,N$$

It immediately follows that

$$N(j|v) = \sum_{x\in A} N(j|j_1,...,j_{N(x|u)})$$

Suppose $P_u(x) = P(x), x \in A$. Then

$$|N(j|v) - nP_{\bar{\rho}}(j)| = |\sum_{x\in A} N(j|j_1,...,j_{N(x|u)}) - \sum_{x\in A} nP(x)p(j|x)|$$

$$\leq \sum_{x\in A} |N(j|j_1,...,j_{N(x|u)} - p(j|x)N(x|u)|$$

$$\leq \sum_{x\in A} \delta\sqrt{N(x|u)p(j|x)(1-p(j|x))}$$

$$= \delta\sqrt{n}\sum_{x\in A}\sqrt{P(x)p(j|x)(1-p(j|x))}$$

$$\leq \delta\sqrt{n}(\sum_x P(x))^{1/2}.(\sum_x p(j|x)(1-p(j|x)))^{1/2}$$

$$\leq \delta\sqrt{an}$$

From this, it follows that

$$v \in T(P_{\bar{\rho}}, n, \delta\sqrt{a})$$

Thus, we have proved that

$$\tilde{E}(n,u,\delta) \leq E(\bar{\rho}^{\otimes n}, \delta\sqrt{a})$$

whenever $P_u(x) = P(x)\forall x \in A$. Now define

$$D_i' = \tilde{E}(n,u_i,\delta)D_i\tilde{E}(n,u_i,\delta), i = 1,2,...,M$$

Then,

$$D_i' \leq \sum_{i=1}^{M} E(\bar{\rho}^{\otimes n},\delta\sqrt{a})D_iE(\bar{\rho}^{\otimes n},\delta\sqrt{a})$$

$$= E(\bar{\rho}^{\otimes n}, a\sqrt{\delta})DE(\bar{\rho}^{\otimes n},\delta\sqrt{a})$$

$$\leq I$$

Now, we observe that

$$\rho(u)E(n,u,\delta) = \bigotimes_{x\in A} \rho(x)^{\otimes N(x|u)} E(\rho(x)^{\otimes N(x|u)},\delta)$$

$$\leq (\Pi_{x\in A} 2^{-N(x|u)S(\rho(x))+\delta\sqrt{N(x|u)})K_1(x)} E(n,u,\delta)$$

where as defined earlier

$$K_1(x) = \sum_i \sqrt{P_{\rho(x)}(i)(1-P_{\rho(x)}(i))log(P_{\rho(x)}(i)}$$

and hence

$$\rho(u)E(n,u,\delta) \leq 2^{-n\sum_x P_u(x)S(\rho(x))+K_5\delta\sqrt{n}} E(n,u,\delta)$$

and if we assume further that $u \in T(P,n,\delta)$, then we get

$$\rho(u)E(n,u,\delta) \leq 2^{-n\sum_x P(x)S(\rho(x))+K_6\delta\sqrt{n}} E(n,u,\delta) - - - (3)$$

since for such u,

$$|P_u(x) - P(x)| \leq \delta/\sqrt{2n}$$

as observed earlier. It follows from (3) and the above lemma that

$$Tr(D_i') \geq 2^{n\sum_x P(x)S(\rho(x))-K_6\delta\sqrt{n}}(Tr(\rho(u_i)D_i') - Tr(\rho(u_i)(1-E(n,u_i,\delta)))$$

Now we have

$$Tr(\rho(u_i)D_i') = Tr(\rho(u_i)D_i) - Tr((\rho(u_i) - \tilde{E}(n,u_i,\delta)\rho(u_i)\tilde{E}(n,u_i,\delta))D_i)$$

$$\geq 1-\epsilon - \| \rho(u_i) - \tilde{E}(n,u_i,\delta)\rho(u_i)\tilde{E}(n,u_i,\delta) \|_1$$

$$\geq 1-\epsilon-\beta Tr(\rho(u_i)(1-\tilde{E}(n,u_i,\delta)))$$

Let π denote any specific rank one projection of the form $|i_1,...,i_n><i_1,...,i_n|$ where $|i_1,...,,i_n>$ are eigenvectors of $\bar{\rho}^{\otimes n}$ obtained by tensoring the orthonormal basis of eigenvectors of $\bar{\rho}$. Then, we have

$$Tr(\rho(u_i)\tilde{E}(n,u_i,\delta)) = \sum_\pi Tr(\rho(u_i)\pi\tilde{E}(n,u_i,\delta)\pi)$$

$$= \sum_\pi Tr(\pi\rho(u_i)\pi\tilde{E}(n,u_i,\delta))$$

$$= Tr(\tilde{\rho}(u_i)\tilde{E}(n,u_i,\delta))$$

$$\geq 1 - N/\delta^2$$

Thus,

$$Tr(\rho(u_i)\tilde{E}(n,u_i,\delta)) \leq N/\delta^2$$

and also

$$Tr(\rho(u_i)E(n,u_i,\delta)) \leq N/\delta^2$$

Thus, we get

$$Tr(D_i') \geq 2^{n\sum_x P(x)S(\rho(x))-K_6\delta\sqrt{n}}(1-\epsilon-\beta N/\delta^2 - N/\delta^2)$$

$$= 2^{n\sum_x P(x)S(\rho(x))-K_6\delta\sqrt{n}}(1-\epsilon-(\beta+1)N/\delta^2)$$

It follows that

$$Tr(\sum_{i=1}^M D_i') \geq M.2^{n\sum_x P(x)S(\rho(x))-K_6\delta\sqrt{n}}(1-\epsilon-(\beta+1)N/\delta^2)$$

on the one hand and on the other,

$$Tr(\sum_{i=1}^M D_i') = \sum_{i=1}^M Tr(\tilde{E}(n,u_i,\delta)D_i\tilde{E}(n,u_i,\delta))$$

We have already noted that if $P = P_u$, then

$$\tilde{E}(n,u,\delta) \leq E(\bar{\rho}^{\otimes n}, \delta\sqrt{a})$$

so if we assume that $P_{u_i} = P\forall i$, then we get

$$Tr(\sum_{i=1}^{M} D_i') \leq Tr(E(\bar{\rho}^{\otimes n}, \delta\sqrt{a})DE(\bar{\rho}^{\otimes n}, \delta\sqrt{a})$$

$$\leq Tr(E(\bar{\rho}^{\otimes n}, \delta\sqrt{a}))$$

$$\leq 2^{nS(\bar{\rho})+K_7\delta\sqrt{an}}$$

and hence we get

$$2^{nS(\bar{\rho})+K_7\sqrt{an}} \geq M.2^{n\sum_x P(x)S(\rho(x))-K_6\delta\sqrt{n}}(1-\epsilon-(\beta+1)N/\delta^2)$$

from which we get the upper bound on M in the special case when $P_{u_i} = P\forall i$:

$$M \leq (1-\epsilon-(\beta+1)N/\delta^2)^{-1}2^{n(S(\bar{\rho})-\sum_x P(x)S(\rho(x)))+(K_6+K_7\sqrt{a})\delta\sqrt{n}}$$

Suppose now that $u_1, ..., u_M, D_1, ..., D_M$ are as in the greedy algorithm. Let Q be any empirical probability distribution on A corresponding to the integer n. This means that there is a sequence $u \in A^n$ such that $Q(x) = N(x|u)/n, x \in A$. It is clear that the number of empirical probability distributions on A corresponding to n cannot exceed $(n+1)^a$. (Each symbol $x \in A$ can occur in a sequence of length n, k times where $k = 0, 1, ..., n$). We now consider the subset F_Q of all integers $1, 2, ..., M$ such that u_i is of empirical type Q, ie, $N(x|u_i)/n = Q(x), \forall x \in A$. Let M_Q denote the cardinality of F_Q. Then it is clear that

$$M = \sum_Q M_Q$$

where the summation is over all empirical distributions Q, there being atmost $(n+1)^a$ of them. By the above inequality, we have since $Tr(\rho(u_i)D_i) \geq 1-\epsilon, D_i \leq E(n, u_i, \delta), \forall i \in F_Q$ and $\sum_{i \in F_Q} D_i \leq I$,

$$M_Q \leq (1-\epsilon-(\beta+1)N/\delta^2)^{-1}2^{n(S(\bar{\rho}_Q)-\sum_x Q(x)S(\rho(x)))+(K_6+K_7\sqrt{a})\delta\sqrt{n}}$$

$$\leq 1-\epsilon-(\beta+1)N/\delta^2)^{-1}2^{nC+(K_6+K_7\sqrt{a})\delta\sqrt{n}}$$

where

$$\rho_Q = \sum_x Q(x)\rho(x)$$

and

$$C = max_P(S(\sum_x P(x)\rho(x)) - \sum_x P(x)S(\rho(x)))$$

the maximum being taken over all probability distributions P on A. Summing this over all empirical distributions Q of length n on A gives us

$$M = \sum_Q M_Q \leq (n+1)^a.(1-\epsilon-(\beta+1)N/\delta^2)^{-1}2^{nC+(K_6+K_7\sqrt{a})\delta\sqrt{n}}$$

which is the desired upper bound on M.

A remark: Now suppose we do not assume $P_{u_i} = P\forall i$. Since $u_i \in T(P, n, \delta)$, we have $|P_{u_i}(x) - P(x)| \leq \delta/\sqrt{2n}\forall i, x$.

For any u, we define $\tilde{E}(n, u, \delta)$ in the same way as above except that we replace P by P_u. Then, by the same arguments as above, we have

$$\tilde{E}(n, u, \delta) \leq E(\bar{\rho}_u^{\otimes n}, \delta\sqrt{a})$$

where

$$\bar{\rho}_u = \sum_{x \in A} P_u(x)\rho(x)$$

We proceed in the same way as above by defining D_i' in terms of the new $\tilde{E}$ leading to the results:

$$D' = \sum_i D_i' = \sum_i \tilde{E}(n, u_i, \delta)D_i\tilde{E}(n, u_i, \delta)$$

$$= E(\bar{\rho}^{\otimes n}, \delta\sqrt{a})DE(\bar{\rho}^{\otimes n}, \delta\sqrt{a}) +$$

$$\sum_i (E_i - \bar{E})D_i(E_i - \bar{E}) + \sum_i (E_i - \bar{E})D_i\bar{E} + \bar{E}D_i(E_i - \bar{E})$$

where

$$E_i = \tilde{E}(n, u_i, \delta), \bar{E} = E(\bar{\rho}^{\otimes n}, \delta\sqrt{a})$$

Thus, in terms of norms,

$$Tr(D') \leq Tr(\bar{E}) + 2.max_i \parallel E_i - \bar{E} \parallel_1^2 + max_i \parallel E_i - E \parallel_1$$

Note that $u_i \in T(P, n, \delta)$. Now if $u \in T(P, n, \delta)$, then P_u is close to P, hence $\bar{\rho}_u$ will be close to $\bar{\rho}$ and so $\tilde{E}(n, u, \delta)$ will be close $\bar{E}$. Thus, E_i will be close to $\bar{E}$ and we would get that $Tr(D')$ cannot exceed $Tr(\bar{E})$ by an order of an exponential of n provided that we are able to show that $\parallel E_i - \bar{E} \parallel^2$ cannot grow faster than an exponential of n^p where $p < 1$). Hence, the desired lower bound of the form $M \leq 2^{n(S(\bar{\rho}) - \sum_x P(x)S(\rho(x)) + K\delta\sqrt{n}}$ would follow. To make this more precise, we observe that $u \in T(P, n, \delta)$ implies

$$|P_u^{\otimes n}(v) - P^{\otimes n}(v)| = |\Pi_{x \in A} P_u(x)^{N(x|v)} - \Pi_{x \in A} P(x)^{N(x|v)}|$$

$$= |e^{\sum_x N(x|v)log(P_u(x))} - e^{\sum_x N(x|v)log(P(x))}|$$

$$=\leq |exp(\sum_x N(x|v)(log(P_u(x)) - log(P(x))) - 1|$$

We can write

$$P_u(x) = P(x) + f(x)/\sqrt{n}, x \in A, |f(x)| \leq 1, \sum_x f(x) = 0$$

Hence,

$$|P_u^{\otimes n}(v) - P^{\otimes n}(v)| \leq$$

$$|exp(\sum_x N(x|v)log(1 + f(x)/P(x)\sqrt{n})) - 1| = |exp(\sum_x N(x|v)(f(x)/P(x)\sqrt{n} - f(x)^2/2P(x)^2n + ...)) - 1|$$

If $v \notin T(P, n, \delta)$, then $N(x|v) < nP(x) + O(\sqrt{n})$ in which case we see that

$$\sum_x N(x|v)(f(x)/P(x)\sqrt{n} - f(x)^2/2P(x)^2n + ...) < O(1/\sqrt{n})$$

since $\sum f(x) = 0$. If $v \in T(P, n, \delta)$, then $N(x|v) = nP(x) + g(x)\sqrt{n}$ where $|g(x)| < 1$. Hence, in this case,

$$\sum_x N(x|v)(f(x)/P(x)\sqrt{n} - f(x)^2/2P(x)^2n + ..) = \sum_x (g(x)f(x)/P(x) - f(x)^2/2P(x)) + O(1/\sqrt{n})$$

$$= c_0 + O(1/\sqrt{n})$$

Other useful inequalities are

$$|log(P_u^{\otimes n}(v)) - log(P^{\otimes n}(v))| =$$

$$= |\sum_x N(x|v)log(P_u(x)) - \sum_x N(x|v)log(P(x))| =$$

$$|\sum_x N(x|v)(log(1 + f(x)/P(x)\sqrt{n})|$$

[20] Restricted quantum gravity in one spatial dimension and one time dimension. The metric is

$$d\tau^2 = (1 + 2U(t,x))dt^2 - (1 + 2V(t,x))dx^2$$

The position fields are $U(t,x)$ and $V(t,x)$ and to find the momentum fields, we must first evaluate the Lagrangian density

$$L = g^{\mu\nu}\sqrt{-g}(\Gamma^\alpha_{\mu\nu}\Gamma^\beta_{\alpha\beta} - \Gamma^\alpha_{\mu\beta}\Gamma^\beta_{\alpha\beta})$$

This Lagrangian density is a function of $U, V, U_{,\mu}, V_{,\mu}$. Define the position fields as U, V and the canonical momentum fields as

$$\pi_U = \frac{\partial L}{\partial U_{,0}}, \pi_V = \frac{\partial L}{\partial V_{,0}}$$

Then, apply the Legendre transformation after solving for $U_{,0}, V_{,0}$ in terms of $U, V, \nabla U, \nabla V$ to get the Hamiltonian density as

$$\mathcal{H}(U, V, \nabla U, \nabla V, \pi_U, \pi_V) = \pi_U U_{,0} + \pi_V V_{,0} - L$$

Then, set up the Schrodinger wave equation

$$(\int H(U(x), V(x), \nabla U(x), \nabla V(x), -i\delta/\delta U(x), -i\delta/\delta V(x))dx)\psi_t(\{U(x), V(x) : x \in \mathbb{R})$$

$$= i\frac{\partial}{\partial t}\psi_t(\{U(x), V(x) : x \in \mathbb{R})$$

References

[1] Rohit Singh, Naman Garg and H.Parthasarathy, "Statistical modeling of transmission lines using stochastic differential equations with random loading effects along the line", Technical report, NSIT, 2016.

[2] Rohit Singh, Naman Garg and H.Parthasarathy, "Quantization of transmission line equations using the GKSL formalism", Technical report, NSIT, 2016.

[3] Naman Garg, H.Parthasarathy and D.K.Upadhyay, "Real time simulation of Hudson-Parthasarathy noisy Schrodinger equation and Belavkin equation", Technical report, NSIT, 2016.

[4] Naman Garg, H.Parthasarathy and D.K.Upadhyay, "Real time control of the the Hudson-Parthasarathy noisy Schrodinger equation with observer obtained from Belavkin's filter, Technical report, NSIT, 2016.

[5] D.W.Stroock and S.R.S.Varadhan, Multidimensional Diffusion Processes, Springer, 1997.

[6] L.D.Landau and E.M.Lifshitz, "The classical theory of fields", Butterworth and Heinemann.

[7] Rohit Singh, Jyotsna Singh and H.Parthasarathy, "Application of wavelets to partial differential equation based modeling of images, Technical Report, NSIT, 2016.

[8] H.Parthasarathy, "Time travel in the special and general theories of relativity and the notions of space and time in quantum gravity, Lecture delivered at the G.B.Pant University, 15th September, 2016.

[9] Rohit Singh, Jyotsna Singh and H.Parthasarathy, "Image parameter estimation using Edgeworth models for non-Gaussian noise and nonlinear filtering theory in discrete time", Technical Report, NSIT, 2016.

[10] Rohit Singh, Naman Garg, Kumar Gautam and H.Parthasarathy, "Design of quantum gates using scattering theory", Technical Report, NSIT, 2016.

[11] Lalit Kumar and H.Parthasarathy, "The magnetic field produced by a non-uniform transmission line carrying current when nonlinear hysteresis and nonlinear capacitive effects are taken into account, Technical report, NSIT, 2016.

[12] K.R.Parthasarathy, "An introduction to quantum stochastic calculus", Birkhauser, 1992.

[13] K.R.Parthasarathy, "Coding theorems of classical and quantum information theory", Hindustan Book Agency.

[14] K.R.Parthasarathy, "Mathematical foundations of quantum mechanics", Hindustan Book Agency.

[15] J.Gough and Koestler, Quantum Filtering in Coherent States".

[16] S.Weinberg, The quantum theory of fields, Vols.1,2,3, Cambridge University Press.

[17] Leonard Schiff, Quantum mechanics, TMH.

[18] P.A.M.Dirac, "The principles of quantum mechanics", Oxford University Press.

[19] S.Weinberg, Gravitation and Cosmology:Principles and applications of the general theory of relativity.

[20] Mandel and Wolf, Optical coherence and quantum optics.

[21] W.O.Amrein, A Hilbert space approach to quantum mechanics, CRC press.

[22] T.Kato, Perturbation theory for linear operators, Springer.